Global Geodetic Observing System

Hans-Peter Plag · Michael Pearlman
Editors

Global Geodetic Observing System

Meeting the Requirements of a Global Society on a Changing Planet in 2020

Springer

Editors
Dr. Hans-Peter Plag
University of Nevada, Reno
Nevada Bureau of Mines &
Geology & Seismological Laboratory
Reno NV 89557
USA

Dr. Michael Pearlman
Harvard-Smithsonian
Center for Astrophysics
60 Garden St.
Cambridge MA 02138
MS 16
USA

ISBN 978-3-642-02686-7 e-ISBN 978-3-642-02687-4
DOI 10.1007/978-3-642-02687-4
Springer Dordrecht Heidelberg London New York

Library of Congress Control Number: 2009932078

Cover design: deblik

Printed on acid-free paper

Springer is part of Springer Science+Business Media (www.springer.com)

Foreword

The Global Geodetic Observing System (GGOS) has been established by the International Association of Geodesy (IAG) in order to integrate the three fundamental areas of geodesy, so as to monitor geodetic parameters and their temporal variations, in a global reference frame with a target relative accuracy of 10^{-9} or better. These areas, often called 'pillars', deal with the determination and evolution of (a) the Earth's geometry (topography, bathymetry, ice surface, sea level), (b) the Earth's rotation and orientation (polar motion, rotation rate, nutation, etc.), and (c) the Earth's gravity field (gravity, geoid). Therefore, Earth Observation on a global scale is at the heart of GGOS's activities, which contributes to Global Change research through the monitoring, as well as the modeling, of dynamic Earth processes such as, for example, mass and angular momentum exchanges, mass transport and ocean circulation, and changes in sea, land and ice surfaces. To achieve such an ambitious goal, GGOS relies on an integrated network of current and future terrestrial, airborne and satellite systems and technologies. These include: various positioning, navigation, remote sensing and dedicated gravity and altimetry satellite missions; global ground networks of VLBI, SLR, DORIS, GNSS and absolute and relative gravity stations; and airborne gravity, mapping and remote sensing systems. The optimal assimilation of such heterogeneous observations into models of geodynamics, oceanography, hydrology, glaciology, and weather and climate, will be done by interdisciplinary teams of researchers from geodesy and other sciences, and through the coordinated work of all IAG Services and Commissions. Naturally, addressing problems of such large scale and complexity requires international effort and commitment. Such initiatives are already underway (GEO, GEOSS), and GGOS represents IAG, and geodesy in general, in all of them, and provides the scientific and infrastructure contribution of geodesy to the Earth sciences.

The science and applications that GGOS addresses have important implications for the well-being of the global society. In an era of economic uncertainty and rapid environmental change it is imperative that action be taken to minimize risks from natural hazards, climate change, sea level rise, etc., to develop forecasting models for oceans and weather, and early warning systems for severe storms, tsunamis, and other hazards, and to manage our natural resources and our environment in a sustainable manner. To understand the Earth processes responsible for the aforementioned hazards requires continuous monitoring campaigns over long periods of time, as well as novel modeling of the observed changes with time. In other words, we can no longer speak of geodesy in three dimensions; we have entered a new and exciting era of four-dimensional geodesy, in which modern geodesy has become an indispensible contributor to the understanding of System-Earth and its evolution in time. IAG is well-positioned and proud to be able to contribute to this international effort through the work of GGOS, and therefore considers GGOS as its flagship Component.

The GGOS 2020 document describes the challenges, science, technology, applications, strategies, future plans and expected contributions of IAG and GGOS to the Earth sciences through the next decade. It contains the collective work over a period of several years of many individuals and organizations too many to list here without whom this volume would not have been possible. The IAG, and I personally, express our sincere gratitude to each one of them. Many thanks are due to the authors of the various chapters and the editors of this volume, and in particular to Hans-Peter Plag, for the countless hours he has devoted to writing, editing and coordinating, and his enthusiastic dedication to the project.

Calgary, February 2009

Prof. Michael G. Sideris
President, International Association of Geodesy

Preface

About this book

Background

This book describes the scientific rationale and the specifications for the Global Geodetic Observing System (GGOS) of the International Association of Geodesy (IAG) in terms of concepts, conventions, infrastructure and services, that would meet future requirements of a global community facing increasingly challenges on a changing planet. With this in mind, the document provides the basis for the further development of GGOS over the next decade and beyond. GGOS is built upon the basis provided by the existing Services and Commissions of IAG and is one of the major IAG components. In order to maximize the benefits to users of the considerable infrastructure and resources available to these Services, the concept for GGOS and the strategy for its development and implementation require careful considerations of the future needs of society for geodetic observations and services.

Improvements to the International Terrestrial Reference Frame (ITRF) and the availability of geodetic observations of changes in Earth's shape, gravity field and rotation over the last few decades have been a major driver of scientific discovery. Further improvement can be expected to lead to more exciting discoveries, particularly in combination with emerging new observation technologies for monitoring the variability of the Earth's gravity field and surface deformations. In a broader sense, the geodetic reference frames and observations have contributed to a transition of many processes in society and are expected to continue to do so. This great potential for scientific progress in support of societal needs associated with an improved geodetic observing system motivated the process that led to this book.

The context for this book is the increasing societal and scientific need for Earth observations, and their dependence on an appropriate geodetic foundation as well as a continuous series of geodetic observations. There is a growing awareness that sustainable development, which is the agreed-upon leading principle and goal of the global community, cannot be achieved without sufficient knowledge about the state, trends and processes in the Earth system. This is manifested in the establishment of the Group on Earth Observations (GEO) with currently about 75 member countries. The main purpose of GEO is to facilitate the implementation of the Global Earth Observation System of Systems (GEOSS), with the vision for this system *to realize a future wherein decisions and actions for the benefit of humankind are informed by coordinated, comprehensive and sustained Earth observations and information* (GEO, 2005a).

Geodesy provides the foundation for most Earth observations as well as crucial observations of changes in the Earth's geometry, gravity field, and rotation, which are all related to mass transport in the Earth system and the system dynamics. Therefore, geodesy is crucial for meeting many of the requirements for observations of global change and observations supporting studies of the Earth system. Providing the basis for precise positioning and navigation, geodesy is also crucially supporting or enabling many activities and processes in a modern society.

Realizing the importance of the geodetic reference frame and the contribution of geodesy to Earth observations, GEO has included a specific Task AR-07-03 "Global geodetic reference frames" in its Work Plan 2007-2009 and and as Sub-Task DA-09-2c in the Work Plan 2009-2011. Understanding the requirements for GGOS is a central goal of this task. The present book provides this input to the GEO Task.

The development of Earth observations takes place in a context where a considerable fraction of the funding for Earth observation infrastructure and research is allocated in response to major natural and anthropogenic disasters without a sufficiently well developed core infrastructure stable over time. Many satellite missions are research-oriented, whereas operational monitoring of many key indicators of the Earth system is insufficiently implemented (GEO, 2005b).

In geodesy, this situation is not much different. Current limitations in funding, often with a lack of appreciation of decision makers of the importance of the geodetic observing system for Earth observations and society at large, has led to the global geodetic community seeking yo provide better products and services based on incremental improvements to the system in an overall framework that severely limits the options for such improvements.

Scope

The advent of the space-geodetic techniques, and the rapid improvement and growth of communication techniques and capacities, has launched a revolution in the field of applied and global geodesy. Moreover, geodetic imaging increasingly gains importance, and the integration of the new techniques and methods into the traditional point-based approach of geodesy poses a major challenge. Therefore, it is timely to assess thoroughly the user requirements for the geodetic observations and products, and based on these requirements to design an optimal future system, which makes use of the maturing space-geodetic techniques as well as emerging imaging techniques. In order to do so, the authors for the contributions collected in this book had to take a fresh approach to the problem, not only with respect to the infrastructure but even more so concerning the underlying concepts, including the conventional approach to geodetic reference frames. Some of the concepts described or proposed here contradict current "best practices" and time will tell whether these new concepts will facilitate significant progress or whether they will have to be modified.

The authors of the contributions collected in the book do not attempt to assess current systems, concepts, products and services, but rather take a new look at the problem of building a geodetic observing system. The starting point is a rigorous review of the societal and scientific problems that require geodetic observations for their solution. This analysis leads to a set of general user requirements. These requirements are then, in a second step, used to derive functional system specifications. A third step focuses on the design of a system that would meet these specifications.

Collectively, the chapters of this book provide:

(1) a description of the scientific and societal problems, as well as practical applications that benefit from geodetic observations, services and products;
(2) a comprehensive overview of the user requirements for geodetic observations and products as derived from a broad range of societal benefit areas and scientific requirements;
(3) the functional specifications for a geodetic observing system capable of meeting the user requirements;
(4) a concept for future realizations of a (terrestrial) reference system able to meet the user requirements;
(5) the design of a system capable of addressing the functional specifications, in terms of conventions, techniques, infrastructure, and data analysis; and
(6) considerations and recommendations for the system implementation.

The anticipated audience

This book is a comprehensive document describing the background rationale for GGOS. It was written by a team of Chapter Lead Authors, each supported by Chapter Writing Teams. Besides including geodetic experts in all relevant fields, the chapter teams also include experts from other fields of Earth sciences and Earth observations. This book serves two purposes: (1) to inform users of Earth observations (in particular, GEO) of the potential of GGOS, and (2) to ensure that the GGOS community is aware of the users' needs and requirements so as to integrate GGOS into GEOSS for maximum mutual benefit. Thus, this book seeks to facilitate communication across several sectoral and discipline boundaries, including those between geodesy and other Earth sciences, between scientists and operational agencies, and between GGOS and GEOSS.

Documents consulted

Geodesy has a long tradition of assessing the requirements of society and of projecting these into future developments of the geodetic techniques and observing systems. This book continues this tradition, and it therefore benefited from a number of reports made available over the last four decades. These reports include, but are not limited to, the "Williamstown Report" (Kaula, 1970), the "Erice Report" (Mueller & Zerbini, 1989), the report on geodesy in 2000 prepared by the U.S. National Research Council in 1990 (Commission on Physical Sciences, Mathematics, and Applications, 1990), the "Coolfont Reports" (NASA, 1991a,b,c), the gravity report by the U.S. National Research Council (Commission on Geosciences & Resources, 1997), the *Living on a Restless Planet* report of the Solid Earth Science Working Group of NASA (Solomon & the Solid Earth Science Working Group, 2002), the report of an InSAR Workshop (Zebker, 2005), and the recent ESA document *The Changing Earth* (Battrick, 2006).

In the frame of the Integrated Global Observing Strategy - Partnership (IGOS-P) and GEO, several reports documented the needs for Earth observations in several societally relevant fields. Examples are the documents of GEO, such as GEO (2005a,b), the IGOS-P Theme reports (e.g., IGOS-P Ocean Theme Team, 2001; Lawford & the Water Theme Team, 2004; Marsh & the Geohazards Theme Team, 2004; Townshend & the IGOL Writing Team, 2004; Key & the IGOS-Cryo Writing Team, 2004), as well as reports produced by the various United Nations (UN) Agencies and programs. The latter include in particular the recent UN Water report (United Nations, 2006).

In a number of recent reports, user requirements for geodetic observations have been considered. Some of these reports are focused on national developments (e.g., Williams et al., 2005), improvements to the current situations (e.g., Plag, 2006a), or single technological aspects (such as Niell et al., 2006). Of direct importance for this book are the documents and publications produced by IAG scientists and teams focusing on GGOS, namely the papers in Rummel et al. (2000) and the GGOS Implementation Plan (Beutler et al., 2005). A considerable number of recent studies concerning relevant Earth system processes and the geodetic observations required to study these processes have been produced. Examples are the UNAVCO report on solid Earth science (UNAVCO, 1998), the German report on mass movements (Ilk et al., 2005), and the U.S. report on InSAR (InSAR Working Group, 2005). In addition to these report, a number of science reports from related fields have been consulted, such as the report on earthquake science by the National Research Council (Board on Earth Sciences and Resources, 2003), the NASA study on a global earthquake satellite system (Raymond et al., 2003), and the National Research Council Decadal Survey (National Research Council, 2007).

Reno, Boston,
March 2009

Hans-Peter Plag
Michael Pearlman

Acknowledgments

This work has been supported by many colleagues in geodesy and various fields of Earth science. The Editors are grateful to the many authors of the chapter writing teams, and in particular the chapter lead authors for their contributions.

The book has gone through several rounds of reviews. Individual reviews of an early draft were provided by Chris Hughes, Norman Miller, Ivan Mueller, Chris Reigber, Fernando Sanos, and Christian Tscherning. An open review in the IAG community resulted in more than 300 individual comments. A final review of parts of the book was carried out by an IAG Panel consisting of Hermann Drewes, Chris Rizos, and Michael Sideris. The Editors extend their sincere thanks to all who invested time and effort into turning this book into a valuable basis for the further development of GGOS.

Financial support for the Lead Editor was provided by NASA (through a contract with the Jet Propulsion Laboratory and in the frame of several research projects), and the Norwegian Mapping Authority.

Parts of the work described in this document were carried out at the Jet Propulsion Laboratory, California Institute of Technology, under contracts with the National Aeronautics and Space Administration.

Executive Summary

H.-P. Plag, B. Lilja Bye, R. Gross, T. A. Herring, M. Pearlman, P. Poli, C. Rizos, D. Sahagian, J. Zumberge

Preamble: Geodesy is the science of determining the geometry, gravity field, and rotation of the Earth, and their evolution in time. Traditionally, geodesy has been serving other sciences and many societal applications, including mapping. With the advent of satellite geodesy and an accuracy improvement of more than three orders of magnitude over the last three decades, geodesy has developed into a science making unique contributions to the study of the Earth system, its inherent dynamics, and its response to climate change, as well as a tool underpinning a wide variety of other remote sensing techniques. Facilitated by the Global Navigation Satellite Systems such as the Global Positioning System, a wide and growing variety of applications associated with positioning and navigation are being developed, particularly in combination with products derived from global geodetic observations. This book describes the requirements for a global observing system to provide products and services with the geodetic accuracy necessary to address important geophysical questions and societal needs, and to provide the robustness and continuity of service which will be required of this system in order to meet future needs.

(Chapter 1) **Living on a dynamic planet – the challenge:** A growing population is living on a dynamic planet, endowed with finite resources and limited capacity to accommodate the impact of the increasingly powerful anthropogenic factor. Sustainable development is crucial for realizing a stable and prosperous future for the anthroposphere, as has been acknowledged by a number of World summits. Although there are many influential factors, a detailed understanding of the Earth system with its major processes and its trends is one of the prerequisites for sustainable development. A deeper understanding cannot be reached without sufficient observations of a large set of quantities of the Earth system. As emphasized by the Earth Observation Summits (EOSs), there is an urgent need for a comprehensive, coordinated and sustained program of Earth observation. Earth observations are not only necessary

for a scientific understanding of the Earth, they are fundamental for most societal activities, ranging from disaster prevention and mitigation, the adequate provision of resources such as energy, water and food, the understanding of climate change, the protection of the biosphere, environment, and human health, to the building and management of a prosperous and sustainable global society.

(Chapter 1) **Geodesy is fundamental in meeting this global challenge:** Geodesy provides the foundation on which all Earth observation systems are built. In this function, geodesy is essential for Earth observation just like the foundation and frame of a house are necessary to keep it stable over time. But modern geodesy does more: it also provides comprehensive observations of changes in the Earth's shape, gravity field and rotation, the so-called "three pillars of geodesy." The principal geodetic quantities associated with these "pillars" are intimately related to mass transport in the fluid envelope of the solid Earth and its interior, as well as the dynamics of the Earth system. Therefore, the geodetic observing system provides essential observation of Earth system processes. It turns out, not surprisingly, that the geodetic observing system is similarly essential for exploring the planets, the solar system, and beyond.

(Chapter 1) **Geodesy is in transition:** The advent of space-geodetic techniques and the rapid improvement of communication technologies and capacities have fundamentally changed, if not revolutionized, geodesy and its methods. While previously point coordinates were given with respect to local or regional reference frames, positions can now be observed with respect to a global reference frame with unprecedented accuracy. Based on these techniques, changes in the Earth's shape, rotation and gravity field are determined with increasing spatial and temporal resolution, increasing accuracy, and with decreasing latency. These observations capture the "fingerprints" of mass movements in the oceans, atmosphere, ice sheets and terrestrial water storage; they provide the "scales" to weigh changes in the mass in the ocean; they allow the determination of the kinematics and strain field of the Earth's surface and the displacement field associated with earthquakes; they provide information on the water content in the atmosphere; and they constitute crucial constraints for all models of mechanical processes in the Earth system.

With the development of the space-geodetic techniques, the scope of the geodetic observing system is rapidly extending from a provider of the reference frame, and the tools for the determination of accurate positions, to a system monitoring the mass transport and the dynamics of the solid Earth and its fluid envelope with unprecedented spatial and temporal resolution and accuracy. Thus, this observing system is in transition from a utility for other geoscientists, to a provider of a consistent set of Earth observations relevant for nearly all societal benefit areas of Earth observations.

Geodesy is a "service science". In the past the "customers" of geodesy mainly came from the surveying and mapping profession; today, however, geodesy also serves the geophysical, oceanographic, atmospheric, and environmental science communities. Thus, it is their user requirements that also influence the development of the geodetic observing system.

(Chapter 2) **International cooperation is essential for geodesy:** Over many years, the international scientific community has managed in a major cooperative effort the establishment and maintenance of a global infrastructure that provides the observational basis for the determination of highly accurate positions anywhere on Earth and in space. This achievement has been facilitated by the International Association of Geodesy (IAG) and is based on the voluntary commitment of national geodetic authorities, space agencies, research institutes, universities, and individuals. Two reference systems are basic in geodesy, namely the celestial reference system and the terrestrial reference system. The International Earth Rotation and Reference Systems Service (IERS) has the responsibility for defining these geometric reference systems, and to realize them through appropriate frames. The International Celestial Reference System (ICRS) is the fundamental basis for the definition of celestial positions, and the International Terrestrial Reference System (ITRS) is the fundamental basis for describing terrestrial positions. These systems are conventional coordinate systems including all conventions for the orientation and origin of the axes, the scale, physical constants, models, and processes to be used in their realization.

The ICRS is realized through the International Celestial Reference Frame (ICRF), which is a set of estimated coordinate positions of extragalactic reference radio sources distributed over the sky. The ITRS, in turn, is realized through the International Terrestrial Reference Frame (ITRF), which is a set of globally distributed points on the solid Earth's surface, for which estimates of coordinate positions and (currently constant) velocities are derived from space-geodetic observations at these points.

Conceptually, the link between ITRS and ICRS is provided by the Earth rotation. Consequently, the ITRF and ICRF are connected through estimates of the Earth rotation parameters, which are also derived and made available through the IERS as so-called Earth Orientation Parameters (EOP) as determined by space-geodetic techniques.

Currently, the ICRF is determined by the technique of Very Long Baseline Interferometry (VLBI). For the determination of the ITRF, a combination of several independent space-geodetic techniques, including VLBI, Satellite Laser Ranging (SLR), Lunar Laser Ranging (LLR), Global Navigation Satellite System (GNSS), and Doppler Orbitography and Radiopositioning Integrated by Satellites (DORIS) is employed. Similarly, the EOPs are derived from a combination of these techniques. For each of these techniques, a technique-specific IAG Service maintains a global network of tracking stations (based on voluntary efforts of many contributors). Each of these techniques has unique advantages as well as disadvantages, and only the combination of the techniques guarantees an accurate and stable reference frame. Therefore, the most important elements for the determination and maintenance of the ITRF are the so-called "core stations", which have at least three of the independent space-geodetic techniques co-located (in addition to absolute and relative gravity observations and tide gauges, where possible). However, globally, there are currently only about 15 of these core stations, while about 40 stations are considered necessary in order to meet the most demanding user requirements.

The GNSSs have developed into the most widely applied technique for positioning (and navigation). The dramatic development of the Global Positioning System (GPS) over the last ten years into an accurate and highly efficient technology for positioning has been facilitated by the work of the International GNSS Service (IGS).

The Global Geodetic Observing System (GGOS) of the IAG is the proposed unifying umbrella for the IAG Services, which integrates the observing systems for changes in the Earth's shape, gravity field, and rotation and improves internal consistency. It links the geodetic services into the global Earth observation systems in order to provide a consistent service to the users. In particular, GGOS aims to ensure that the geodetic products and tools respond to increasingly more demanding user requirements.

Much of the international cooperation originates from regional and national organizations, which not only facilitate the dissemination of the global developments into the regions, but also are influential in motivating national bodies to contribute to international geodetic activities. Today, the ITRF, and the products and services that give access to the ITRF anywhere and anytime, are crucial for many economic and scientific applications. They have become so integrated in many applications that they are often taken for granted, as an integral part of the societal infrastructure freely available to everybody. However, without the international cooperation in geodesy, this global reference frame could not be maintained at its current level of accuracy and accessibility. Considering the nature of the voluntary commitment of many contributors on which GGOS is based, the incomplete spatial coverage of the ground-based networks, and the complementarity of the geodetic techniques, national decisions to discontinue geodetic infrastructure such as the operation of ground stations, or to withdraw support for specific techniques, can have severe consequences for GGOS and its products, in particular the ITRF.

(Chapter 3) **The development of the geodetic observing system needs research:** Maintaining a terrestrial reference frame at the level that allows, for example, the determination of global sea level changes at the sub-millimeter per year level, pre-, co- and postseismic displacement fields associated with large earthquakes at the sub-centimeter level, timely early warnings for earthquakes, tsunamis, landslides, and volcanic eruptions, as well as the monitoring of mass transport in the Earth system at the few gigatons level, requires a comprehensive Earth system approach.

Currently, geodesy is facing an increasing demand from science, the Earth observation community, and society at large for improved services, observations and products. Most of these requirements are in terms of improved accuracy (in particular, instantaneous accuracy), better reliability (including addressing the issue of liability), and improved access to the reference frame. The IAG and GGOS are aware of the enormous challenges implied by the demand to improve the accuracy from an average level of close to 10^{-9} (i.e., 1 ppb of the Earth's radius) to an instantaneous level (with daily or higher temporal resolution) of 10^{-10}, as required in order to meet emerging user requirements. In fact, GGOS faces two types of scientific and technological challenges, namely an "internal" challenge and an "external" challenge. The "internal" challenge to geodesy is concerned with developing GGOS and the geodetic technologies in order to meet the demanding requirements in terms of reference frame accuracy and availability, as well as the spatial and temporal resolution and accuracy of the observations. In many cases, it is not so much the measurements from a single technique that ultimately limits accuracy, but rather the ability to attribute signals to specific sources, and to model these. Therefore, meeting this challenge requires integration of techniques and models. This challenge is a central theme for research and development inside IAG for the future. The "external" challenge is related to geodesy's contribution to Earth system monitoring and science. The signals induced by global change in the Earth's shape, gravity field and rotation are small (on the order of parts-per-billion of the quantities) and embedded in often larger variations not caused by global change. Besides measuring the geodetic quantities with an accuracy considerably better than the signals, identifying and extracting the global change signals also requires the modeling of all known processes in an Earth system model taking into account the interactions between the various Earth system components. This challenge requires geodesy to interact with all Earth sciences and to accommodate the terrestrial processes in data processing and modeling.

(Chapter 4) **The benefits of the global and national geodetic infrastructure are enormous:** A very accurate and stable global geodetic reference frame, such as the ITRF, is indispensable for Earth observation, science and the functioning of a modern society. In such a frame, coordinates can be attached to points and objects (e.g., an airplane, a measuring sensor, a mark in the ground) and their movements over time can be described (e.g., the position of a point on the Earth's surface before, during, and after an earthquake). The benefits of the ITRF and the global geodetic infrastructure are wide-ranging. GGOS and, in particular, the key product ITRF:

- contribute substantially, directly or indirectly, to many economic activities and to the global wealth;
- allow for the exploitation of the space-geodetic technologies for a wide range of practical and scientific applications;
- provide a foundation on which today's national and regional reference frames are built and link these frames to each other;
- allow the interrelation of all geo-referenced data to be described in the same frame, thus facilitating full interoperability of geo-related databases and services;
- support governmental and intergovernmental priorities and international activities, such as sustainable development, climate change, the Global Earth Observation System of Systems (GEOSS), the Intergovernmental Panel on Climate Change (IPCC), and the United Nations (UN);
- provide a mechanism in many countries, including developing ones, for national participation in important global programs aimed at a better understanding of the Earth system, its climate, global geodynamics, geohazards, etc., and the mitigation of the impact of natural and anthropogenic hazards on society; and

- provide a mechanism for participation of the private sector and research institutes in international projects and activities, particularly in the field of technology development.

(Chapter 4) **The societal prospects of space geodesy:** The technological development facilitated through the new space-geodetic techniques for navigation and positioning poses challenges by creating new requirements for accessibility, accuracy and long-term stability. The rapid development of satellite-based precise point positioning techniques, which allow the determination of very accurate position anytime and anywhere on the planet, enables a wide range of position-related applications. The new geodetic technologies are leading to fundamental changes not only in all areas of navigation and transport, but also for application in process control (e.g., farming, construction, mining, resource management), construction and monitoring of infrastructure (e.g., off-shore platforms, reservoirs dams, bridges, and other large civil structures), surveying and mapping (including off-shore), and Earth observation. Geodetic techniques are crucial for the assessment of geohazards and anthropogenic hazards, and they will play a pivotal role in early warning systems of such hazards and disasters. The outcomes include increased security, a better use of resources, and progress towards sustainable development.

A well-defined and accessible reference frame, together with high-speed communications and advanced data processing, enables modern societies to operate in a very cost efficient manner, and hence create a basis for higher standards of living. National studies have shown that a number of major areas in national economies depend to a large part (up to 40%) on their geodetic infrastructure and services. Taking into account the fact that most national reference frames are fully dependent on the global infrastructure and frames, any degradation of the global infrastructure may have serious consequences for national economies.

The availability of a global geodetic reference frame such as ITRF and the tools to determine precise point coordinates anytime and anywhere on Earth have a profound effect on almost all areas of society. Since the ITRF is accessible anywhere on the planet, it improves access to an important technological resource, particularly in developing countries. Therefore, it is an important contribution compatible with the principle of sustainable development demanding equal access to resources for all.

(Chapter 5) **Towards a geodetic Earth system service:** Changes in the Earth's shape, gravity field, and rotation are inherently related to the dynamics of and mass transport in the Earth system. With the rapid progress of the geodetic observation techniques, an integrated GGOS constitutes the basis for an Earth system service that provides information on the state of and trends in the Earth system with respect to relocation of mass, deformations of the Earth's surface, and changes in the Earth's dynamics.

Mass transport on time scales up to decades takes place mainly in the fluid envelope of the solid Earth, where water transport is three orders of magnitudes larger than any other type of mass transport. Thus, information on the fluxes in the global water cycle, including the ice sheets and glaciers, oceans, and terrestrial hydrosphere can be provided with unprecedented spatial and temporal resolution and accuracy, particularly for global and regional scale changes. This information is crucial to understanding the impact of global change on the water cycle, in particular the ice sheets, sea level, and large terrestrial water catchments,

Surface displacements are related to both mass relocations on and above the solid Earth's surface and geodynamic processes within the solid Earth. Surface displacements are caused, for example, by earthquakes, tectonic processes, magma flow in the crust, and anthropogenic ground water changes. Thus, information on surface displacement provides a basis for, for example, scientific studies of geohazards, hazard assessment, early warning, and resource management.

For times scales of up to decades, changes in the dynamics of the Earth system, particularly its rotational dynamics, are brought about to a large extent by changes in the climate system. The solid Earth, oceans, and atmosphere continuously exchange angular momentum, and changes in the mean circulation of the atmosphere and ocean affect the rotation of the solid Earth. Mass redistribution on the Earth's surface, for example, through melting of ice sheets, deform the solid Earth and, as a result, also change the rotation. Earth rotation is affected by these processes in an integral way, and thus is an ideal parameter to assess the overall state of the system.

***(Chapter 5)* Geodetic observations and products are crucial for maximizing the benefits of Earth observation:** Geodesy provides the foundation for a global geodetic reference frame such as the ITRF that can be used by all Earth observing systems to monitor atmosphere, ocean, and other resources, and which relates the measurements to a globally consistent reference frame. Without a sufficiently accurate and stable ITRF, the benefit of Earth observations for most of the nine Societal Benefit Areas (SBAs) identified by the EOSs would be significantly reduced. Monitoring quantities relevant to geohazards, the global water cycle, climate, weather, energy, and even health, depends on a ready and reliable access to an accurate global geodetic reference frame. Today, only the ITRF meets these requirements of most applications. Therefore, a the ITRF is crucial for realizing GEO's vision for GEOSS, i.e. *a future wherein decisions can be based on sufficient information for the benefit of humankind.*

Geodesy supports Earth system observation, modeling, interpretation, and prediction in general. Some of the tools of geodesy, in particular GNSS, already yield routine observations of the atmosphere, such as the water vapor fields in the lower troposphere, the mass fields in the stratosphere, and the electron content fields in the ionosphere. The raw GNSS measurements are inherently calibrated with respect to atomic clocks. There are no other observations of the Earth's global atmosphere that can claim such a recurrent, atomic calibration. In that respect, geodesy could further help track climate change. On the modeling and prediction issues, geodesy could support the development of Earth system circulation models for the fluid envelope of the Earth with space- and time-varying gravity fields.

Despite considerable progress over the last two decades, mainly due to technological improvements, the quality of the reference frame has been hampered by fluctuations in institutional support and contributions. In particular, infrastructure central to the long-term stability of the reference frame, such as SLR stations and VLBI antennas, have been retired without replacements; a development potentially leading to a degradation of the ITRF accuracy. In the near future, satellite missions central for monitoring ice sheets, sea level, and the global water cycle will cease to operate, and follow-on operational missions must be planned now.

Unanticipated impacts of global change can be very costly in terms of life and property. However, unnecessary mitigation can be costly, too. A good example is provided by the anticipated sea level changes widely acknowledged as a slowly developing hazard with potentially disastrous consequences. Mitigation of the sea level rise impact is a long-term process which requires a planning and implementation time scale of the order of decades. Mitigation of sea level rise impact is extremely expensive and risky: too little will cause severe impact, too much will put unnecessary demands on national and regional economies. Therefore, decisions must be based on solidly founded sea level scenarios in order to minimize the risk associated with misjudgment (in either direction). Considering the typical life time of coastal infrastructure of 100 to 200 years, the sea level scenarios have to cover at least one hundred years. Crucial information required to improve the understanding of sea level and ice sheet changes, and to set up future sea level scenarios comes from Earth observation systems. Satellite altimeters, satellite gravity missions, GNSS satellites, tide gauges and other *in situ* techniques are all necessary components of the "sea level observing system". However, with all these components in place, the observations cannot provide the required fidelity if not linked to a stable global reference frame. Without this frame, past and present changes in ice sheets and sea level cannot be sufficiently quantified and understood, and plausible future scenarios of regional and local sea level cannot be provided to society as a basis for informed planning.

(Chapter 6) **Geodesy is essential for exploring the planets, solar system and beyond:** Planetary geodesy, radio science, interferometry (including imaging VLBI, astrometric VLBI, and Earth-space VLBI), and interplanetary navigation all require accurate terrestrial and celestial reference frames well linked together by Earth rotation observations for making and interpreting their measurements. The performance of the GGOS is not a limiting factor for these applications. However, in order to meet demanding future requirements, it will be important to develop GGOS such that the terrestrial and celestial reference frames and the Earth rotation parameters meet these requirements.

(Chapter 7) **User requirements for geodetic observations and products are demanding:** The current scientific and societal user requirements are demanding in terms of accuracy, resolution, latency and reliability, and the requirements are expected to increase in the future. The GGOS products must have sufficient accuracy, temporal and spatial resolution, and latency to meet these requirements. The most demanding users of the terrestrial reference frame in terms of accuracy and long-term stability are most likely the scientific studies of sea level change caused by climate change. In order to have a frame at least an order of magnitude more accurate than the signal to be monitored, the terrestrial reference frame should be accurate at a level of 1 mm and be stable at a level of 0.1 mm/yr. The most demanding applications of the geoid are likely to be the determination of the mean sea surface topography for oceanic general circulation models, and the GNSS determination of the height of surface points at the millimeter level. These applications require the static geoid to be accurate at a level of 1 mm and to be stable at a level of 0.1 mm/yr; consistent with the accuracy and stability of the terrestrial reference frame. The most demanding application in terms of accuracy and latency of EOPs and their consistency with the terrestrial and celestial reference frames is likely to be the tracking and navigation of interplanetary spacecraft. This application is capability-driven and requires the most accurate EOPs that can be determined, realizing that those determined in near real-time are somewhat less accurate than those determined with a delay of a couple of weeks. Quantitatively, an accuracy at a level of 1 mm for the EOPs should be achieved. For the time variable geoid, the monitoring of the water cycle at sub-regional to global scales appears to be the most demanding applications requiring the geoid variations to be monitored accurate to 1 mm, stable to 0.1 mm/yr, with a spatial resolution of 50 km and a time resolution of 10 days.

(Chapter 8) **Towards a modern geodetic reference frame:** A modern geodetic reference frame supporting precise point positioning consists of:

- a highly-accurate, global geodetic reference frame based on a sufficient number of multi-technique tracking stations;
- a service providing satellite orbits and clocks as well as Earth rotation parameters of high quality and long-term consistency in this global reference frame;
- a highly-accurate model of the gravity field (in particular, the geoid) and its changes;
- a well-determined tie between the geometric and gravimetric reference frames; and
- a velocity model that allows the determination of time-variable transformations between the global reference frame and national reference frames.

On a national level, the classical geodetic reference frames are still typically reliant on relative positioning. However, it is anticipated that increasingly for many applications a transition to precise point positioning will take place in many countries. A core element for this transition will be a reference frame service providing access to the reference frame anywhere on Earth, including the ocean surface, with a high instantaneous accuracy.

A deficiency of the current terrestrial reference frame is that it is only defined for relatively few points (of the order of 500) on the Earth's land surface. For all other points, no 'reference motion' is available hampering the identification of anomalous motion. Therefore, it is proposed to augment the current reference polyhedron with a dynamic Earth reference model. This model, in principle, will provide infinite spatial and temporal resolution for geometry and gravity, and thus establishes a reference frame accessible anywhere on Earth (and above) at any time. The dynamic Earth reference model will combine geometry, gravity and rotation into one consistent model. However, implementing this model poses significant scientific challenges, which will define a central theme for geodesy over the next decade.

(Chapter 9) **Infrastructure for geodetic Earth system monitoring:** GGOS is based on a combination of terrestrial, airborne, and spaceborne techniques, each with unique characteristics and contributions, and a layered infrastructure ranging for the ground-based networks to artificial satellites, infrastructure on the Moon, and quasars. Parts of the infrastructure are still in the form of research facilities, while other parts are fully operational.

The global ground-based infrastructure comprises not only the global *in situ* networks of several geometric and gravimetric techniques, but also the numerous data centers, analysis centers, and web-based services, that are required to determine and maintain the reference frames as well as to make them accessible for a wide range of users and their applications. Despite a large international effort, most networks are still characterized by spatially uneven distributions, and hence have large gaps in coverage. For some techniques, such as SLR, spatial gaps are large and place significant limitations on the achievable accuracy. Of particular importance are stations where several techniques are co-located, thus allowing the integration of the products of techniques into one coherent frame. Of the order of 40 evenly distributed core stations, i.e., stations with three or more space-geodetic techniques co-located, are required; however, currently there is a severe gap over the southern hemisphere. Without closing this gap, many of the most demanding user requirements will not be met.

The satellite component contributing to GGOS includes low Earth orbiting satellites (e.g., dedicated gravity missions and altimeters), dedicated laser-ranging satellites (e.g., LAGEOS), and GNSS satellites. The former provide observations related to mass transport and displacements of the solid Earth, ice, and ocean surfaces. Mission continuity is a key infrastructure issue.

The dedicated laser-ranging satellites are crucial for the connection of the reference frame origin to the center of mass of the Earth system, a mandatory requirement for studies of global processes. These satellites have very long lifetimes, but their number is very small.

The signals from the GNSS satellites provide the basis for the "work horse" in GGOS. With currently about 400 tracking stations in more than 80 countries, this "work horse" allows for an accurate monitoring of the global reference frame and for access to the frame anytime and anywhere on Earth. Without the freely available signals of GPS, the impressive development of geodesy over the last two decades would have been impossible.

Today, infrastructure on the Moon consists of retro-reflectors for LLR.

VLBI utilizes radio signals emitted by quasars, and contributes unique observations that are especially important for the monitoring of Earth rotation, which provides the link between ICRF and ITRF. In fact, VLBI is the only space-geodetic technique capable of simultaneously monitoring ITRF, ICRS and Earth rotation. Furthermore, unlike the other space-geodetic techniques, VLBI provides a unique ITRF scale, traceable directly to the speed of light, which is essential to various long-term monitoring goals of GGOS, including changes in global hydrology and sea level rise.

Observations with terrestrial gravimeters, both absolute and relative, provide the basis for studies of many geophysical phenomena, including (but not limited to) free oscillations of the Earth, solid Earth and ocean tides, surface loading, changes in ice sheets, and sea level changes. Absolute gravimetry, combined with geometric techniques, is a terrestrial technique supporting SLR in constraining the tie between the reference frame origin and the center of mass of the Earth system.

In total, an estimated 500 person years per year are provided on the basis of voluntary commitment by national operational and research institutes to maintain the ground-based networks, the data centers, analysis centers and user interfaces. Not included in this estimate are the resources required to support the satellite missions and the GNSS satellites themselves.

(Chapter 9) **For a full exploitation of the potential, an operational core component is needed:** Currently, GGOS and the IAG Services are based on the voluntary commitments of many national authorities, institutions, and individuals. Moreover, GGOS, to a large extent, is still science-driven. As a consequence, the observing system keeps changing due to technological developments and scientific priorities, as well as national political decisions. The impact of fluctuations in the regional coverage of the terrestrial component can be severe, often dependent on national priorities or funding availability. A high redundancy is needed to compensate for these fluctuations. Technological progress leads to changes that are not always properly coordinated. Satellite missions are even more science-driven than the other components of GGOS, and discontinuation of important observation programs has happened in the past, and unfortunately are likely to continue to happen in the future. Funding for the global geodetic infrastructure depends on the national decisions and priorities in many countries, and this implies considerable volatility, sometimes threatening the proper maintenance of the reference frames and of the IAG Services themselves. All of these factors lead to temporal inhomogeneities in the system, its observations, and, most importantly, the geodetic reference frames. At the same time, as a consequence of the growing demands for geo-referencing in a wide range of applications, issues are raised concerning the reliability and continuity of the geodetic products, as well as liability of the service and data providers. Therefore, in order to fully exploit the potential of geodesy and to develop GGOS into an Earth system service, a fully operational core infrastructure is needed. Considering the scale of GGOS, such a core will require an approach based on intergovernmental agreements, implying firm commitments by the contributing nations. GGOS therefore has started a dialog at the international level, in particular within GEO, in order to develop an intergovernmental framework for these activities.

(Chapter 10) **Implementation of GGOS needs a multi-faceted organizational framework:** GGOS is based on the IAG Commissions, Inter-Commission Committees, and the Services of IAG. In order to maintain GGOS in the future, the technique-specific and the combination services must continue their work using state-of-the-art observational and analysis tools, with GGOS providing the overarching strategy and organizational framework. In particular, GGOS will have to ensure the coordination of the multi-technique network (including the data flow), it

will have to maintain the standards and conventions necessary to ensure consistency across the components contributing to GGOS, and it will have to develop a plan for an uninterrupted sequence of geodesy-related space missions. GGOS will need to be embedded within the framework of global Earth observation currently represented by GEO, the surveying and navigation communities, and the science community. GGOS will have to serve as an interface to all these stakeholders in GGOS as well as society at large. An on-going dialog of GGOS with its stakeholders, including the funding agencies, the space agencies, and relevant UN agencies, with the goal to ensure long-term stability of GGOS, and to secure long-term funding for GGOS, will be central for a successful implementation of GGOS.

Contents

List of Figures

List of Tables

List of Contributors

Editors

Hans-Peter Plag
Nevada Bureau of Mines and Geology, and Seismological Laboratory, University of Nevada, Reno, NV, USA, e-mail: hpplag@unr.edu

Michael Pearlman
Harvard-Smithsonian Center for Astrophysics, Cambridge, MA, USA, e-mail: mpearlman@cfa.harvard.edu

Chapter Lead Authors:

Gerhard Beutler (Chapter 10)
Gerhard, Astronomical Institute, University of Bern, Switzerland

Richard S. Gross (Chapter 7)
Jet Propulsion Laboratory, California Institute of Technology, Pasadena, USA

Thomas A. Herring (Chapter 8)
Massachusetts Institute of Technology, Cambridge, MA, USA

Hans-Peter Plag (Exec. Sum., Chapters 1, 2, and 11)
Nevada Bureau of Mines and Geology, and Seismological Laboratory, University of Nevada, Reno, USA

Chris Rizos (Chapter 4)
School of Surveying & Spatial Information Systems, The University of New South Wales, Sydney, Australia

Markus Rothacher (Chapter 9)
ETH, Zurich, Switzerland

Reiner Rummel (Chapter 3)
Institute for Astronomy and Physical Geodesy, Technische Universität München, Munich, Germany

Dork Sahagian (Chapter 5)
Environmental Initiative at Lehigh University, Lehigh University, Bethlehem, PA, USA

James F. Zumberge (Chapter 6)
Jet Propulsion Laboratory, Pasadena, CA, USA

Contributing authors

Doug Alsdorf (Chapter 5)
School of Earth Sciences, Ohio State University, Columbus, Ohio, USA

Zuheir Altamimi (Chapters 2, 8)
Institut Géographique National Laboratoire de Recherche en Géodésie (LAREG) Champs-sur-Marne, France

Dirk Behrend (Chapter 9)
NVI, Inc./Goddard Space Flight Center, Greenbelt, MD, USA

Srinivas Bettadpur (Chapters 2)
Center for Space Research, The University of Texas at Austin, Austin, Texas, USA

Gerhard Beutler (Chapters 1, 2, 3, 7, 9, 11)
Astronomical Institute, University of Bern, Bern, Switzerland

Georg Beyerle (Chapter 2)
GeoForschungsZentrum, Potsdam, Germany

James S. Border (Chapter 6)
Jet Propulsion Laboratory, California Institute of Technology, Pasadena, CA, USA

Dorota Brzezinska (Chapter 4)
Department of Civil and Environmental Engineering and Geodetic Science, The Ohio State University, Columbus, OH, USA

Anny Cazenave (Chapter 2)
Laboratoire d'Etudes en Géophysique et Océanographie Spatiales, LEGOS-CNES, Observatoire Midi-Pyrénées, Toulouse, France

David Crossley (Chapter 2)
Earth and Atmospheric Sciences, Saint Louis University, St. Louis, MO, USA

Veronique Dehant (Chapters 3, 6)
Royal Observatory of Belgium, Brussels, Belgium

Andrea Donnellan (Chapters 2, 9)
Jet Propulsion Laboratory, California Institute of Technology, Pasadena, CA, USA

William M. Folkner (Chapter 6)
Jet Propulsion Laboratory, California Institute of Technology, Pasadena, CA, USA

Rene Forsberg (Chapters 2, 4)

DTU Space, National Space Institute, Danish Technical University, Copenhagen, Denmark

Richard S. Gross (Exec. Sum., Chapters 1, 2, 3, 11)
Jet Propulsion Laboratory, California Institute of Technology, Pasadena, CA, USA

Thomas A. Herring (Exec. Sum., Chapters 1, 11)
Massachusetts Institute of Technology, Cambridge, MA, USA

Jacques Hinderer (Chapters 2, 9)
Ecole et Observatoire des Sciences de la Terre, Institut de Physique du Globe de Strasbourg, Strasbourg, France

Karl Heinz Ilk (Chapter 3)
Institute of Theoretical Geodesy, University of Bonn, Bonn, Germany

Gary Johnston (Chapter 4)
Geoscience Australia, Canberra, Australia

Dayton L. Jones (Chapter 6)
Jet Propulsion Laboratory, California Institute of Technology, Pasadena, CA, USA

Steve Kenyon (Chapter 4)
National Geospatial-Intelligence Agency, Arnold, MO, USA

Attila Komjathy (Chapter 2)
Jet Propulsion Laboratory, California Institute of Technology, Pasadena, CA, USA

Cornelius Kremmer (Chapter 5)
Nevada Bureau of Mines and Geology, and Seismological Laboratory, University of Nevada, Reno, NV, USA

Bente Lilja Bye (Exec. Sum.)
Norwegian Mapping and Cadastre Authority, Hønefoss, Norway

Chopo Ma (Chapters 2, 9)
Goddard Space Flight Center, Greenbelt, MD, USA

Anthony J. Mannucci (Chapter 2)
Jet Propulsion Laboratory, California Institute of Technology, Pasadena, CA, USA

Thomas Martin-Mur (Chapter 6)
Jet Propulsion Laboratory, California Institute of Technology, Pasadena, CA, USA

John M. Melack (Chapter 5)
Bren School of Environmental Science and Management, and Department of Ecology, Evolution and Marine Biology, University of California, Santa Barbara, CA, USA

Ruth Neilan (Chapter 10)
Jet Propulsion Laboratory, California Institute of Technology, Pasadena, CA, USA

Carey Noll (Chapters 2, 9)

Goddard Space Flight Center, Greenbelt, MD, USA

Axel Nothnagel (Chapter 2)
Geodetic Institute, Universität Bonn, Bonn, Germany.

Jürgen Oberst (Chapters 6, 9)
German Aerospace Center, Institute of Planetary Research, Berlin, Germany

Erricos C. Pavlis (Chapter 2)
Joint Center for Earth Systems Technology/University of Maryland, Baltimore County, Baltimore, MD, USA

Michael R. Pearlman (Exec. Sum., Chapters 1, 2, 5, 9, 10)
Harvard-Smithsonian Center for Astrophysics, Cambridge, MA, USA

Hans-Peter Plag (Chapters 3, 5, 7, 8, 9, 10)
Nevada Bureau of Mines and Geology, and Seismological Laboratory, University of Nevada, Reno, NV, USA

Paul Poli (Exec. Sum., Chapters 2, 3, 5, 8, 11)
Météo France, Centre National de Recherches Météorologiques (CNRS/GAME), CNRM/GMAP, Toulouse, France (now at ECMWF, Reading, U.K.)

Stephen J. Reid (Chapter 5)
Environmental Initiative, Lehigh University, Bethlehem, PA, USA

Bernd Richter (Chapter 9)
BKG, Frankfurt, Germany

Chris Rizos (Exec. Sum., Chapters 1, 11)
School of Surveying & Spatial Information Systems, The University of New South Wales, Sydney, Australia

Matthew Rodell (Chapter 5)
Hydrological Sciences Branch, NASA's Goddard Space Flight Center, Greenbelt, MD, USA

Markus Rothacher (Chapters 3, 10, 11)
ETH Zurich, Switzerland

Reiner Rummel (Chapters 1, 10, 11)
Institute for Astronomy and Physical Geodesy, Technische Universität München, Munich, Germany

Dork Sahagian (Exec. Sum., Chapters 1, 11)
Environmental Initiative at Lehigh University, Lehigh University, Bethlehem, PA, USA

Uli Schreiber (Chapter 2)
Forschungseinrichtung Satellitengeodäsie, Technical University of Munich, Fundamentalstation Wettzell, Kötzting Germany

Tilo Schöne (Chapter 9)

GeoForschungsZentrum, Potsdam, Germany

Ken Senior (Chapter 2)
U.S. Naval Observatory, Washington, DC, USA

Dru Smith (Chapter 4)
NOAA, National Geodetic Survey, Silver Spring, MD, USA

Seth Stein (Chapter 3)
Department of Earth and Planetary Sciences, Northwestern University, Evanston, IL, USA

Gilles Tavernier (Chapter 9)
CNES, Toulouse, France

Robert Thomas (Chapters 3, 5)
USA

James G. Williams (Chapter 6)
Jet Propulsion Laboratory, California Institute of Technology, Pasadena, CA, USA

Philip L. Woodworth (Chapters 2, 3, 5, 9)
Proudman Oceanographic Laboratory, Liverpool, U.K.

Xiaoping Wu (Chapter 6)
Jet Propulsion Laboratory, California Institute of Technology, Pasadena, CA, USA

Susanna Zerbini (Chapters 2, 3)
Department of Physics, Sector of Geophysics, University of Bologna, Bologna, Italy

Victor Zlotnicki (Chapter 3)
Jet Propulsion Laboratory, California Institute of Technology, Pasadena, CA, USA

Cinzia Zuffada (Chapter 2)
Jet Propulsion Laboratory, California Institute of Technology, Pasadena, CA, USA

James Zumberge (Exec. Sum., Chapters 1, 11)
Jet Propulsion Laboratory, California Institute of Technology, Pasadena, CA, USA

Reviewers

Chris Hughes
Proudman Oceanographic Laboratory, Liverpool, U.K.

Norman Miller
Berkeley National Laboratory and University of California, Berkeley, CA, USA

Ivan I. Mueller
USA

Christoph Reigber
Germany

Fernado Sanso
University of Milan, Milan, Italy

Christian Tscherning
Department of Geophysics, University of Copenhagen, Copenhagen, Denmark

Chapter 1
Introduction

H.-P. Plag, G. Beutler, R. Gross, T. A. Herring, C. Rizos, R. Rummel, D. Sahagian, J. Zumberge

1.1 The challenge: living on a changing, dynamic planet

Earth is a restless planet (Solomon & the Solid Earth Science Working Group, 2002). With its atmosphere, oceans, ice cover, land surfaces and its interior, it is subject to a large variety of dynamic processes operating on a wide range of spatial and temporal scales, driven by large interior as well as exterior forces. Many areas of the Earth's surface are exposed to natural hazards caused by dynamic processes in the solid Earth, the atmosphere and the oceans. Earthquakes, tsunamis, volcano eruptions, tectonic deformations, landslides, deglaciation, sea level rise, floods, desertification, storms, storm surges, global warming and many more are well known phenomena that are expressions of the dynamics of our restless planet. In modern times these processes are influenced, as well, by anthropogenic effects; to what extent is still largely unknown.

Earth is a finite planet. Resources such as clean water, arable land, flora and fauna, minerals, and energy are limited. Probably even more importantly, the capacity of the Earth system to maintain a delicate equilibrium under increasing anthropogenic pressure is limited.

A growing population has to cope with this restless, and finite, planet. On the one hand, settlements are encroaching into areas of high risks from natural hazards with major infrastructure being built in locations with high risks of large earthquakes, volcanic eruptions, storm surges, tsunamis, landslides and flooding, thus increasing the vulnerability of society. Increasingly, critical infrastructure is destroyed in natural disasters, affecting the economy on national and global levels, and displacing large populations, with severe social implications. On the other hand, the growing demands for access to food, water, materials, and space put stress on the finite resources of the planet. The anthroposphere has grown into a powerful force rapidly transforming the Earth's surface layers (as documented, e.g., by Turner II et al., 1990) and capable of changing major processes, including those of the climate system. However, humanity has not reached the necessary understanding to actually

H.-P. Plag, M. Pearlman (eds.), *Global Geodetic Observing System*,
DOI 10.1007/978-3-642-02687-4_1,

wield this power. Earth system processes, whether natural or modified by humans, affect our lives and the lives of future generations: decisions made today will influence the well-being of future generations. In order to minimize the anthropogenic impact on Earth system processes and in order to preserve resources for future generations, a better understanding of Earth system processes is required.

Reaching a condition of "sustainable development" has been recognized as a necessary (albeit not sufficient) prerequisite for living on a restless planet with finite resources, and with a limited capacity to accommodate the impact of the increasingly powerful anthropogenic factor. A number of World Summits have acknowledged that a sustainable development is mandatory for realizing a stable and prosperous future for the anthroposphere. Although there are many other influential factors, understanding the Earth system, its major processes and its trends, is one of the prerequisites for the success of the quest for sustainable development. Major decisions determining our future will have to be based on a much deeper understanding of this complex system.

A deeper understanding of the Earth system cannot be achieved without sufficient observations of a large set of quantities characteristic of Earth system processes. As emphasized by the Earth Observation Summits (EOSs), there is an urgent need for comprehensive Earth observations (see the documents in the Appendices of GEO, 2005b). Earth observations are not only necessary for a scientific understanding of the Earth, they are fundamental for most societal areas ranging from disaster prevention and mitigation, the provision of resources (such as energy, water and food), improving our understanding of climate change, the protection of the biosphere, environment, and human health, and ultimately to the building and management of a prosperous global society.

1.2 The potential: geodesy's contribution to a global society

Geodesy is the discipline that deals with the measurement and representation (geometry, physics, temporal variations) of the Earth and other celestial bodies (Sideris, 2007). The "three pillars" of geodesy are the Earth's time-dependent geometric shape, gravitational field, and rotation (Figure 1.1). Today, along with these pillars a number of related quantities are observed with terrestrial and space-geodetic techniques using a combination of spaceborne and airborne sensors and *in situ* networks (see Chapter 2). With its observational means, geodesy has the potential to determine, unambiguously and with utmost precision, the geometric shape of land, ice, and ocean surfaces as a global function of space and time. Since the dense web of microwave radiation used for geodetic positioning passes through the atmosphere, its interaction with the atmosphere yields important weather parameter information. The geometric methods when combined with global gravity information and the geoid, allow us to infer mass anomalies, mass transport phenomena and mass exchange in the Earth's system. Finally, the variations in Earth rotation reflect mass transport in the Earth system and the exchange of angular momentum among its components.

Observations of the Earth's variable shape, gravity field, and rotation provide the basis for the realization of the reference systems that are required in order to assign (time-dependent) coordinates to points and objects, and to describe the motion of the Earth in space (Figure 1.1). For this purpose, two reference systems are intrinsic in geodesy, namely the celestial reference system and the terrestrial reference system, which are dynamically linked to each other by the Earth's rotation. The two most accurate reference systems currently available are the International Celestial Reference System (ICRS) and the International Terrestrial Reference System (ITRS)(see Section 2.2 for more details), which are defined by the International Earth Rotation and Reference Systems Service (IERS). These systems are conventional coordinate systems that include all conventions for the orientation and origin of the axes, the scale, and the physical constants, models, and processes to be used in their realization. Based on observations, these systems can be realized through their corresponding "reference frames". The frame corresponding to the ICRS is the International Celestial Reference Frame (ICRF), which is a set of estimated positions of extragalactic reference radio sources. The frame corresponding to the ITRS is the International Terrestrial Reference Frame (ITRF), which is a set of estimated positions and velocities of globally distributed reference marks on the solid Earth's surface. These two frames are linked to each other by estimates of the Earth Orientation Parameterss (EOPs). ICRS, ITRF and the EOPs are provided by IERS.

Today, the internationally coordinated geodetic observations of the global geodetic station networks provide a continuous monitoring of the ITRF. This well-defined, long-term stable, highly-accurate, and easily accessible reference frame is the basis for all precise positioning on and near the Earth's surface. It is the indispensable foundation for all sustainable Earth observations, *in situ*, as well as airborne and spaceborne. Furthermore the ITRF underpins all geo-referenced data used by society for so many uses. At the most foundational level the ITRF rigorously supports the Spatial Data Infrastructure (SDI). The SDI is a model of all geo-referenced data that consists of many layers, all connected to the geodesy layer which is the realization of the ITRF at national and regional (and increasingly the international) scale. The other layers of the SDI are like elements of a "house", built on strong foundations, and include map and image data of the physical surface of the Earth, its terrain, waterways, forests, vegetation and habitats; transport and built infrastructure such as roads, railways, and other structures; cadastral land boundaries; political boundaries; and many others. These layers of digital geo-referenced data are crucial for many activities, ranging from mapping, construction, land development, natural resource management and conservation, navigation - in fact all decision-making that has a geo-related component.

Historically, geodesy was limited to determining the shape of the Earth, its gravity field, and its rotation including their changes over time. With modern instrumentation and analytical techniques, the scope of geodesy has extended to include the causes of the observed changes, i.e., the dynamics of and mass transport within the Earth system. With this broader scope, new pathways emerge in which geodesy can

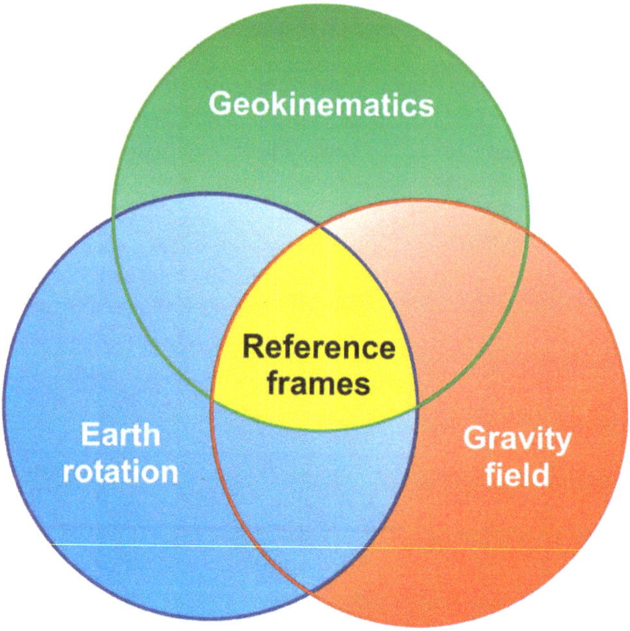

Fig. 1.1. Constituents of an integrated geodetic monitoring system. The "three pillars" of geodesy provide the conceptual and observational basis for the reference frames required for Earth observation. These three pillars are intrinsically linked to each other as they provide different observation related to the same Earth system processes.

contribute to the scientific understanding of the Earth system as well as the development, functioning, and security of society in general.

To a large extent, geodesy is a "service science". In the past, the main "customers" of geodesy came from the surveying and mapping profession, while today geodesy serves all Earth science, including the geophysical, oceanographic, atmospheric, and environmental science communities. Consequently, today the development of the geodetic observing system is guided by the user requirements of a much broader "customer" base.

With the "three pillars", geodesy precisely observes and consistently monitors mass movement in the Earth system and its associated dynamics:

- **Geokinematics:** measuring the geometric shape of the Earth's surface (solid Earth, ice and oceans) and its kinematics and variations, on global to local spatial scales, and at time scales from rapid to secular;
- **Earth rotation:** monitoring the variations of the Earth's rotation as an indicator of all angular momentum exchange inside, on or above the solid Earth, as well as of the torques acting on the solid Earth (including those due to the Sun and Moon); and
- **Gravity field:** determining and monitoring the Earth's gravity field and inferring the underlying mass redistributions in the solid Earth, liquid core, atmosphere, oceans, hydrosphere, and cryosphere.

Ultimately, all geodetic observations are affected by the same physical Earth system processes. Thus, geodesy provides a unique framework for monitoring and ultimately understanding the Earth system as a whole. Modern space-geodetic techniques are well suited for observing phenomena on global to regional scales, and thus are an important complement to traditional *in situ* observation systems.

The rapid development of space-geodetic techniques (see Chapter 2) also enables auxiliary applications that utilize the atmospheric disturbance of geodetic measurements (ionosphere, troposphere, magnetic field) for non-geodetic applications. Atmospheric disturbances formerly were the natural factor limiting the accuracy of geodetic measurements. Now this "noise" is increasingly being recognized as "signal", and the distortions of microwave signals propagating through the atmosphere can be "inverted" for atmospheric parameters and utilized for numerical weather prediction (e.g., Jerrett & Nash, 2001; Elgered et al., 2005), climate studies, and studies in atmospheric physics.

A major driver for the development of the geodetic observing system is the progress of science. In addition, technological advances with improved sensors, networks, and communications, the impact of nanotechnology, and the development of new and improved observing systems (for example, Interferometric Synthetic Aperture Radar (InSAR), LIght Detection And Ranging (LIDAR) and all remote sensing missions, including Gravity Recovery and Climate Experiment (GRACE), Gravity field and steady-state Ocean Circulation Explorer (GOCE), and future satellite missions) are key drivers. As pointed out above, the mounting pressure of environmental changes and the associated societal needs demand improved Earth observations which in turn put increasing demands on the geodetic observing system. Issue such as hazards monitoring and understanding of global change, the exponential growth of, and need for, geo-spatial information, and the complexity and scale of the global problems that cannot be solved by a single science require a well developed geodetic observing system. Geodetic expertise is therefore increasingly needed, and valued, by other sciences (Sideris, 2007).

With this development, geodesy faces several challenges (Sideris, 2007), namely: (1) inter-disciplinarity is required in order to contribute to collaborative solutions to problems, to allow for an optimal assimilation of a wide spectrum of observations into inter-disciplinary models, and to enable to interpretation and separability of the various signals; (2) development of a framework for a four-dimensional geodesy is required, in which temporal variations in the shape of the Earth and its gravity field are fully accounted for, long-term observation campaigns and archiving are planned with the 4-D nature of the system in mind, and an accuracy level for geometric and gravimetric quantities of much better than 10^{-9} (approaching 10^{-12}) is achieved; and (3) the recognition of what geodesy is and who benefits from needs to be communicated through appropriate outreach, and geodesy, in particular the International Association of Geodesy (IAG), faces the challenge of how best to promote the geodetic contributions to science and society at large.

Many scientific applications depend on a detailed knowledge of the Earth's shape, its gravity field and rotation (see Chapter 3), and in the past geodesy has (with ever-increasing accuracy) provided the necessary observations. The relatively recent advent of space-geodetic techniques has brought about a rapid development in global geodesy. The relative precision of the measurements is approaching the very impressive level of 1 parts per billion (ppb) or even better. Today, geodetic techniques permit the measurement of changes in the geometry of the Earth's surface with an accuracy of millimeters over distances of several thousand km.

Over the last one and a half decades, the global geodetic networks have provided an increasingly detailed picture of the kinematics of points on the Earth's surface and the temporal variations in the Earth's shape. Among other applications, the observations have been used to determine improved models of the secular horizontal velocity field (e.g., Kreemer & Holt, 2001; Kierulf et al., 2002; Kreemer et al., 2003), to derive seasonal variations in the terrestrial hydrosphere (e.g., Blewitt et al., 2001), to study seasonal loading (e.g., Dong et al., 2002), to invert for mass motion (e.g., Wu et al., 2003), and to improve the modeling of the seasonal term in polar motion (Gross et al., 2004). Geodetic techniques provide the means to observe surface deformations on volcanoes (e.g., Lu et al., 2000; Lanari et al., 2002; Bonforte & Puglisi, 2003), in unstable areas (e.g., Ferretti et al., 2004), associated with earthquakes and fault motion (e.g., Banerjee et al., 2005; Vigny et al., 2005; Kreemer et al., 2006b), or subsidence caused by anthropogenic activities such as groundwater extraction (e.g., Strozzi et al., 2002). Current developments indicate that geodetic observing techniques will be able to determine the magnitude of large earthquakes in near-real time and thus help mitigate the problem of low initial magnitudes estimated by seismic techniques (Blewitt et al., 2006b).

Spaceborne sensor systems play an important role in Global Change studies. With satellites it is feasible to observe Earth system processes globally, uniformly and with relatively rapid repetition rates. Nevertheless, the results are still inconclusive, as evidenced by the ongoing debate about global warming (see, e.g., Hogan, 2005, and the references therein).

If the geodetic observations and products can be provided on a global scale with a precision at or below the 1 ppb level, consistently, and with sufficient stability over decades, geodesy can make very important contributions to our understanding of the state and dynamics of System Earth (see Chapter 5). A prerequisite for exploiting the full potential of geodesy for Earth observation, Earth system monitoring, and many practical applications, is a sophisticated integration of all geodetic techniques (spaceborne, airborne, marine and terrestrial), processing models and geophysical background models into one system model. This integration will permit – as part of global change research – the assessment of surface deformation processes and the quantification of mass anomalies and mass transport inside the individual components, and mass exchange between the components of the Earth's system. These quantities serve as input to the study of the physics of the solid Earth, ice sheets and glaciers, hydrosphere and atmosphere. They are of particular value for the study of complex phenomena such as glacial isostatic adjustment, the evolution of tectonic stress patterns, sea level rise (and fall), the hydrological cycle, transport processes in the oceans, and the dynamics and physics of the atmosphere (troposphere and ionosphere).

Geodesy is crucial not only for Earth observation and science, but today it is also indispensable for many activities in a modern society. Traditionally, geodesy has served society by providing reference frames for a wide range of practical applications from regional to global navigation on land, sea, and in the air, construction of infrastructure, to the determination of reliable boundaries of real estate properties. Reference frames were, however, national or regional in scope, adequate for the determination of coordinates relative to a network of reference points. Thus, determination of precise coordinates required simultaneous measurements at several points. Today, the Global Navigation Satellite System (GNSS) also provides access to precise point coordinates in a global reference frame anytime and anywhere on the Earth's surface with centimeter-level accuracy, without requiring additional measurements on nearby reference points.

On the user side, such technological developments have stimulated new applications demanding even greater accuracy and improved access to geodetically-determined positions. On local to regional scales, applications such as land surveying, monitoring of critical infrastructure, prevention and mitigation of impacts of environmental hazards, and numerous technical applications require more or less instantaneous access to a reliable reference frame with centimeter-level accuracy or better. Already today, the economic benefit of the geodetic reference frame is enormous (see, e.g., Williams et al., 2005), and as more and more societal applications become depended on precise positioning this is very likely going to increase. In particular, the emerging combination of broadband communications, geo-databases and easily accessible accurate positioning can be expected to facilitate the development of many new applications and services (see Chapter 4), which will transform society and lead to an increasing dependence on the geodetic foundation, i.e., the terrestrial geodetic reference frame and tools for easy access to this frame.

1.3 The observing system: the current development of the Global Geodetic Observing System

The international cooperation fostered by the IAG has led to the establishment of the IAG Services, which provide increasingly valuable observations and products not only to the scientific community but also for a wide range of non-scientific applications. The IAG has therefore taken the first steps towards the implementation of the Global Geodetic Observing System (GGOS). GGOS was created as an IAG Project during the IUGG meeting in 2003 in Sapporo, Japan. After the first two years devoted to the definition of the internal organizational structure of GGOS and its relationship with external organizations, the Executive Committee of the IAG at its meetings in August 2005 in Cairns, Australia, decided to progress the Project into the implementation phase. Finally, at the IUGG meeting in 2007 in Perugia, Italy, the IAG elevated GGOS to the status of a full Component of IAG as the Observing System of IAG.

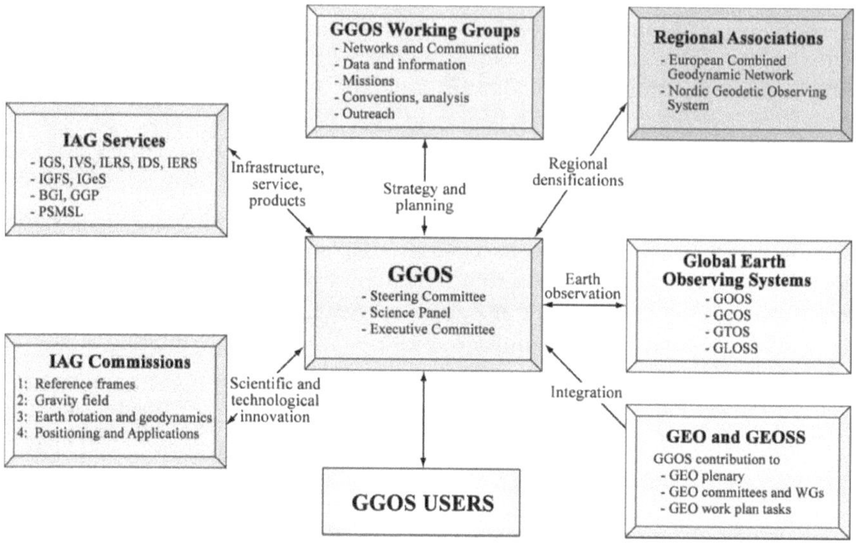

Fig. 1.2. Organizational links and relationships of GGOS. GGOS is being built on the scientific support from the IAG Commissions and the infrastructure of the IAG Services. GGOS integrates the work of the Services through a number of GGOS Working Groups and provides coordination and advice through its Committees. GGOS links these entities to the main programs in Earth observations, and provides a unique interface for GGOS users to the geodetic services. Modified from Plag (2006a).

GGOS as an organization is being built on the existing IAG Services as a unifying umbrella. Figure 1.2 shows the current organizational structure of GGOS with its Committees, Panels and Working Groups, the links to the IAG Services and Commissions, regional organizations, and to the outside world. In particular, the large international programs such as the Group on Earth Observations (GEO), which is implementing the Global Earth Observation System of Systems (GEOSS), and the relevant United Nations programs (see Chapter 5 for more details of these programs). GGOS provides the links between the IAG Services and the main programs in Earth observations and Earth science. It constitutes a unique interface for many (although not all) users of the geodetic services. GGOS adds a new quality and dimension to Earth system research by combining the geodetic techniques into one observing system of highest accuracy in a well-defined and reproducible global terrestrial frame. The observing system, in order to meet its objectives, has to combine the highest measurement precision with spatial and temporal consistency that is maintained over decades. The research needed to achieve these goals will influence the agenda of the IAG Commissions and the GGOS Working Groups.

According to the IAG By-Laws, GGOS *works with the IAG Services and Commissions to provide the geodetic infrastructure necessary for the monitoring of the Earth system and global change research.* The vision for GGOS implicit in this statement is to empower Earth science to extend our knowledge and understanding of Earth system processes, to monitor ongoing changes, and to increase our capability to predict the future behavior of the Earth system. The mission of GGOS embedded in the statement is to facilitate collaboration of the IAG Services and Commissions, and other stakeholders in the Earth science and Earth observation communities, to provide scientific advice and coordination that will enable the IAG Services to develop products that can meet the requirements of global change research, and to improve the accessibility of geodetic observations and products for a wide range of users. The IAG Services benefit from GGOS as a framework for communication, coordination, and scientific advice necessary to develop improved or new products with increased accuracy, consistency, resolution, and stability. IAG benefits from GGOS as an agent for improved visibility of geodesy's contribution to the Earth sciences and to society in general. The users benefit from GGOS as a single interface to the global geodetic observation system not only for access to products but also to voice their needs. Society benefits from GGOS as a utility supporting Earth science and global Earth observation systems as a basis for informed decisions.

GGOS as an observing system utilizes the existing and future infrastructure provided by the IAG Services. It will provide consistent observations of the spatial and temporal changes in the shape and gravity field of the Earth, as well as the temporal variations of the Earth's rotation (Figure 1.1). In particular, GGOS will provide on a global scale and in relation to one reference system a means to determine the spatial and temporal changes in the shape of the solid Earth, oceans, ice cover and land surfaces. In other words, it will provide a global picture of the surface kinematics of our planet. It will provide, in addition, estimates of mass anomalies, mass transport and mass exchange within the Earth system. Surface kinematics and mass transport together are the key to the determination of global mass balance, and an important contribution to the understanding of the energy budget of our planet (e.g., Rummel et al., 2002, 2005; Drewes, 2006). Moreover, the system will provide the observations that are needed to determine and maintain a terrestrial reference frame of higher accuracy and greater temporal stability than what is available today (Beutler et al., 2005).

GGOS as a system will exploit (and try to extend) the current constellation of satellite missions relevant to this goal, and missions planned for the next two decades, by integrating them into one observing system. The foundation for this integration are the existing global ground networks of tracking stations for the space-geodetic techniques: Very Long Baseline Interferometry (VLBI), Satellite Laser Ranging (SLR), Lunar Laser Ranging (LLR), GNSS, and Doppler Orbitography and Radiopositioning Integrated by Satellites (DORIS). GGOS will integrate these tracking networks with terrestrial gravity networks. GGOS will complement the space segment and global ground network with airborne and terrestrial campaigns that serve the purpose of calibration and validation, regional densification, and refinement. Assimilation of these observations into models of weather, climate, oceans, hydrology, ice and solid Earth processes will fundamentally enhance the understanding of the role of surface changes and in the dynamics of our planet. Furthermore, through the analysis of the dense web of microwave radiation connecting the GNSS satellites with Low Earth Orbiters (LEO) and with the Earth's surface a powerful new technique emerges for probing the atmosphere's composition.

It is clear that GGOS has two very distinct aspects: (1) the "organization GGOS" consisting of components such as the Steering Committee, Science Panel, Bureaus, Working Groups, etc., and (2) the "observating system GGOS" comprising the infrastructure of many different instrument types, satellite missions, and data and analysis centers. While GGOS as an organization is establishing its structure from essentially new entities, the observational infrastructure for GGOS as the observing system is being largely provided by the IAG Services. Most of this book addresses issues related to the observing system aspect of GGOS, while the organizational aspect is considered mainly in Chapter 10.

The challenge for geodesy in terms of Earth system monitoring is well summarized by Chao (2003), who states: "*After three decades and three orders of magnitude of advances, space geodesy is poised for prime time in observing the integrated mass transports that take place in the Earth system, from high atmosphere to the deep interior of the core. As such space geodesy has become a new remote sensing tool, in monitoring climatic and geophysical changes with ever increasing sensitivity and resolution.*

The transport of mass and energy are key processes that determine the dynamics of our Earth system. The Earth system can be conveniently viewed through its components, so-called geophysical fluids – the atmosphere, hydrosphere, cryosphere, biosphere, lithosphere, and the deep interior of mantle and cores. All geophysical fluids undergo a host of mass transports for various reasons, external as well as internal. Studying these processes is undoubtedly one of the most interdisciplinary field in all of Earth sciences. However, mass transport has not received due attentions." Meeting the challenge of developing the geodetic observing system into a mass transport and dynamics observing system is a primary motivation for this book.

GGOS (the observing system) faces two types of scientific and technological challenges, namely an "internal" challenge and an "external" challenge (see Chapter 3). The "internal" challenge to geodesy is to develop GGOS and the geodetic technologies so that they meet the demanding user requirements in terms of reference frame accuracy and availability, as well as in terms of spatial and temporal resolution and accuracy of the geodetic observations. Developing an observing system capable of measuring variations in the Earth's shape, gravity field, and rotation with an accuracy and consistency of 0.1 to 1 ppb, with high spatial and temporal resolution, and increasingly low time latency, is a very demanding task. Accommodating the transition of new technologies as they evolve in parallel to maintaining an operational system is part of this challenge. The "external" challenge is associated with the integration of the "three pillars" into a system providing information on mass transport, surface deformations, and dynamics of the Earth. The Earth is a complex system with physical, chemical and biological processes interacting on spatial scales from micrometers to global and temporal scales from seconds to billions of years. Therefore, addressing the "external" challenge requires a "whole Earth" approach harnessing the expertise of all fields of Earth science.

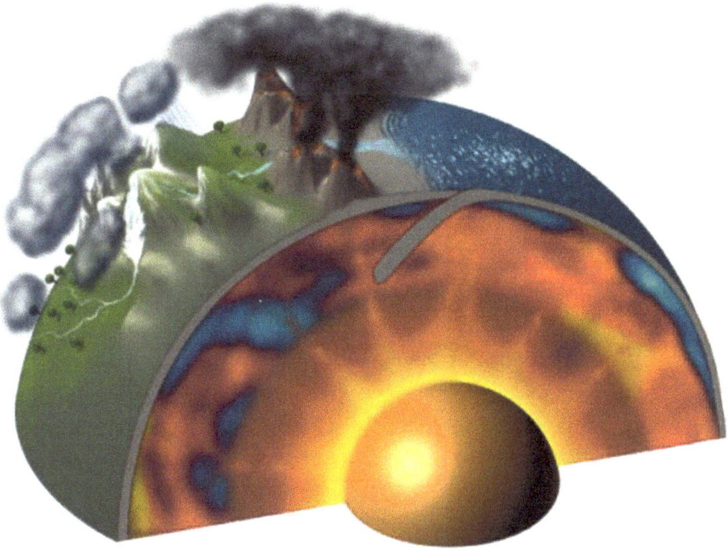

Fig. 1.3. The dynamic Earth. The interaction of solid Earth, hydrosphere, and atmosphere processes has created a highly complex system. From Solomon & the Solid Earth Science Working Group (2002).

The "internal" challenge provides GGOS with a central theme for research and development inside IAG. This book is a first step in sketching a roadmap for this central theme that will lead to a fully defined implementation plan.

The biggest challenge for geodesy, however, may arise from recent developments in global Earth observation. Stimulated by the international quest for sustainable development and the resulting demand for information on the current state and future evolution of the Earth system (GEO, 2005a), the need for comprehensive Earth observations is acknowledged in extensive programs of the United Nations, the European Union, and the international community, culminating in the establishment of the GEO at the EOS-III on 16 February 2005 in Brussels, Belgium. GEO has the task of implementing according to the Ten-Year Implementation Plan (TYIP) endorsed by EOS-III (GEO, 2005a, see also Section 5.1). This TYIP is likely to guide the development of global Earth observation programs over the next decade. The challenge is therefore to integrate GGOS as an organization into the context of Earth observation and society, and to develop GGOS as an observing system in accordance with the strategies and methodologies of the global observing systems for the mutual benefit of all. Earth observation and society at large will benefit from the availability of geodetic observations and products, and GGOS will benefit from an improved visibility and acknowledgment of the valuable service it provides.

In order to facilitate the integration of GGOS into GEOSS, IAG is a Participating Organization in GEO and is represented there by the GGOS organization. GGOS is also a contributing system to the GEOSS, which is implemented by GEO. GGOS was a Partner of Integrated Global Observing Strategy Partnership (IGOS-P)

(Plag et al., 2006a) and continues to contribute to several of the GEO Community of Practices (COPs) that developed out of IGOS-P Themes. Moreover, steps are being taken to strengthen joint initiatives with government organizations and international bodies. These initiatives will enhance the visibility of geodetic activities in the context of Earth sciences, Earth observation and practical applications (Plag, 2006b).

1.4 The strategy: where to go from here

Identifying the requirements for observations and products of geodesy for a wide range of scientific and societal applications is an important prerequisite for defining a set of functional specifications of a geodetic observing system that would be able to serve some or all these applications. Compiling a comprehensive set of URs for geodetic observations and products and deriving functional specifications for a global geodetic observing system is one of the two major goals of this book. The other goal is to specify, based on the functional requirements, the system design of a future GGOS and to define the steps towards the implementation of this GGOS.

In Chapter 2 we first give an introduction to the "ways and means of geodesy" in general, and global geodesy in particular. Emphasis is on the introduction of modern geodetic techniques and methods, but the achievements and current contributions are briefly reported. This Chapter sets the stage for what is currently available and achievable.

Chapters 3 to 6 review the requirements for geodetic observation, products, and services for scientific investigations, monitoring the Earth system, maintaining a modern society, and exploring the planets and the solar system, respectively. In Chapter 3, the open scientific questions concerning the solid Earth, atmosphere, hydrosphere, and cryosphere and their interactions are reviewed with emphasis on how geodetic observations could contribute to providing answers to these fundamental questions. Chapter 4 looks at the many activities in a modern society that depend on or benefit from geodetic observations and products, such as navigation, surveying, mapping, construction, process control, and outdoor activities, and discusses the requirements particularly in terms of access to coordinates in a well-defined and well-maintained reference frame. Chapter 5 starts with the requirements of the key societal benefit areas of Earth observation (see Table 5.1 in Chapter 5 on page 155) as identified by the EOS-II, listed in the Reference Document for the TYIP for GEOSS (GEO, 2005b). These essentially qualitative requirements are then further developed into a set of quantitative requirements. Geodesy is not only essential for many applications on Earth but it also provides the basis for studying and exploring the planets and the solar system. These requirements are addressed in Chapter 6.

In Chapter 7 the results of the previous chapters are used to compile a comprehensive set of quantitative requirements linking the different requirements to applications and users. Based on this set, functional specifications for an observing system are derived.

Two global geodetic references systems, one rotating with the solid Earth, and the other one fixed in space, are fundamental concepts for geodetic theories, models and observations, and their realization through corresponding reference frames is a key task of global geodetic activities. Both the reference systems and the frames are governed by conventions not only concerning the axes and origin of the reference system but also the observations, constants, analysis methods, and models used in their realization. Chapter 8 reviews the current approach and develops it further so that future reference frames will meet the requirements defined in the previous chapters.

Chapter 9 addresses the design of the GGOS that is able to meet the functional specifications set out in Chapter 7. In this Chapter, use is made of the full set of available techniques including a consideration of the ground-based, airborne and spaceborne components.

The current GGOS is based on the voluntary commitment of many contributing countries, organizations, institutions and individuals. This situation leads to fluctuations in available resources, and therefore requires a high degree of redundancy in order to ensure a sufficient geodetic infrastructure. This infrastructure is central to the provision of a reference frame meeting the requirements of both scientific and non-scientific applications as well as for the contribution to international programs and activities directed towards global Earth observation. Chapter 10 describes steps necessary for the implementation of the system defined in Chapter 9 taking into account the available infrastructure as well as the current organizational and funding situation. With respect to the organizational background, the Chapter considers alternative approaches, including an intergovernmental one.

Finally, Chapter 11 provides recommendations for the development of GGOS, the implementation of its proposed components, and its future organization. Recommendations are given for improving the framework conditions, the infrastructure, the products, and the organizational background for global geodesy as a multinational endeavor.

Chapter 2
The goals, achievements, and tools of modern geodesy

H.-P. Plag, Z. Altamimi, S. Bettadpur, G. Beutler, G. Beyerle, A. Cazenave, D. Crossley, A. Donnellan, R. Forsberg, R. Gross, J. Hinderer, A. Komjathy, C. Ma, A. J. Mannucci, C. Noll, A. Nothnagel, E. C. Pavlis, M. Pearlman, P. Poli, U. Schreiber, K. Senior, P. L. Woodworth, S. Zerbini, C. Zuffada

2.1 Introduction

Friedrich Robert Helmert (1843-1917) defined geodesy as the science "of measurements and mappings of the Earth's surface". Over time, this definition of geodesy has been extended, mainly as a consequence of technological developments allowing geodesy to observe the Earth on global scales with high accuracy. Today, geodesy is the science of determining the geometry, gravity field, and rotation of the Earth and their evolution in time. This understanding of modern geodesy has led to the definition of the "three pillars of geodesy", namely (1) Geokinematics, (2) Earth Rotation and (3) the Gravity Field (see Figure 1.1 on page 4). These three pillars are intrinsically linked to each other, and they jointly change as a consequence of dynamical processes in the Earth system as a whole. The changes in Earth's shape (including the surface of the water and ice bodies), i.e. the geokinematics, are the result of dynamic processes in the solid Earth and its fluid envelope, affecting mass distribution and angular momentum, and thus changing the gravity field and Earth rotation.

Traditionally, geodesy has been a service science, providing an important utility to other sciences and many applications. This aspect has remained unchanged, and a principal tool and output of geodesy is a reference frame allowing the determination of the position of points relative to each other. But geodesy has developed into a science that can no longer satisfy this service aspect without encompassing and monitoring the whole Earth system, its kinematic and dynamics. As an additional benefit, geodesy is increasingly forced not only to "measure" the geokinematics, gravity field, and rotation, but also to "model" these quantities on the basis of mass transport and dynamics.

The instruments (or measurement tools) are of crucial importance in geodesy. They in essence define the scope of the problems, which may be addressed by geodesy. Before the advent of the space age the geometrical aspects were studied mainly by measuring angles and time (time-tagging of the observations). In the best

H.-P. Plag, M. Pearlman (eds.), *Global Geodetic Observing System*,
DOI 10.1007/978-3-642-02687-4_2, © Springer-Verlag Berlin Heidelberg 2009

case, angles were measured with sub-arcsecond accuracy, and time with an accuracy of a few microseconds. The angles define a unit vector from the observer to the observed object (a terrestrial target or a celestial object such as stars, the Moon, etc.) at particular epochs. When observing celestial objects, the classical observation technique is called *astrometry*. For time measurement one made the distinction between the *astronomical clocks* (defined, for example, by Earth rotation or, alternatively, by the motion of the Moon and/or planets) and the man-made mechanical clocks. Accuracy and long-term stability of the astronomical clocks could never be reached by mechanical clocks. They were, however, of crucial importance for solving practical problems in navigation (like the problem of determining the longitude at sea or the longitude difference between sites on different continents) and, of course, for interpolating the astronomical time. Gravity was studied by measuring the zenith (actually nadir) direction (i.e., the unit vector along which gravity acts) in a well-defined geometric reference frame and/or by measuring the absolute value of the gravity vector. Both measurement types are heavily affected by the mass distribution in the environment of the measuring instruments, which makes the interpretation of their contribution to global gravity field determination problematic.

The advent of the space age (marked by the launch of the first artificial Earth satellite on October 4, 1957) together with the development of atomic clocks (first realized by crystal oscillators in the 1950s, then by atomic clocks like, for example, the hydrogen masers) to precisely measure epochs and time intervals initiated an extremely rapid development of novel observation techniques and, associated with that, scientific opportunities, which revolutionized the entire field of geodesy. It became in particular possible to

1. connect different continents by simultaneously observing high orbiting, bright satellites from sites located on different continents using astrometry;
2. measure distances through the measurement of the propagation time of short light pulses between an observatory on the Earth's surface and an artificial Earth satellite;
3. exploit the signals emitted by stable oscillators onboard navigation satellites and recorded by receivers on the Earth surface or in the near-Earth space to determine the time development of the distance between the satellite emitting the signal and the receiver(s) recording it;
4. correlate the signals emitted by Quasars (radio galaxies "at rest" in the inertial space) and received by two radio telescopes to establish the distance difference between the telescopes, as seen from the Quasars at the measurement epochs;
5. use the trajectories of artificial Earth satellites to determine the Earth's gravity field;
6. use atomic time to study the rotation of the Earth and the motion and rotation of other objects in our planetary system.

The first of the above items initiated the concept of modern terrestrial reference systems and frames, with the frames being the realization of the systems. Items 2-4 represent "new" observation techniques, which in essence ruled out astrometry and replaced it by the measurement of distances or distance differences. A somewhat simplistic order of magnitude calculation shows that this step resulted in a gain in accuracy of about 2-3 orders of magnitude: A typical error of 0.1"-1.0" in the astrometric position implies a tangential error at a typical distance of 1000 km to a satellite of about 0.5-5 m, whereas the new observation techniques typically measure distances with accuracies of about 1-5 mm. This gain gives access to a whole suite of new problems, which can now be addressed by modern geodesy. Items 1 to 3 are so-called satellite-geodetic techniques. Items 1 to 4 are also referred to as space-geodetic techniques.

Item 5 allows us to study the Earth's global gravity field in detail. By modeling the satellite orbits as solutions of the equations of motion, which contain the parameters describing the Earth's gravity field, and by using the satellite geodetic observations in particular of the Laser satellites, and, more recently, of LEOs equipped with GNSS receivers, it became possible to determine the Earth's global gravity field already before the end of the 20th century in astonishing detail. A quantum jump in accuracy and resolution is being achieved with the suite of dedicated space missions Challenging Minisatellite Payload (CHAMP), GRACE, and, most recently, GOCE, which were deployed in the first decade of the 21st century.

Item 6 marks an important change of paradigm in geodesy and fundamental astronomy: Instead of using the Earth rotation and lunar/planetary motion to define and realize time, in particular Universal Time (UT), it is now possible to study the Earth's rotation and planetary motion as a function of atomic time (or Coordinated Universal Time (UTC), which is today derived uniquely from atomic time). This aspect is of particular importance for the problems associated with the second pillar of modern geodesy, namely the study and monitoring of Earth rotation.

In accordance with the development of measuring techniques, the concepts of the Earth system, including the solid Earth, changed, influenced by both geodetic observations and a better understanding of the Earth system and its main processes. For a long time, geodetic concepts were based on a static view of the solid Earth, and terrestrial reference frames were based on fixed coordinates of points on the Earth's surface. Over the last five decades, the development in our understanding of the solid Earth and the total Earth system has made it clear that the solid Earth's surface undergoes continuous deformations, changing the relative position of all points on a wide range of time scales. The invention and rapid improvement of the space-geodetic technologies have provided a wealth of observations documenting the surface deformations, irregularities in the Earth's movement in space and the extent of mass movements in the Earth's system. At the same time, scientific and societal applications pose increasing requirements on the accuracy and reliability of positioning as well as navigation. A detailed review of requirements for geodetic observations and products in Earth observations, scientific studies, and societal applications (see Chapters 3 to 7) demonstrates that in terms of precise point positioning, the requirements in terms of accuracy are on the order of centimeter for real-time or low-latency application, 1 cm or better on daily time scales, a few mm/yr for intraseasonal time scales, and of the order of 0.1 mm/yr on interannual to secular time scales. Thus, relative to the size of the Earth, a general accuracy requirement for geodetic observations and products of the order of 10^{-9} or less can be stated.

Considering the characteristics of the spatio-temporal variations in Earth's shape, rotation and gravity field, the task to establish a reference frame with an accuracy at or below the 1 ppb level is a demanding and scientifically challenging endeavor. The Earth's surface is constantly deformed by internal and external processes including earthquakes, Earth tides, surface loading (present and past) caused by the atmosphere, hydrosphere, and cryosphere, sediment loading, and mantle convection. All these processes have to be accounted for at a level well below the targeted accuracy of 1 ppb. This requires geodesy to interact with other geosciences and to take an Earth system approach, which considers the effects of external forcing, atmosphere, ocean, terrestrial hydrosphere, and cryosphere on the solid Earth. Consequently, the realization and maintenance of reliable reference frames on local, regional and global scales as well as the provision of techniques for high-precision positioning has received growing attention within geodesy and in Earth science in general.

Considering the importance of the geodetic reference frames, in Section 2.2, the concepts for reference systems and their realization through reference frames is introduced and the two main geodetic reference systems are described. This sets the stage for a more detailed discussion of the "three pillars of geodesy" and their interrelations in Section 2.3 and the current state-of-art of the observing system in each of the three pillars in the Sections 2.4 to 2.6, respectively. Subsequently, Section 2.7 addresses a central issue for geodesy, that is access to accurate time. Section 2.8 briefly describes measures taken to ensure consistency between the geodetic observations and identifies key open questions. Finally, Section 2.9 introduces auxiliary applications of geodetic and related observations, which increasingly are developed adding a multi-application aspect to geodesy and opening new fields of research.

2.2 Geodetic reference systems and frames

As pointed out earlier in this book, a principal goal of geodesy is to provide the means to assign coordinates to points as a function of time. Position and movement are not absolute quantities and depend on the reference frame to which they are referred. In particular, observations of any celestial body, be it natural or artificial, or of a point in the Earth system, can be used to describe the motion of this body only if the observations can be referred to a well defined coordinate system. In an ideal world, such a system could be defined through three coordinate axes, the origin, and a scale, with the axis either being fixed in space or having a known movement with respect to something else that is fixed. In the real world, the provision of an accessible coordinate system requires far more definition, which comprises a reference system.

In the context of space geodesy, making use of natural and artificial celestial objects, there is a need for both the Celestial Reference System (CRS) and a Terrestrial Reference System (TRS). The CRS, which is fixed in space, is required to describe the motions of galaxies, stars, the sun, planets including the Earth itself, the Moon and the satellites of other planets, and artificial satellites. Observations of points on the Earth's surface or related to the Earth's surface are often easier to relate to movements of these points if they are referred to a TRS with axes fixed in some way to the solid Earth and moving with the Earth in space.

It is of obvious practical advantage to agree upon one definition for each of the celestial and terrestrial reference systems. This has led to the adoption of conventional celestial and terrestrial reference systems (CCRS and CTRS, respectively). A conventional reference system includes the specification of the origin, the direction of the axes (orientation in space), and the scale of the system in an appropriate way. However, more is needed in order to complete the system such as conventions (see Box 1 for an definition of geodetic conventions) on physical constants.

Box 1: Geodetic conventions and standards.

Convention In the context of this report, convention refers to an agreement between groups, especially an international agreement, that is slightly less formal than a treaty. Conventions are for example the agreed-upon way to transform from inertial frame to terrestrial frame, splitting three unique angles that connect the two frames to a set of conventionally defined sub-group of angles (polar motion, Earth rotation, nutation and precession). In geodesy, conventions often regulate ways to process data in order to ensure comparability of the resulting products. In many cases, standards (see below) adopted by e.g. IAG/IUGG become part of conventions. For example, standards that define the "refractivity" of the atmosphere at various wavelengths adopted by IAG/IUGG are later used in "conventional" approaches of the determination of the propagation delays through the atmosphere.

Standard In geodesy, a standard refers to an authorized model (normally authorized by IAG or IUGG or other international bodies recognized by IAG/IUGG) used to define a unit of measurement. Examples of standards are the definition of the meter, the speed of light, and similar physical constants.

Another issue is to gain access to such a reference system. Modern conventional celestial and terrestrial reference systems in fact are realized through coordinates of a set of points and objects determined from observations analyzed with appropriate mathematical and physical models. Such a realization of a reference system is denoted as reference frame. In practice, the realization of a reference system through such a frame requires continuous monitoring of the points or objects. Given the nature of the problem, any realization also requires the specification of additional boundary conditions that the reference frame should fulfill. Moreover, models used to analyze the observations and to correct for disturbances in the coordinates of the points and objects are an integral part of the realization, and therefore have to be included in the convention specifying the reference system and its realization through a frame.

It is not always clear whether the boundary conditions and models are considered as part of the conventional reference system, part of the reference frame realizing the system or the subject of an additional convention. There is certainly a trade-off between the completeness of the conventions specifying the reference system and the need to change the reference system when models or constants improve.

Figure 2.1 gives an overview illustrating the conventional reference systems and their realizations presently adopted by the relevant international scientific unions. The two fundamental systems accepted by the relevant international scientific bodies are the ICRS and the ITRS, which are realized by IERS through the ICRF and the

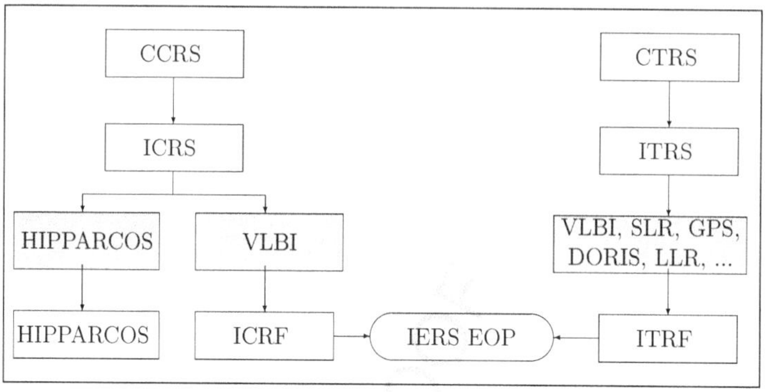

Fig. 2.1. Overview of current conventional reference systems and their realizations. The current Conventional Celestial Reference System (CCRS) adopted by the IAU is the ICRS. In the radio-wavelength, this system is realized as ICRF through VLBI measurements of extragalactical objects and as such maintained by the IERS. At optical wavelengths, the observations made with the HIPPARCOS satellite allowed the materialization of the ICRS through the HIPPARCOS stellar frame.The tie between the HIPPARCOS and the ICRS is determined to a high degree of accuracy (Kovalevsky, 1997). The current Conventional Terrestrial Reference System (CTRS) accepted by IUGG is the ITRS, which is realized through the ITRFmaintained by the IERS. The tie between the ICRF and ITRF is provided by the IERS' EOP. These describe the orientation of the Celestial Ephemeris Pole (CEP) in the terrestrial and celestial systems through the polar coordinates x and y and the nutation offsets $d\psi$ and $d\varepsilon$, respectively, and the orientation of the Earth around this axis through UT1-TAI as function of time. From Plag (2006a).

ITRF, respectively. The IERS is a service under the joint auspice of IAG and IAU, and for the ICRF, both organizations take responsibility.

These two frames are linked to each other through the Earth rotation. Today, IERS provides parameters related to Earth's rotation under the name of EOPs.

The ICRS is defined and maintained by the IERS. It was adopted by the IAU and the IUGG as the primary celestial reference system, replacing its optical predecessors based on fundamental star catalogs (see Box 2). The observation and analysis aspects related to the realization of the ICRS through the ICRF are today coordinated by the International VLBI Service for Geodesy and Astrometry (IVS).

The ITRS is also defined and maintained by the IERS. It is adopted by IAG and IUGG as the primary terrestrial reference system, in particular, for Earth science applications. Unlike the ICRS, the realization of the ITRS through the ITRF is based on a combination of results from several space geodetic techniques, and local survey measurements between reference points of geodetic instruments (so-called local ties) co-located at the same sites. The combination is coordinated by the IERS, while the observational aspects for each individual technique involved are coordinated by technique-specific Services. Co-location sites (where two or more instruments are operating in close vicinity), are key elements in the ITRF combinations. While any individual space geodesy technique (VLBI, SLR, DORIS, GNSS) is able to provide necessary information for the ITRF, only the combination of the independent tech-

Box 2: The ICRF

The ensemble of distant extragalactic objects constitutes a quasi-inertial reference frame in which the motion and orientation of the Earth can be measured. In practice, this frame is accessed from the Earth through VLBI observations of compact radio sources, for the most part quasars. The red shifts of these quasars are large enough that their physical transverse movement cannot be detected by current radio or optical techniques, and the objects can be treated conceptually as fixed points in the sky. The International Astronomical Union (IAU) recognized the utility and accuracy of the extragalactic celestial reference frame by adopting the ICRF effective 1 January 1998. ICRF-Extension.2 is a catalog of some 700 radio source positions (Fey et al., 2004, see also Figure 9.5). The positions and errors of the 212 "defining" sources of the ICRF define (realize) the axes and precision of the ICRS on which all celestial positions are now placed. While the right ascension origin and pole of the ICRF are consistent with the previous FK5 stellar frame within the much larger errors of FK5, the concept of the ICRS/ICRF is fundamentally different in several respects. The defining objects of the ICRF have no real proper motions, and the axes of the ICRS are decoupled from the equator, the ecliptic and any particular epoch.

The quasars and other compact radio sources that are included in the ICRF have point-like optical images. Their red shifts indicate great distances so their emissions must be powered by processes different from stars and galaxies, most probably mass inflow onto massive black holes. At the resolution of geodetic/astrometric VLBI using S-band (2 GHz) and X-band (8 GHz), the objects are generally not point-like but have some structure that can also change with time. Such structure changes can be seen as changes in position up to 1 milliarcsecond. The brightest extragalactic radio sources in fact have too much detectable structure to be good astrometric objects. By balancing the competing criteria of source strength, compactness and constancy of structure and position, a set of $\sim$100 geodetic sources has been selected for routine geodetic VLBI observations while the remainder of the ICRF improves the distribution and density over the sky (see Figure 9.5). It should be noted that the small number of VLBI stations in the Southern Hemisphere causes the ICRF to be weaker in all aspects in the southern sky. The quasars in the ICRF emit relatively strongly at microwave frequencies while the great majority of quasars are much weaker or radio-quiet.

The ICRF now constitutes the fundamental celestial frame for all astrometric and geodetic purposes. This includes both planetary ephemerides and satellite orbits. The former have been related to the ICRF by specialized VLBI observations of transmitters on planets and spacecraft as well as from locations of VLBI stations. Satellite orbit determination requires accurate measurements of the actual rotation angle of the Earth UT1-UTC as a priori information since the rotation of the orbit nodes cannot be modeled over a long period. VLBI observations of GNSS satellites should be feasible in the future as the observing bandwidth for geodetic VLBI is extended. Such observations would directly connect the satellite frames to the ICRF. The motion of the Earth's axis in space, precession and nutation, is also observed using the ICRF. These measurements provide information about the structure of the Earth as it responds to the torques of the Sun, Moon and planets.

The ICRF is essential to geodesy as it is the frame for measuring EOP and the ultimate frame for satellite orbits. The ICRF is also the basis for astrometry. In this regard the ICRF has different realizations at various wavelengths, the microwave VLBI realization being the most accurate at this time. The astrometric satellite GAIA is scheduled for launch in late 2011 and has the potential for generating an optical extragalactic realization with an order of magnitude better precision and two orders of magnitude more objects. Other space missions may refine the positions and proper motions of the brightest stars with corresponding improvement of star tracking for satellite orientation. For most geodetic purposes, however, these improvements will not be applicable since no correspondingly precise ground-based observing system exists. An accurate microwave realization for geodetic VLBI will still be needed.

niques allows for the complete determination of ITRF (origin, scale and orientation). In principle, the particular strengths of one observing method can compensate for weaknesses in others if the combination is properly constructed, suitable weights are found, and accurate local ties in co-location sites are available.

The conventions for both the ITRS and ICRS and their realizations are detailed in the IERS Conventions (e.g., McCarthy & Petit, 2004). As accuracy requirements evolve and technical and modeling capabilities increase, these conventions are modified and developed under the auspice of IERS in a continuous process with support from the broad geodetic science community.

In the conventions, the motion of the reference points in ITRF currently is described by a linear model, thus reducing the information necessary to determine the motion of the reference points relative to their coordinates at a reference epoch and a constant velocity. This representation is no longer appropriate to accommodate possible future user requirements to have access to the actual instantaneous point position over the Earth surface and new representation and models are being discussed (see Chapter 8).

The coordinates and constant velocities of the points that define a particular reference frame depend on the points, techniques, models, and analysis tools used in the determination of these quantities. Therefore, for any given reference system, there can be a multitude of reference frames realizing the system at various degrees of accuracy. For global terrestrial reference frames, the ITRS is increasingly used as the underlying system, thus gaining importance for practical applications. For example, the U.S. Government and the European Commission agreed to align the reference frames of the Global Positioning System (GPS) and GALILEO as close as possible to ITRS (European Commission, 2004). In practice, this goal is achieved by aligning the GNSS reference frames to the ITRF, which is the most accurate realization of ITRS. The reference frame of the positioning services provided by GPS, is the most recent realization of the World Geodetic System 1984 (WGS 84) (e.g., Assistant Secretary of Defence for Command, Control, Communication, and Intelligence, 2001). As a consequence, this realization of WGS84 is today closely aligned to ITRF and in fact supported by ITRF.

ITRF is currently the most accurate realization of ITRS (Altamimi et al., 2002, 2007). The ITRF is updated regularly with the most recent versions being ITRF97, ITRF2000, and ITRF2005. In geodetic analyses of observations of different groups using different techniques and different software packages, coordinates agree to the centimeter level. Secular trends determined from long GPS records using different analysis approaches may disagree on the order of 1 to 2 mm/yr, but most of these discrepancies are due to the approach used to align the solution to ITRF. A significant bias may result from a potential secular translation of the Reference Frame Origin (RFO) with respect to the Center of Mass of the whole Earth system (CM). Recent studies estimate the bias to be of order ± 2 mm/yr (e.g., Ray et al., 2004; Morel & Willis, 2005; Plag, 2006b; Plag et al., 2007a), depending on the geographical location.

The translation of the RFO with respect to the CM introduces particularly large uncertainties in sea level studies. Taking the effect on vertical velocities of the secular translation between ITRF2000 and ITRF2005 (Figure 2.2) as an indication of the uncertainty in the tie of the RFO to the CM, the effect on global sea level trend estimates is of the order 0.2 to 0.3 mm/yr. Consequently, not only maintenance but also improvement of the ITRF as the essential architecture for almost all geodetic measurements is a crucial requirement for sea level studies.

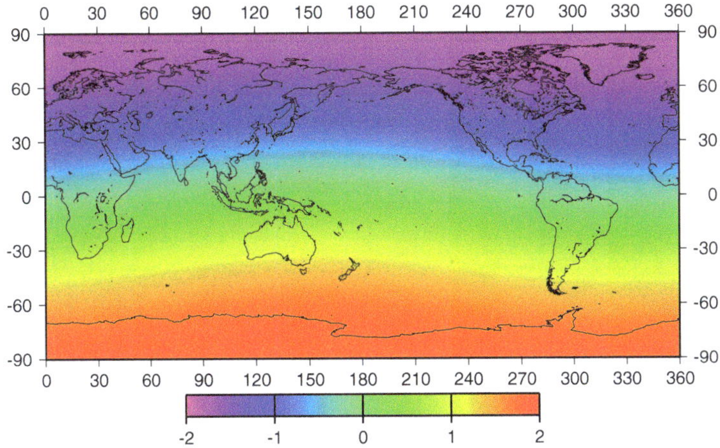

Fig. 2.2. Effect of secular translation between ITRF2000 and ITRF2005 on vertical rates. The vertical rates are for a secular translation velocity of $\mathbf{d} = (-0.2, 0.1, -1.8)$ mm/yr as given on http://itrf.ensg.ign.fr/ITRF_solutions/2005/.

2.3 The tools and products of modern geodesy

Today, the toolbox of geodesy comprises a number of space-geodetic and terrestrial techniques, which together allow for detailed observations of the "three pillars of geodesy" on a wide range of spatial and temporal scales (Figure 2.3). With a mix of terrestrial, airborne, and spaceborne techniques, geodesy today determines and monitors changes in Earth's shape, gravitational field and rotation with unprecedented accuracy, resolution (temporal as well as spatial), and long-term stability (Table 2.1). At the same time, geodetic observation technologies are in constant development with new technologies extending the observation capabilities almost continuously in terms of accuracy, spatial and temporal coverage and resolution, parameters observed, latency and quality. Together, these observations provide the basis to determine and monitor the ITRF and ICRF as the metrological basis for all Earth observations. Equally important, the observations themselves are directly related to mass transport and dynamics in the Earth system. Thus, the geodetic measurements form the basis for Earth system observations in the true meaning of these words. Beutler et al. (1999) suggested a development towards an interdisciplinary service in support of Earth sciences for the IGS. With the establishment of GGOS,

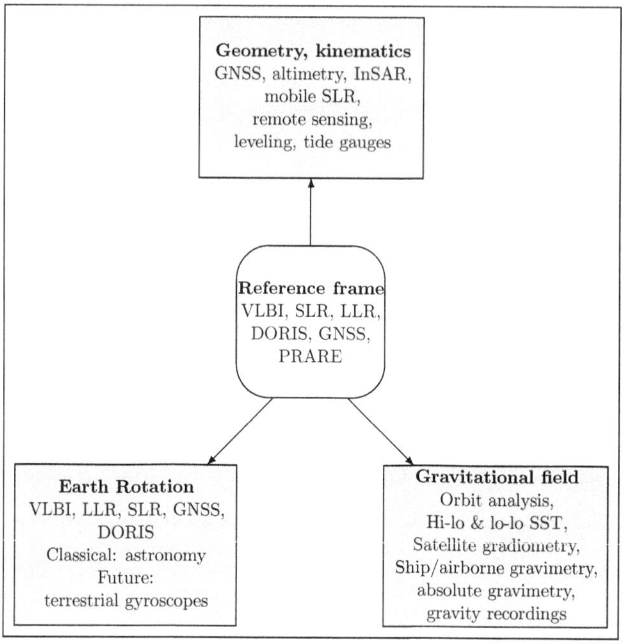

Fig. 2.3. The "three pillars of geodesy" and their techniques. Today, the space-geodetic techniques and dedicated satellite missions are crucial in the determination and monitoring of geokinematics, Earth's rotation and the gravity field. Together, these observations provide the basis to determine the geodetic reference frames with high accuracy, spatial resolution and temporal stability. From Plag (2006a), modified from Rummel (2000). For acronyms, see the list in Appendix 11.

IAG has extended this concept of an observing system and service for Earth system sciences to the whole of geodesy.

From the discussion of the reference systems and frames in the previous section it is obvious that there is an intimate relationship between the three pillars of geodesy and the reference systems and frames (Figure 2.3). For geokinematics and Earth rotation, the relationship works both ways: The reference systems are required for positioning purposes (terrestrial and celestial) and for studying Earth rotation, and monitoring through the space geodetic techniques is necessary to realize the two frames and the (time-dependent) transformation between them.

The ICRF, the ITRF, and the EOPs are needed to derive a gravity field, which is consistent with the ICRF, the ITRF, and the corresponding EOPs. Therefore, one might think at first that the gravity field is not necessary to define and realize the geometric reference systems. However, in order to realize the ITRF, observations made by the satellite geodetic techniques (SLR, GNSS, DORIS) are needed. For these techniques, a gravitational reference system and frame (including a gravity field representation and the parameters associated with it, and the geoid, the mean equipotential surface "near sea level", which may be derived from the gravity field representation) is required as well and cannot be separately determined from the geometrical frames. The problems are obviously inseparable when dealing with the

Table 2.1. The Global Geodetic Observing System (GGOS). For acronyms, see the list in Appendix 11.

Component	Objective	Techniques	Responsible
I. Geokinematics (size, shape, kinematics, deformation)	Shape and temporal variations of land/ice/ocean surface (plates, intra-plates, volcanos, earthquakes, glaciers, ocean variability, sea level)	Altimetry, InSAR, GNSS-cluster, VLBI, SLR, DORIS, imaging techniques, leveling, tide gauges	International and national projects, space missions, IGS, IAS, future InSAR service
II. Earth Rotation (nutation, precession, polar motion, variations in LOD)	Integrated effect of changes in angular momentum and moment of inertia tensor (mass changes in atmosphere, cryosphere, oceans, solid Earth, core/mantle; momentum exchange between Earth system components)	Classical astronomy, VLBI, LLR, SLR, GNSS, DORIS, under development: terrestrial gyroscopes	International geodetic and astronomical community (IERS, IGS, IVS, ILRS, IDS)
III. Gravity field	Geoid, Earth's static gravitational potential, temporal variations induced by solid Earth processes and mass transport in the global water cycle.	Terrestrial gravimetry (absolute and relative), airborne gravimetry, satellite orbits, dedicated satellite missions (CHAMP, GRACE, GOCE)	International geophysical and geodetic community (GGP, IGFS, IGeS, BGI)
IV. Terrestrial Frame	Global cluster of fiducial point, determined at mm to cm level	VLBI, GNSS, SLR, LLR, DORIS, time keeping/transfer, absolute gravimetry, gravity recording	International geodetic community (IERS with support of IVS, ILRS, IGS, and IDS)

definition in the geometry and gravity domains (origin, orientation, scale of the geometric networks, low degree and order terms of the Earth's gravity field).

This consistency between geometric and gravitational products is important today, it will be of greatest relevance in the future for the understanding of the mass transport and the exchange of angular momentum between the Earth's constituents, in particular between solid Earth, atmosphere, and oceans. The aspect of consistency is also of greatest importance for all studies related to global change, sea level variation, and to the monitoring of ocean currents. Only if consistency on the 10^{-9} level or better between all reference frames it achieved, will it be possible to perform meaningful research in the areas mentioned.

In the narrowest possible sense, geodesy has the tasks to define the geometric and gravitational reference systems, and to establish the celestial, terrestrial, and gravitational reference frames. Moreover geodesy has to provide the transformation between the terrestrial and celestial reference frames. These key tasks would be relatively simple to accomplish on a rigid Earth without hydrosphere and atmosphere. However, in the real Earth environment already the definition of the terrestrial and gravitational reference systems is a challenge. The corresponding reference frames can only be established by permanent monitoring based on a polyhedron of terrestrial geodetic observing sites, and of space missions.

This ambitious and expensive geodetic monitoring is necessary and its result, properly time-tagged and mutually consistent, is a stringent requirement in a broad field of scientific and societal applications. There is strong science justification for these geodetic products as a prerequisite (see Chapter 3). Also, some tasks of societal relevance may only be addressed if this permanent geodetic monitoring is available (see Chapter 4), and monitoring of the Earth system, including for example sea level and ice sheet variations, would not be possible without it (see Chapter 5).

The following three sections give an overview of the current status of the global geodetic observing system relevant to the three pillars. Many (but not all) items or activities, which will be mentioned in these section below, are coordinated by entities working under the auspices of IAG. IAG has been in the "monitoring business" since the late 19th century, when the International Latitude Service (ILS) was created to monitor polar motion. More recently the IAG created technique-specific Services to coordinate observation and analysis for the new space geodetic techniques. Also, on the level of IUGG and IAU the IERS was given the charter mentioned above and is coordinating related activities. These Services, which will be mentioned below, are important building blocks of the GGOS.

2.4 Observing Earth geometry and kinematic

2.4.1 Overview

Changes in the Earth's shape are measured with a mix of ground- and space-based techniques. These techniques can be separated into two broad classes:

(1) space-geodetic tracking techniques that monitor the deformation of a polyhedron (points) defined by ground-based networks of tracking stations which either passively utilize signals from satellites (GNSS) or stellar objects (VBLI) or actively send out signals to satellites (SLR and DORIS); and

(2) air- or spaceborne remote sensing techniques that send signals from airplanes or satellites to the Earth's surface and utilize the reflections to map the surface.

The space-geodetic tracking methods provide time series of point movements with high temporal resolution and high accuracy. Tracking stations are normally placed on the land surface. Remote-sensing techniques in general have much lower temporal resolution but provide information with potentially high spatial resolution and much better coverage, including the surface of oceans, lakes, and ice sheets. Beside altimeters, the remote-sensing techniques also include the imaging techniques (such as InSAR and LIDAR), which provide high-resolution images of a surface and its temporal changes.

2.4.2 *Space-geodetic tracking techniques*

The geometric space-geodetic techniques in general consist of a ground-based component of fixed stations from which the motion of satellites or astronomical objects (moon, quasars) are tracked with electromagnetic waves (including visible light). These stations can be passive in the sense that they do not emit signals but "only" receive signals from remote sources (GNSS, VLBI) or active (SLR, LLR, DORIS).

Common to all these methods is that the data analysis requires a good *a priori* station motion model describing in particular any variation with periods shorter than the analysis interval (for example, 1 day). For the methods based on range measurements, effects on the satellite also need to be accommodated. Coordinate time series resulting from space-geodetic analyses therefore are generally residuals with respect to the station motion model and other modeled effects.

VLBI: Very Long Baseline Interferometry is a space-geodetic technique based on radio astronomy and developed in the 1970s. A radio interferometer consists of a pair of directional antennas (radio telescopes) receiving radio signals from sources in a targeted radio frequency band. The signals from the two receivers are cross-correlated (multiplied and accumulated) to produce a cross-correlation "fringe pattern".

VLBI uses networks of radio antennas typically 20-30 meters in diameter (Figure 2.4) to observe radio signals from extragalactic objects (quasars). Quasars are at such great distances from Earth that they provide fixed points in the sky. Their transverse physical motion cannot be detected with any existing observing system. A radio signal from a quasar passing a VLBI station is received and recorded digitally with very precise time provided by a hydrogen maser. The same signal will travel an additional distance $c\tau$ before arriving at the second VLBI station, where c is the speed of light and τ is the time difference of the signal arriving at the first and second station (Figure 2.5). The distance $c\tau$ depends on the length of the baseline between the stations and its orientation with respect to the direction to the quasar (e.g., Lambeck, 1988; Robertson, 1991; Sovers et al., 1998). The time delay between the arrival times at the two stations can be determined with a precision equivalent to a few millimeters using purpose-built hardware correlators.

The global network of 40 VLBI stations (see Figure 2.6) is today coordinated by the IVS. A typical VLBI session currently includes eight stations observing about 60 quasars several times in a time period of 24 hours. Overlapping networks and network sessions of up to 20 stations connect the 40 global VLBI stations. The time delays from each baseline in the network are used to estimate the station positions with precision of < 1 cm, and the (relative) station velocities can be measured by observations over several years.

Currently, VLBI is the only non-satellite geodetic technique contributing to the IERS. Its unique and fundamental contributions to geodesy and astronomy are (1) the ICRF, the realization of the ICRS, (2) UT1-UTC (apart from leap seconds the difference of universal time as realized by Earth rotation and the atomic time, respectively, see Section 2.7), (3) long-term stability of nutation, and (4) the scale

Fig. 2.4. 32-meter VLBI antenna in Tsukuba, Japan.

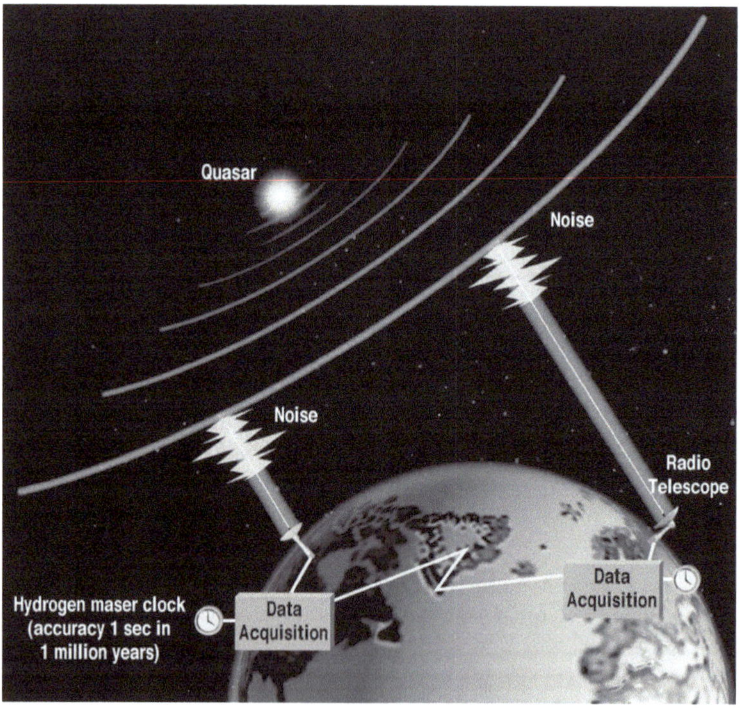

Fig. 2.5. Principle of very long baseline interferometry.

of the ITRF (together with SLR). VLBI also contributes with station positions and velocities to the establishment of the ITRF and is able to provide several further geodynamical, astronomical, or meteorological parameters.

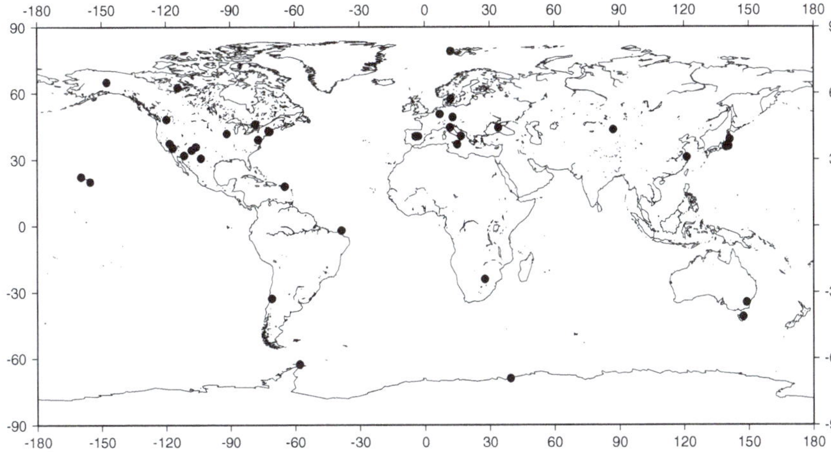

Fig. 2.6. Locations of the 40 VLBI stations in the global network of the IVS.

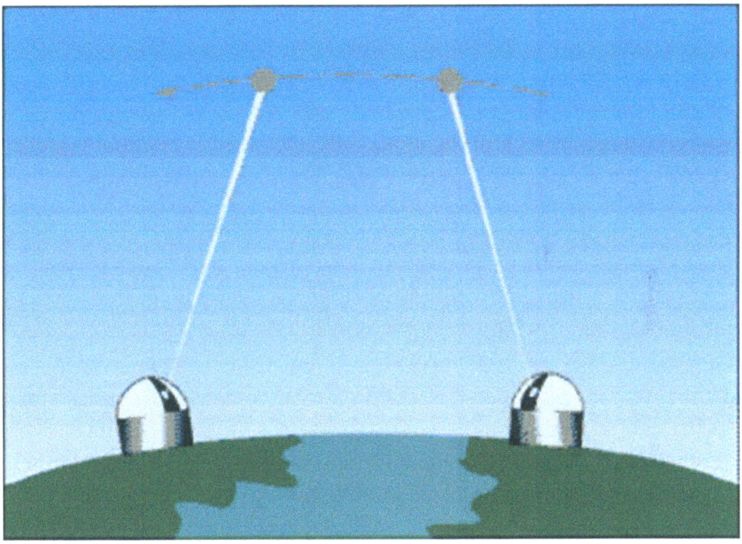

Fig. 2.7. Principle of satellite laser ranging.

SLR and LLR: SLR and LLR use very short laser pulses and fast electronics to measure the instantaneous round trip time of flight of the pulses to satellites equipped with special retroreflectors (Figure 2.7) and to retroreflectors on the Moon (Figure 2.9), respectively. This provides range measurements of a few millimeter precision which are accumulated to help define the terrestrial reference frame and to support Precision Orbit Determination (POD) for active spaceborne Earth sensing missions and studies of lunar science and fundamental physics. The fundamental targets for the reference frame are the LAser GEOdynamics Satellite (LAGEOS)-1 and -2 (Figure 2.8), whose spherical shape and high mass-to-area ratios provide long-term orbital stability for measuring the dynamics of the Earth.

The basic range measurement is sensitive to any geophysical process that changes the distance between the satellite and the observing station, such as displacements of the satellite due to perturbations of the Earth's gravitational field, motions of the observing station due to tidal displacements or plate tectonics, or a change in the orientation of the Earth (which changes the location of the observing station with respect to the satellite). These and other geophysical processes must be modeled when fitting the satellite's orbit to the range measurements as obtained at a number of globally distributed tracking stations. Adjustments to the *a priori* models used for these effects can then be obtained during the orbit determination procedure, thereby enabling, for example, the determination of station positions and Earth orientation parameters (Smith et al., 1985, 1990, 1994; Tapley et al., 1985, 1993).

The technique of LLR is similar to that of SLR except that the laser retro-reflector is located on the Moon instead of on an artificial satellite (Mulholland, 1980; Lambeck, 1988; Williams et al., 1993; Dickey et al., 1994; Shelus, 2001). LLR is technically more challenging than SLR because of the need to detect the much weaker signal that is returned from the Moon. Larger, more powerful laser systems with more sophisticated signal detection systems need to be employed in LLR; consequently, there are far fewer stations that range to the Moon than range to artificial satellites (see Figure 2.10).

The international network of about 40 SLR and two LLR stations (Figure 2.10) currently tracks on a daily basis about 30 satellites ranging in altitude from 400 km to 22,000 km and four retroreflectors on the Moon. As a Service of the IAG and a component of GGOS, laser ranging activities are coordinated by the International Laser Ranging Service (ILRS), which develops standards and specifications necessary for product consistency, sets priorities and tracking strategies, oversees data operations, and provides quality control and a user interface.

These laser ranging activities support programs primarily in geodetic, geophysical, and lunar research activities. The ILRS currently provides the IERS with weekly solutions for station coordinates and EOPs for the monitoring of the ITRF, contributing exclusively the definition and time-varying motion of its origin (with respect to the CM), and in combination with VLBI, its scale. Other contributions include the estimation of static and time-varying components (harmonic coefficients) of Earth's gravity field; accurate satellite ephemerides for POD and validation of altimetry (for satellites such as ICESat, shown in Figure 2.11), relativistic and satellite dynamics tests; and Lunar ephemeris for relativity studies and lunar libration for lunar interior studies. SLR, as a backup system, has also provided POD for missions whose primary tracking systems failed (e.g., GFO-1, ERS-1, Meteor-3M, etc.). Prior to the launch of CHAMP in the 2000, knowledge of the Earth's gravity field was almost uniquely based on SLR and terrestrial gravity measurements. SLR is an essential calibration technique for the GNSS technique and for the new space missions

Fig. 2.8. The LAGEOS-1 satellite (identical to LAGEOS-2). Dedicated laser ranging satellites have a long-term orbital stability because of their spherical shape and high mass-to-area ratio.

Fig. 2.9. Laser reflector on the Moon (image courtesy of NASA).

CHAMP, GRACE, and GOCE. The ILRS is now preparing to support space missions to the planetary system with optical transponders.

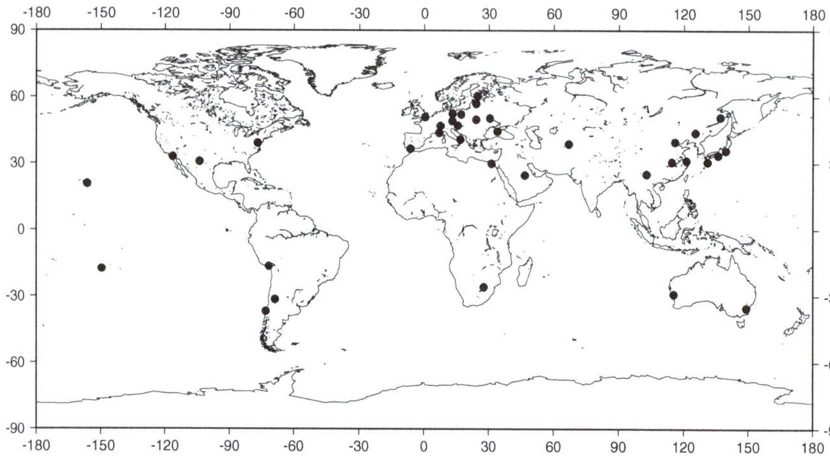

Fig. 2.10. Locations of 38 SLR stations in the tracking network of the ILRS.

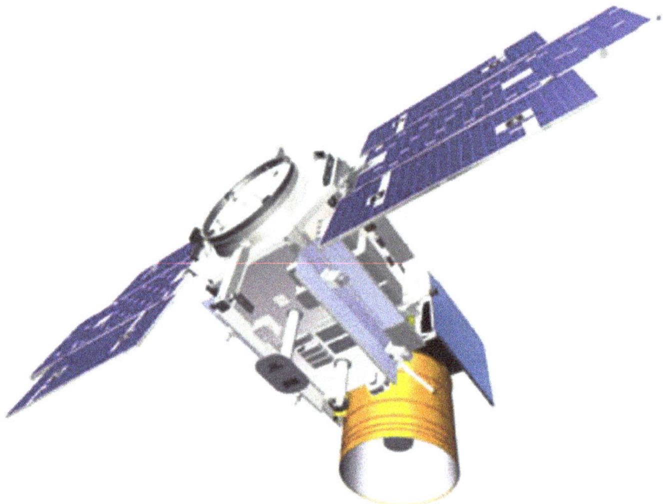

Fig. 2.11. ICESat Satellite (image courtesy of ICESat Science Team).

GNSS: Today's GNSS are the successors of the so-called Doppler systems. They are based (1) on about 30 to 45 satellites emitting microwave signals on at least two carriers, and (2) an unlimited number of receivers capable of tracking the signals of all satellites simultaneously in view (usually between 4 and 12). Today's GNSSs occupy the so-called Medium Earth Orbit (MEO)-belt. The satellites orbit the Earth in heights around 20000 km and complete one revolution within approximately half a day. The U.S. GPS with nominally 24 satellites (see Figure 2.12), uniformly distributed in six orbital planes, which are in turn inclined by 55° with respect to and separated by 60° in the equator, is the best known and most widely used GNSS. The Russian Global Navigation Satellite System (GLONASS) with nominally 24 satel-

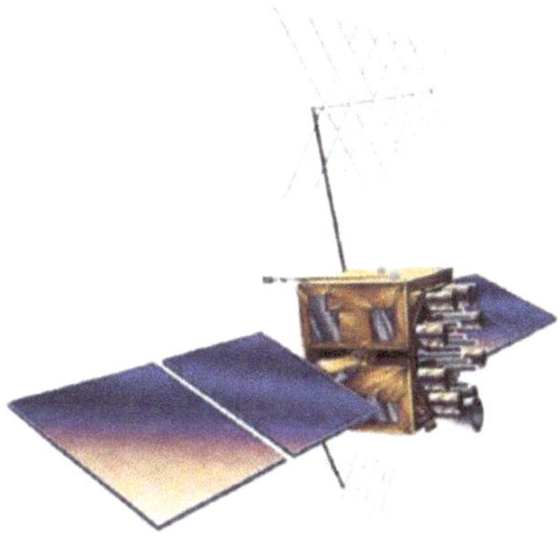

Fig. 2.12. GPS satellite. From http://www.af.mil/shared/media/factsheet/gps.jpg.

lites (Figure 2.13) in three orbital planes inclined by 63 degrees with respect to the equator and separated by 120 degrees in the equator, is currently not fully available (in January 2007 only nine satellites were fully operational). The plan is to achieve Full Operational Capability (FOC) by 2010. The first experimental satellite of the European GALILEO system (GIOVE-A, Figure 2.15) was launched on December 28, 2005 and early in May 2007, this satellite successfully transmitted its first navigation message, containing the information needed by user receivers to calculate their position. GALILEO is planned to reach FOC in 2012. By then, this GNSS is projected to have 30 satellites positioned in three circular MEO planes (Figure 2.15).

Fig. 2.13. Illustration of GLONASS satellite.

The microwave band (the L-band) of the electromagnetic spectrum allows for the weather-independent use of the systems, the two carriers for the elimination of

Fig. 2.14. Artist's impression of the first experimental GALILEO satellite GIOVE-A. From http://esamultimedia.esa.int/images/galileo/GSTB_satellite/.

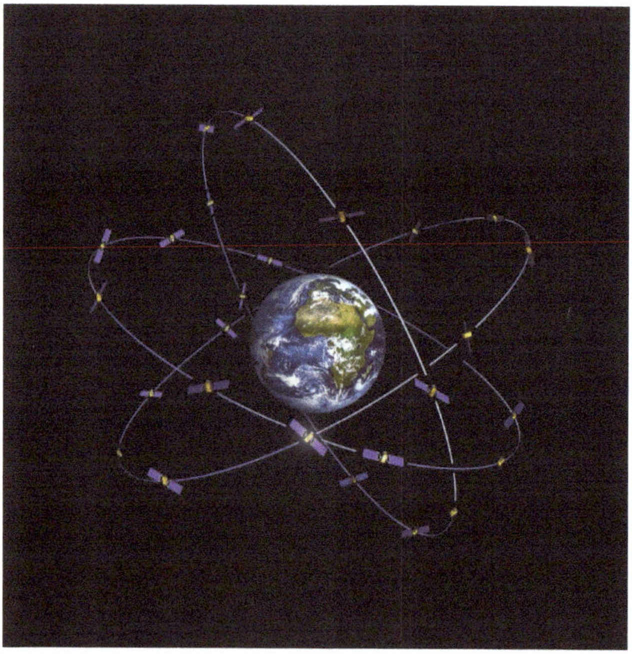

Fig. 2.15. Artist's impression of the complete GALILEO constellation of thirty satellites orbiting in three planes. The three MEO planes are at an inclination of 56° with respect to the equatorial planes, resulting in a good coverage up to a latitude of 75°. From http://esamultimedia.esa.int/images/navigation/.

the ionospheric refraction. The quasi-simultaneity of the observations of different GNSS satellites allows for the elimination (or significant mitigation) of the synchronization errors of the receiver clock with respect to GPS system time (Beutler et al., 2004).

The GNSS were/are deployed primarily for navigation — which is by definition a real time task. They may, however, also be used for science and other positioning applications requiring high accuracy. In this case the observable of choice are not the signals (also called code) modulated on the carrier waves, but the reconstructed carrier itself. The analysis is usually done in the post-processing (but also increasingly in the real-time) mode. This carrier phase observable may be reconstructed with mm-accuracy, which in turns allows for mm-accurate relative positioning, provided not only the receiver clock corrections are estimated from the observations, but the satellite clock corrections, as well. Alternatively to the estimation of the clock errors one may also form the so-called double difference observation (the between-satellite-difference of two between-receiver-difference observations, all observations assumed to be simultaneous).

For science, the following quantities may be determined on a daily basis from a global network of well monumented, permanently operating tracking receivers (the ground tracking network):

- GNSS geocentric satellite positions for the entire day (accurate to few cm)
- GNSS satellite clock corrections (accurate to a few ten picoseconds)
- Mean receiver coordinates per day (accurate to a few mm)
- Position of the Earth's rotation axis on the Earth's surface (polar wobble) (daily estimates, accurate to few mm)
- Length of day (daily estimates, accurate to a few microseconds)
- Tropospheric zenith delays for all stations (which in turn allow it to estimate the total water vapor content over the station - provided station pressure and temperature are recorded as well) with high time resolution.
- Global models/maps of mean electron content (two hours time resolution)
- time and (in particular) frequency transfer between time laboratories (sub-nanosecond accuracy)

It is in essence this catalog of quantities, which is determined every day by the IGS since January 1, 1994 (see `http://igscb.jpl.nasa.gov/`). Since 2003 not only the GPS, but also GLONASS observations are used routinely to derive the official IGS products. The IGS products are based on a weighted combination of the IGS Analysis Centers, generated by the IGS Analysis Coordinator (at least) on a daily basis.

The series of IGS station coordinates is in turn used by the IERS to realize the ITRF together with the corresponding results of the other space-geodetic techniques VLBI, SLR/LLR, and DORIS. The large number of IGS sites (currently more than 400, see Figure 2.16) provides easy access to the ITRF for the user community - going far beyond science.

The IGS series of Earth rotation parameters (see Section 2.5) are also used by the IERS to issue the official transformation parameters between the ICRF and the ITRF. The full set of transformation parameters contains in addition to the above

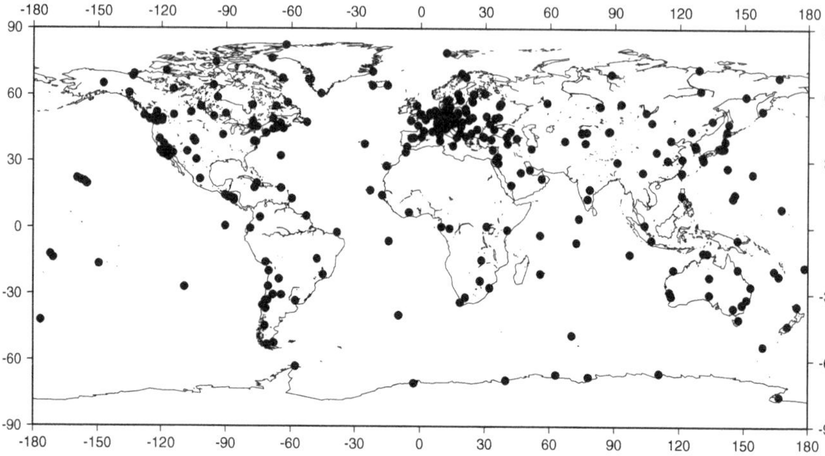

Fig. 2.16. Locations of the more than 400 GNSS stations of the global tracking network of the IGS as of December 2006.

mentioned items UT1-UTC and the nutation angles. These latter quantities can only be provided accurately by VLBI.

In summary one may state that the GNSSs are the workhorses of space geodesy. They provide the basis for numerous applications in geodesy and surveying (virtually all national first order networks refer to the ITRF and are realized using the IGS products) and in the wider area of Earth sciences (in particular atmosphere and ocean sciences, meteorology, and climatology).

DORIS: The DORIS system was designed and developed by the Centre National d'Etudes Spatiales (CNES), in partnership with the Groupe de Recherche de Géodésie Spatiale (GRGS) and the Institut Géographique Nationale (IGN), for precise orbit determination of altimeter missions, and consequently also for geodetic ground-station positioning (Crétaux et al., 1998; Soudarin & Crétaux, 2006). Like GNSS, DORIS is a satellite geodetic technique based on microwave signals, however DORIS is an uplink system from ground stations to spacecraft (Jayles et al., 2006). DORIS beacons transmit on two frequencies, namely 2036.25 Mhz, and 401.25 Mhz. The DORIS system consists of a ground segment, the network of beacons, as well as a space segment, the user satellites, a subset of which actually contribute to the determination of IERS products such as station positioning, and Earth orientation. One characteristic of the DORIS system that is unique with respect to the other space geodetic techniques is the much more homogeneous station distribution. It is the only space geodetic technique with a balanced station distribution in both the Northern and Southern Hemispheres. In addition, another important characteristic is the relative stability of the sites and their longevity with relatively few antenna changes over time (Jayles et al., 2006; Fagard, 2006).

The DORIS network (see Figure 2.17) consists of 50 to 60 stations around the world. The beacons assure an 80% coverage of user satellite orbits near 800 km al-

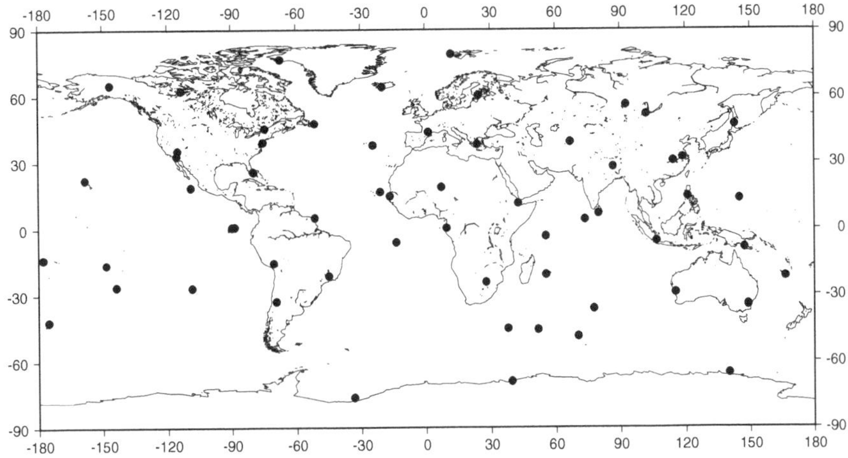

Fig. 2.17. Locations of the 56 DORIS stations of the tracking network of the IDS.

titude, and a 95% coverage of user satellite orbits at 1335 km altitude. Each ground beacon is equipped with a dual frequency transmitter, an Ultra-Stable Oscillator (USO) delivering the reference frequency with a stability of $5.0 \cdot 10^{-13}$ over 10 to 100 seconds, an omni-directional dual-frequency antenna, a a battery pack providing backup power to the beacon during electricity outages, and a meteorological package providing in situ measurements used for the tropospheric correction. Functionally, each beacon may play one of several roles: (1) broadcast upload transmission, (2) time and frequency reference station, (3) time correction beacon, and (4) positioning station. Late in 2005, three stations played the role of master beacons, whose clocks are tied to atomic clocks, and whose delays are estimated with respect to TAI atomic time by time/frequency experts (Jayles et al., 2006). The master beacons serve as time and frequency references for the DORIS system, and handle uploading of data and commands needed by the DORIS spacecraft receivers.

Ground station requirements include the following: (1) The transmitting beacon and its backup power supply must be in a room with moderate temperature with continuous power available; (2) The antenna must be installed outside with a clear sky view above 10° elevation; (3) The local host agency must be willing to carry out minor verifications and adjustments; (4) The frequencies transmitted by DORIS should not interfere with existing receivers in the same area (Fagard, 2006).

An important activity was initiated in 1999 to improve the DORIS system through improvements of the monumentation, installation of new antennae and support structures, and other measures to ensure the stability of the DORIS antenna reference point to within 1 cm over ten years (Fagard, 2006). For example, one goal is that in so far as is possible, ground antennae are now mounted on a concrete pillar deeply anchored into the ground, or on a rigid tower on a deep concrete foundation (see Figure 2.18). As of 2006, 35 of 56 stations are now considered 'excellent' compared to only 10% in 2000 (Tavernier & et al., 2006). The improvements in the network quality are dramatically visible in the residuals (RMS of fit) of DORIS data for the SPOT-2 and TOPEX/Poseidon satellites with the RMS declining from 0.55 mm/s in 1993 to 0.45 mm/s in 2005.

Fig. 2.18. Illustration of two DORIS stations. The stations are Rothera, Antarctica (top), and Thule, Greenland (bottom).

As of 2006, there were co-locations between DORIS antennae and other active IERS techniques at 38 of the 56 permanent DORIS stations. These co-locations are distributed as follows: with GPS at 37 sites, with SLR at 9 sites, and with VLBI at 7 sites. Among these stations, fifteen sites have three co-locations (8 for GPS+SLR+DORIS, and 7 for GPS+VLBI+DORIS). Only two sites worldwide have co-locations with four techniques (GPS+VLBI+SLR+DORIS): Hartebeesthoek and Greenbelt. During the effort of network renovation, some stations were specifically displaced to satisfy both operational constraints and to increase the number of co-locations between DORIS and other geodetic techniques. These include Jiufeng (replacing Purple Mountain), Male (replacing Colombo), Miami (replacing Richmond), Santa Cruz (replacing Galapagos), and Monument Peak (replacing Goldstone) (Fagard, 2006). Since the 1990's, with the growing interest in monitoring changes in global mean sea level, DORIS stations have been increasingly sited near tide gauges. As of 2006, 21 DORIS tide-gauge co-locations were available (Fagard, 2006).

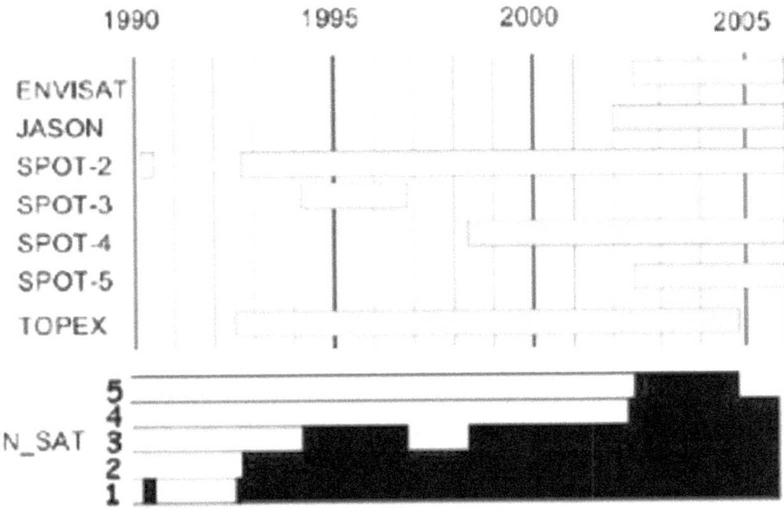

Fig. 2.19. DORIS data available at the IDS Data Centers as of January 2006. With kind permission from Springer Science + Business Media: Tavernier et al. (2006).

The space segment of the DORIS system consists of receivers on user satellites in low Earth orbit. Four satellites have carried first-generation DORIS receivers: TOPEX/Poseidon, SPOT-2, SPOT-3, and SPOT-4. A new second-generation dual channel DORIS receiver was developed in the 1990's. This receiver has been carried on ENVISAT, and a miniaturized version on Jason-1 and SPOT-5 (Tavernier & et al., 2006). Figure 2.19 summarizes the availability of the DORIS data at the data centers of the International DORIS Service (IDS). A DORIS receiver (together with a GPS receiver and a laser retroflector array) is also included on Jason-2 (launched in 2008). Future plans include DORIS receivers on Cryosat-2 (launch in 2009), SARAL/Altika (joint with CNES and the Indian Space Research Organization, launch in 2009-2010). Other possible future DORIS user satellites include Jason-3, HY-2A (a proposed altimeter mission with CNES and the Chinese National Space Agency), and SENTINEL (European Space Agency satellite(s) dedicated to Earth remote sensing).

A major product of the DORIS system are the precise orbits for the user satellites. For satellite altimeter missions such as TOPEX/Poseidon, Jason-1 and ENVISAT, the precise orbits are the key to satisfying the mission objectives of accurately mapping the ocean topography on a routine basis, and determining variations in global mean sea level. The precise orbits have an accuracy of 1-3 cm in the radial direction. DORIS also enables the delivery of routine altimetric science products with latencies of several days. For Jason-1, these rapid delivery orbit products (for use on the IGDR or Interim Geophysical Data Record) have an accuracy of a few cm in the radial component. DORIS near-real-time products will also be available within a few hours on Jason-2. These orbits are expected to have a 10 cm accuracy in the radial component (Jayles et al., 2006). For missions such as Jason-2 and Cryosat-2, a new geodetic bulletin will be available on-board, providing latitude and longitude of the sub-satellite point, and altitude of the satellite over the geoid. The altimeter will use this information in its tracking loops.

The IDS offers routine delivery of ground station positions and Earth orientation based on analysis of the DORIS data in the form of weekly SINEX files. Three analysis centers contributed these SINEX time series to the ITRF2005 solution (Altamimi et al., 2006). The quality of the positioning was evaluated in the construction of the ITRF2005 solution (Altamini et al., 2006). The weighted RMS of the individual weekly time-series combinations can be used as an indication of the positioning quality. The effect of the addition of the large number of satellites in 2002, and the effect of the network improvement project starting in 2000 are clearly visible (see Figure 2.20). Positioning quality with four satellites (post-2002) is 1 to 1.5 cm WRMS. We note that the ITRF2005 DORIS contribution did not include the contribution of Jason, as the USO on the spacecraft experiences a disruption due to periodic passage through the South Atlantic Anomaly (Willis et al., 2004). A correction model has been developed which can partially mitigate the effect (Lemoine & Capdeville, 2006) in the DORIS data. Future DORIS spacecraft USO's will be annealed to prevent this radiation-induced perturbation and resultant data degradation.

Since DORIS is a dual-frequency system, it also measures the ionosphere content along the slant range from the DORIS satellite (800 or 1335 km altitude) to each DORIS ground station. The sampling path is quite different from GPS, whose path from ground station to receiver stretches 20,000 km from Earth. No routine DORIS ionosphere product is delivered as of 2006, however DORIS data were used to validate the ionosphere correction on TOPEX and compute corrections for the Poseidon altimeter (Jayles et al., 2006).

2.4.3 Altimetry

Satellite radar altimetry provides height measurements of the instantaneous surface (sea, ice, or open water on land) with respect to a fixed reference (typically a conventional reference ellipsoid embedded in a global reference frame): the onboard radar altimeter transmits a short pulse of microwave radiation with known power towards the nadir. Part of the incident radiation reflects back to the altimeter. Measurement of the round-trip radar signal travel time provides the height of the satellite (alti-

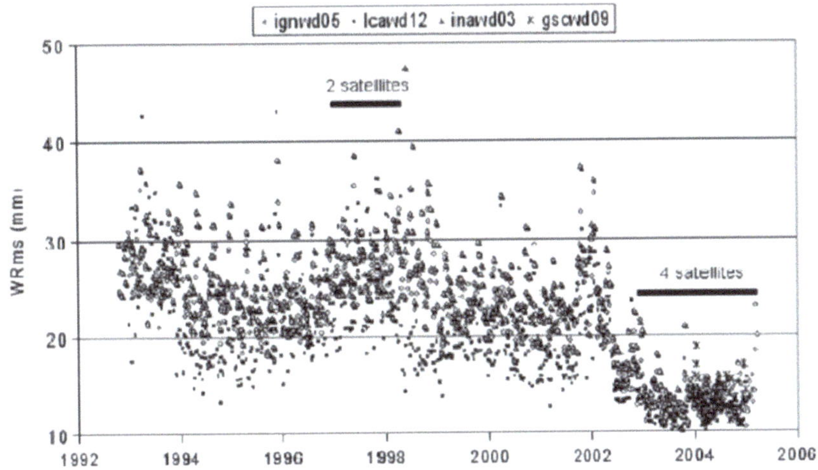

Fig. 2.20. Weighted RMS of the individual weekly time-series combinations from Tavernier & et al. (2006). The results are depicted for four DORIS analysis centers. Note the sensitivity to the number of available satellites, and the effect of the rejuvenation of the network (2000-2005).

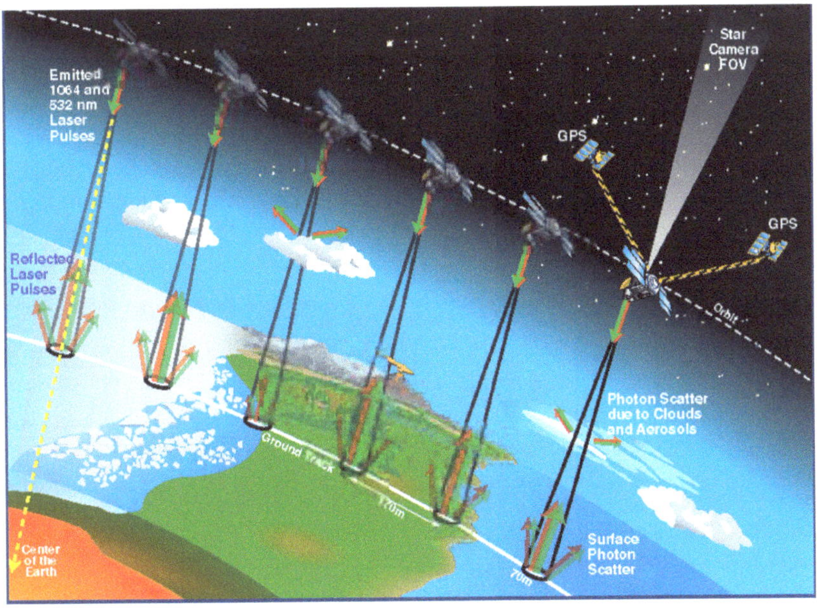

Fig. 2.21. Principle of satellite altimetry. From http://icesat.gsfc.nasa.gov.

metric range) above the instantaneous sea/land water/ice surface. Its difference with the satellite altitude above the reference ellipsoid (computed through precise orbit determination, a long-tested approach in space geodesy) gives sea/land water/ice surface height measurements with respect to the reference (see Figure 2.21).

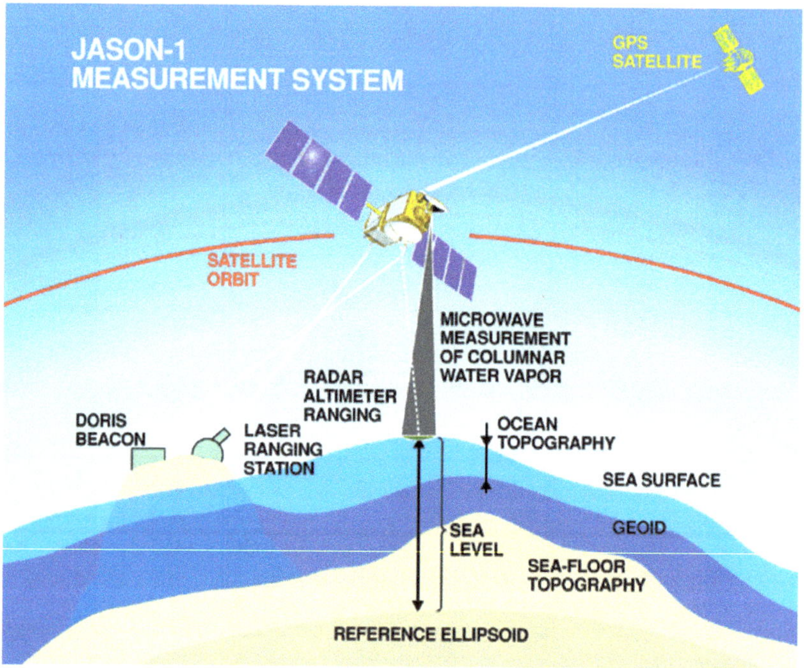

Fig. 2.22. The Jason-1 satellite altimetry mission. Courtesy NASA/JPL.

The range from the satellite to mean sea level must be corrected for various components of the atmospheric refraction and biases between the mean electromagnetic scattering surface and mean sea level at the air-sea interface. A number of corrections due to a number of geophysical effects must also be taken into account.

Table 2.2. Satellite gravity and altimeter mission products to determine mass transport and mass distribution in a multi-disciplinary environment.

Mission	Type	Mission Duration
CHAMP	Gravity, magnetic field, atmosphere	2000–2008
GRACE	Gravity (static and temporal), atmosphere	2002–2010
GOCE	Gravity (stationary, high-resolution)	2009–2011
TOPEX/Poseidon	Ocean altimetry	1992–2005
Jason-1	Ocean altimetry	2001–2008
Jason-2	Ocean altimetry	2008–2015
ICESat	Ice altimetry	2003–2008
CryoSat-2	Ice altimetry	2009–2013
ERS-2	Altimetry, SAR/InSAR, climate, environment	1995–2007
ENVISAT	Altimetry, SAR/InSAR, climate, environment	2002-2008
TerraSAR-X	SAR, InSAR, atmosphere	2006–2011
SWARM	Magentic field	2009–2014

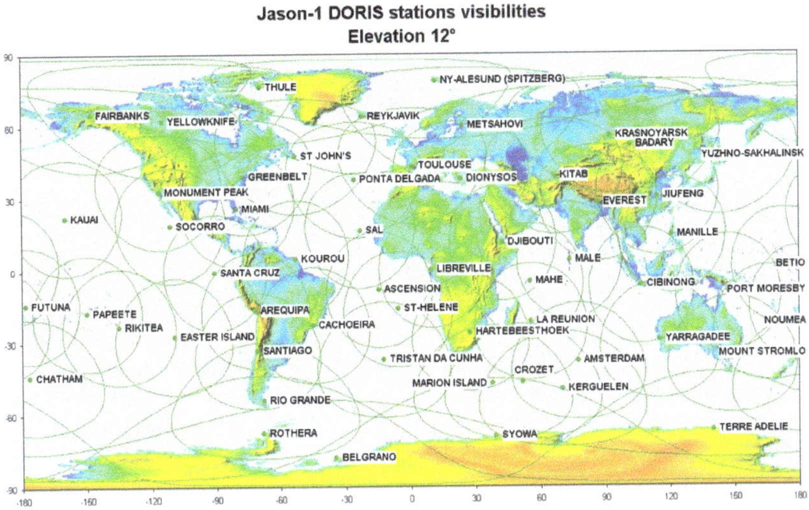

Fig. 2.23. Jason-1 and DORIS. The map shows the visibility of the JASON satellite to each DORIS ground station.

State of the art satellite radar altimetry has more than 2 decades of heritage: GEOS-3 (1975), SEASAT (1978), Geosat (1985- 1989), ERS-1 (1991-1996), TOPEX-/Poseidon (1992-2006), ERS-2 (since 1995), GFO (since 2000), ENVISAT (since 2002) and Jason-1 (since 2001, see Figure 2.22). Over the years, technological improvements (especially for TOPEX/Poseidon, ENVISAT and Jason-1) have decreased considerably the instrumental noise for a point-to-point measurement. Moreover, thanks to a concerted effort in precise modeling of the geophysical and environmental corrections as well as in precise orbit computation, the total rms measurement accuracy of altimetry-based sea surface height is currently about 1-2 cm for a single measurement (e.g., for Jason-1).

Although developed for oceanographic purposes, early altimetry missions have mainly served to map the marine geoid globally with high precision and resolution, leading to considerable achievements in several areas of marine geophysics (e.g., marine tectonics, mechanical and thermal structure of the oceanic lithosphere, seafloor topography, etc.). With the launch of the TOPEX/Poseidon mission in the early 1990s, the precision of sea surface height measurements improved by a factor 10 or more, allowing for the first time precise determination of the temporal variability of the ocean surface, with numerous applications in oceanography. Major results have been obtained on surface currents and the ocean dynamic topography, ocean seasons, El Nino, ocean heat content, sea level rise, ocean tides, waves, etc. Satellite radar altimetry (together with laser altimetry, e.g., the IceSat mission) have also proved very useful for measure the change in elevation of the ice sheets (hence their mass balance in response to global warming) and more recently the water level of lakes, rivers and floodplains on land.

Sea level measured relative to the geoid (the fundamental level surface which will be determined to good accuracy by space geodetic missions such as GOCE in the next few years, see Section 2.6.5), provides the "sea surface topography" which allows estimation of ocean transports, and contributes ultimately to an understanding of climate change (Johannessen et al., 2003).

2.4.4 GNSS scatterometry and reflectometry

In the past few years the potential of GNSS signal reflections for ocean altimetry and remote sensing of sea surface roughness has generated considerable interest. The Passive Reflectometry and Interferometry System (PARIS) was the first concept proposed for ocean altimetry using GNSS L-band signals (Martin-Neira, 1993). Within the PARIS scheme direct and ocean-reflected signals are detected by spaceborne receivers and altimetric height information is extracted from the delay in arrival times of the reflected in relation to the direct signals (Figure 2.24). In the following years altimetric heights with accuracies below 5 cm were determined in a number of airborne and ground-based experiments using special purpose GNSS receiver instrumentation (GNSS reflectometry, e.g., Garrison & Katzberg, 2000; Treuhaft et al., 2001).

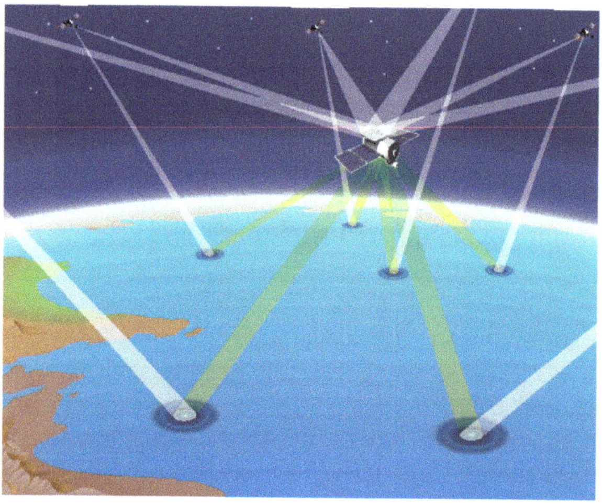

Fig. 2.24. Use of reflected GNSS signals for altimetric measurements. An Earth-orbiting instrument uses direct GNSS signals for precise positioning, but also receives reflected signals to make several simultaneous bistatic altimetric measurements.

In addition, the shape of the code correlation as a function of time delay and Doppler frequency and its dependency on the reflecting surface's slope characteristics can be used to infer the sea surface roughness (GNSS scatterometry, Garrison et al., 1998). Using parameterizations relating the observed roughness to the surface wind vector GNSS scatterometry allows for the remote detection of wind speed and wind direction as well (e.g., Katzberg et al., 2001; Germain et al., 2004).

First spaceborne observations of GNSS signal reflections are reported by Pavelyev et al. (1996) and Lowe et al. (2002). Later, signatures of coherent GPS reflections at grazing incidence angles were found in radio occultation data observed by the GPS/MET, CHAMP and SAC-C satellites (Beyerle & Hocke, 2001; Hajj et al., 2004). More recently, the GNSS scatterometry experiment aboard the United Kindom's Desaster Monitoring Constellation (UK-DMC) satellite successfully demonstrated the feasibility of sea surface state remote sensing from low Earth orbit (Gleason et al., 2005). In the future, satellite constellations furnished with GNSS scatterometry and reflectometry instruments could contribute to the long-term observations of ocean topography as well as constitute essential elements of early warning systems for catastrophic tsunami events.

Science with GPS reflected and scattered signals: This section discusses an emerging technique for Earth remote sensing based on detecting a GNSS signal after it is reflected off the Earth surface, to measure surface topography and roughness at high spatial resolution and rapid temporal coverage. The weak reflected signals require a high-gain multi-beam steerable antenna. Because this technique is promising in terms of spatial and temporal resolution, we devote here some space to the discussion of its potential in a few major areas.

Global Ocean Altimetry: The Oceans, and their interactions with the atmosphere and the lithosphere, play a significant role in Earth's climate. Understanding climate variability implies quantifying all the significant processes that contribute to climate and its changes. One such process, mesoscale ocean eddies, analogous to atmospheric storms, represents one of the dominant global climate errors (see HOT_SWG 2001 for a review); they are essential to understanding ocean circulation on all scales and are an important contribution to the carbon cycle.

On the regional scale, eddies can induce local upwelling and enhance biological production. In the equatorial Pacific, eddies associated with the tropical instability waves can increase the supply of iron and silicate to the euphotic zone resulting in enhancement of the biological productivity (Barber et al., 1996). On the global scale, mesoscale eddies play an important role in the overall transport of heat and momentum. Numerical model simulations with and without the inclusion of mesoscale eddies show a 30% difference in the equator-to-pole heat transport over the Atlantic Ocean (Smith et al., 2000). Ocean eddies have a typical spatial scale on the order of 10 to 100 km and a temporal scale from days to weeks. The sea level signal associated with mesoscale eddies is usually 10 cm or more.

At present, quantifying the role of mesoscale eddies in the ocean circulation and therefore climate variability is limited because their spatio-temporal structures are not resolved by the conventional remote-sensing techniques. Observations of sea surface temperature (e.g., those from Advanced Very High Resolution Radiometers) are frequently contaminated by clouds in the atmosphere. The conventional satellite radar altimeter measures the sea surface height at high spatial resolutions along its ground track (e.g., 7-km for TOPEX/Jason). However, the cross-track distance is usually quite large. For a 10-day repeat orbit with TOPEX/Jason, the cross-track distance is more than 300-km at the equator. Another limiting factor is the long repeat cycle of a given satellite, e.g., 10 days for TOPEX/Jason, 17 days for the Geosat-Follow-On (GFO), and 35 days for Earth Remote Sensings (ERSs). Additionally, some barotropic (i.e., vertically uniform) waves with a periodicity of 20 days or less can be aliased into the 10-day sea level map produced by the TOPEX/Jason data. Hence, there is a need for high spatial and temporal resolution altimetry.

High-resolution ocean altimetric measurements will allow oceanographers to compute high-order quantities like vorticity and eddy fluxes, which will be used to study the interactions between the eddy fields and the time-mean flow. Several important science questions can be addressed by such a high-resolution data. For example, what is the role of mesoscale (ocean) eddies in the large-scale ocean circulation and climate variability? What is the impact of mesoscale eddies on the biological production and therefore the global carbon cycle? If mesoscale eddies are important in modulating the large-scale ocean circulation and climate, there is a need to resolve (or parameterize) ocean eddies in the Earth System Model (coupled atmosphere-ocean-land) for climate prediction purposes.

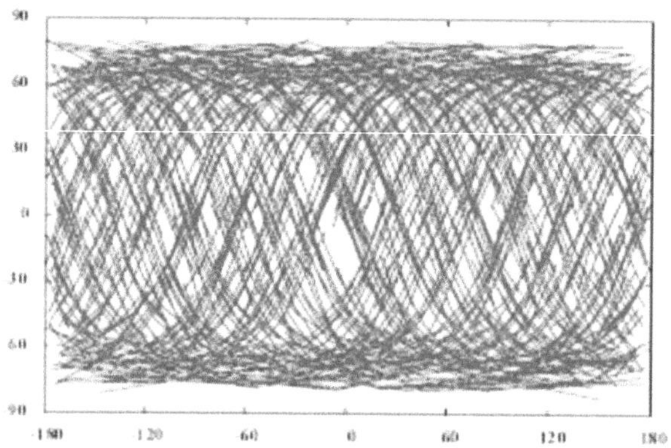

Fig. 2.25. Reflection point loci for one receiver at 400 km, assuming its antenna beam can capture all available reflections, per day. Horizontal axis is longitude, vertical axis is latitude.

Traditional altimetry is limited to looking in the (nominal) nadir direction and obtaining one height observation at a time below the altimeter, following very nearly repeatable tracks passing over the same point every ten days. The concept of wide-swath ocean altimetry improves the coverage and spatial resolution of traditional altimetry by filling the gaps between satellite tracks. However, the wide-swath ocean altimetry uses the same ground tracks of TOPEX/Jason repeating every 10 days. By contrast, a GPS receiver in low-Earth orbit (LEO) with an antenna pointed toward the Earth's surface can, in principle, track about 10 GPS reflections simultaneously, therefore providing a coverage that is an order of magnitude denser than nadir-viewing altimeters. For example, the reflection ground tracks of one single satellite at the altitude of 400 km would cover the Earth nearly uniformly in just 1 day, with at most about 75 km across-track separation, as shown in Figure 2.25. Such dense coverage can be translated into a higher temporal and spatial resolution than that of TOPEX/Jason or the proposed wide swath coverage, thereby providing the ability to recover certain ocean topography features or processes that are precluded with traditional altimeters.

Ocean Surface Statistics and Wind Retrieval: GNSS reflections from the ocean can be used to infer statistical properties of the surface, namely the slope distribution of sea-surface gravity waves, with high spatial and temporal resolution. Such measurements would likely be made concurrent with altimetric measurements (see Ocean Altimetry section above) because the measurement techniques are quite similar. The primary observable is the Mean-Squared Slope (MSS), and recent studies (Germain et al., 2004) have shown a 2D directional-MSS can be obtained. The MSS field provides useful input to ocean-atmosphere coupling phenomena such as surface breaking waves and gas exchange. For example, CO_2 flux measurements may be derived from MSS. With additional assumptions, wind speed or wind vector retrievals can also be obtained from MSS measurements (Garrison et al., 1998; Komjathy et al., 2000; Cardellach et al., 2003; Zuffada et al., 2003). Finally, MSS measurements may clarify the relationship between surface-height dynamics and wind-driven surface velocities (Chelton et al., 2004).

Analysis of the GPS reflection waveform also provides an estimate of the wind speed and direction. While scatterometers such as QuikScat or SeaWinds provide near global coverage in one day, the observations are not necessarily co-located in time and space with the GPS altimetry observations. Instead, GPS reflections provide a unique set of co-located sea surface height and wind observations with near-global daily coverage and with resolution suitable for studying mesoscale features. Accurate sea surface height retrieval requires simultaneous measurements of ocean vector winds. The accuracy of GPS wind measurements is about 2 m/sec for wind speeds ranging from 3 to 15 m/sec, comparable to the traditional radar scatterometer. Thus, the GPS-measured ocean winds will complement the existing radar scatterometer wind observations and, in the context of sea surface height measurement, will provide the needed data set to retrieve the sea surface height with high accuracy. It is anticipated that the GPS altimetry will improve our current capability in two important ways: 1) High-spatial-resolution ocean topography and 2) Improved temporal resolution through rapid coverage. Another possible application of very rapid coverage of the ocean is the monitoring of fast moving barotropic waves that propagate across ocean basins too quickly to be seen by the Jason 10-day repeat cycle.

Ice Science: Detection of GPS reflections at low or grazing angles has the advantage of being coherent and, when combined with the direct signal, provides interferometric fringes from which a very precise estimation of bi-static path delay (down to sub-centimeter) can be detected. In the presence of strong L1 and L2 signals to calibrate the ionosphere, this can be translated into accurate height surface measurements at the specular reflection point.

Recent analysis (Cardellach et al., 2004) used this interferometric signal, detected with the CHAMP radio occultation experiment, to demonstrate a surface height precision of 0.7 m after 0.2 s of integration with a reflection angle of $< 1^\circ$ (i.e., 89° incident angle). The GRSPI instrument will allow the detection of the coherently reflected signal at a higher elevation angle reducing the error in inferred ice surface height to less than 10 cm.

Global observations of sea ice, ice sheets, ice caps, glaciers and their surrounding seas, are paramount in order to determine their mass balance, contributions to sea level change, global circulation and climate change. In fact, model simulations and recent observations suggest that the ice-covered regions of the Earth are the most sensitive to climate change. In the polar region the combination of atmospheric, cryospheric, and oceanographic processes have a large influence on the global climate. Unfortunately, these climatic processes are poorly understood, principally because of a dearth of observations for diagnosing the processes and validating numerical models.

Changes in ice thickness are an indicator of climate change in the polar region as a result of heat exchanges between ocean and atmosphere, and are themselves a primary driver of climate change through the effect of these heat fluxes on atmospheric circulation patterns and the strong positive planetary albedo feedback provided by changes in sea ice, snow cover and melt water. Given the multi-beam bi-static reflections of GPS, a GPS cryospheric sensing system can provide a substantially denser and more rapid coverage than traditional ice altimetry instruments and allow the determination of seasonal and annual variations in sea-ice and land-ice thickness.

Soil Moisture: Soil moisture is an important part of the land hydrology cycle, where it represents the immediate store of infiltrating rainfall, before it either evapotranspires or contributes to groundwater recharge. When the soil gets too dry, plant transpiration drops because the water is becoming increasingly bound to the soil particles. Conditions where soil is too dry to maintain reliable plant growth is referred to as agricultural drought, and is a particular focus of land management. Soil moisture may be measured in situ with different instruments, such as Time Domain Reflectometry (TDR), neutron probe, capacitance probe, etc. but no global remote sensing measurements are currently available. The potential for measuring soil moisture with GPS has been explored through some ground-based and airborne experiments over smooth terrain, led by the University of Colorado in Boulder and NASA Langley Research Center. Theoretical models show that moist soils generate strong reflective layers at the GPS frequencies, due to high gradients in dielectric constant. It was experimentally observed that variations in the reflected signal are uniquely related to changes in the dielectric permittivity, and therefore, to soil moisture because roughness of the area with low grass remains constant. More work is needed to assess requirements, including antenna gains, for potential GPS-based systems for global soil moisture measurements.

Traceability Matrix for Ocean Observations: The 2007 NRC Decadal Report (National Research Council, 2007) stresses that future directions for Earth science at NASA/NOAA will focus on achievement of a national strategy for the Earth Sciences that balances international economic competitiveness, protection of life and property, and stewardship of the planet for this and future generations. Based on the need for climate measurements identified in the report, JPL promoted a study (Sherwood et al., 2006) to explore the science benefits of maintaining GPS receivers on all satellites in orbit for climate science. This is a particularly timely topic since there are currently ten GPS-science capable satellites (COSMIC 1-6, MetOp1, CHAMP, SAC-C, GRACE).

The study performed a series of simulations to determine the science return that could be achieved with varying sizes of GPS receiver constellations. This study can be used to consider the advantages of including GPS science receivers on future satellites as dedicated constellations or constellations of opportunity. For ocean science, we assumed each satellite would be equipped with a Toga receiver (now in development under NASA's Instrument Incubator Program, see Table 2.3), and a steerable 20-dB gain antenna with field of view capable of intercepting all available reflections.

Table 2.3. Instrument characteristics of TOGA receiver. The parameters are for multilag processing and 20 dB antenna gain.

Integration time	Height prec. (cm)	Footprint (km)
1 sec	Near nadir, 5	Along track, < 10
	Near grazing, 25	Cross track, < 10

To evaluate the needed size of receiver constellations as a function of the ocean science capabilities, simulations were performed using the following assumptions: a) the current GPS constellation as available as transmitters and b) reflection-capable receivers are available on constellations of 6, 18, and 37 LEO satellites, respectively. In the first case only the orbits of the COSMIC constellation were chosen, whereas the third case simulates the situation where all existing NASA satellites (assuming their orbits are representative of future satellites) are equipped with GPS receivers capable of tracking and processing reflections. The intermediate case assumes that an additional set of twelve LEO satellites, chosen randomly among the existing NASA satellites, have been added to the COSMIC set. The characteristics of a single measurement are summarized in Table 2.3.

Table 2.4. GPS ocean reflections science questions.

No.	Science Question
1	Can we measure sea ice surface topography (freeboard), to determine sea ice thickness and mass balance?
2	Can we measure wind for a) improved vertical mixing at the mesoscale; b) monitoring and prediction of severe weather systems; c) high resolution wind forcing and attendant coastal ocean response (e.g., local upwelling)?
3	Can we measure the sea surface topography with sufficient spatial and more importantly temporal resolutions to monitor the evolution of mesoscale ocean eddies and coastal oceans?

The traceability matrix summarizing the flow down from science questions (Table 2.4) to observations' requirements and constellation size is presented in Table 2.5. Two science areas have been addressed: ice-free sea surface topography and sea ice topography and mass. Correspondingly, the observational requirements are mapped into latitudinal bins, cell sizes and revisit times. For each case, the percentage of cells that records at least one (in some cases more) reflection is reported.

Table 2.5. Traceability matrix from science questions to observation requirements for GPS ocean reflections measurements.

Cellsize (km)	Science Question	Latitude Bin	Time Scale	Precision (cm)	Constellation size 6	18	37
2	1	> 60	15 days	$5 \approx 10$	$< 78\%$	$< 90\%$	$< 95\%$
	1		30 days		$< 95\%$	100%	100%
	2	All lats	6 hours	N/A	-	-	-
5	2	All lats	6 hours	N/A	-	$< 20\%$	$< 25\%$
	3		1 day	$2 \approx 10 < 23\%$	$< 52\%$	$< 75\%$	
	3	$-60 < x < 60$	1 day	$2 \approx 10$	$< 52\%$	$< 78\%$	$< 90\%$
10	3	$-60 < x < 60$	1 day	$2 \approx 10$	$< 63\%$	$< 94\%$	100%
	3		5 days		$< 99\%$	100%	100%
25	3	$-60 < x < 60$	1 day	$2 \approx 10$	$< 95\%$	100%	100%
	3		5 days		100%	100%	100%

The table quantifies coverage, and required precision. It is very difficult to establish how the precision requirement is met. In fact, this depends on the reflection angle, as reported in Table 1, for the individual measurement as well as on the number of reflections in a given cell and time. The required precision is met with the highest confidence for the situation of 25 x 25 km cell size, both 1 and 5 days repeat cycles. By contrast, the simulations clearly show inadequate coverage for the situation of 5 x 5 km cell size (and below), 6 hours repeat cycle. It is noted that if the constellation of transmitters increases while the number and orbit of the receivers is held constant, the number of measurements in any given cell increases commensurately, thus improving the precision. The coverage is not expected to improve dramatically, since it is ultimately determined by the number and position of the receivers.

2.4.5 Geodetic imaging techniques

InSAR: The processing of Synthetic Aperture Radar (SAR) images using the InSAR techniques has demonstrated the potential to revolutionize deformation monitoring from spaceborne platforms. As opposed to conventional point-level positioning techniques, InSAR gives deformation information for extended areas (up to a few hundred km across). In this sense InSAR truly is a remote sensing technique. It can provide spatially smooth three-dimensional maps of surface change, including that from earthquakes, volcanoes, ice sheets, glaciers, fluid extraction, and landslides.

InSAR for geodetic applications is a method by which radar signals are radiated from a moving platform and are reflected back to the antenna from the surface of the Earth. The intensity and phase of the reflected signal are measured. In order to measure topography, two antennas separated in space are used to measure phase differences between the two antennas from a radar signal reflected from one point on the Earth's surface (Figure 2.26, top picture). The Shuttle Radar Topography

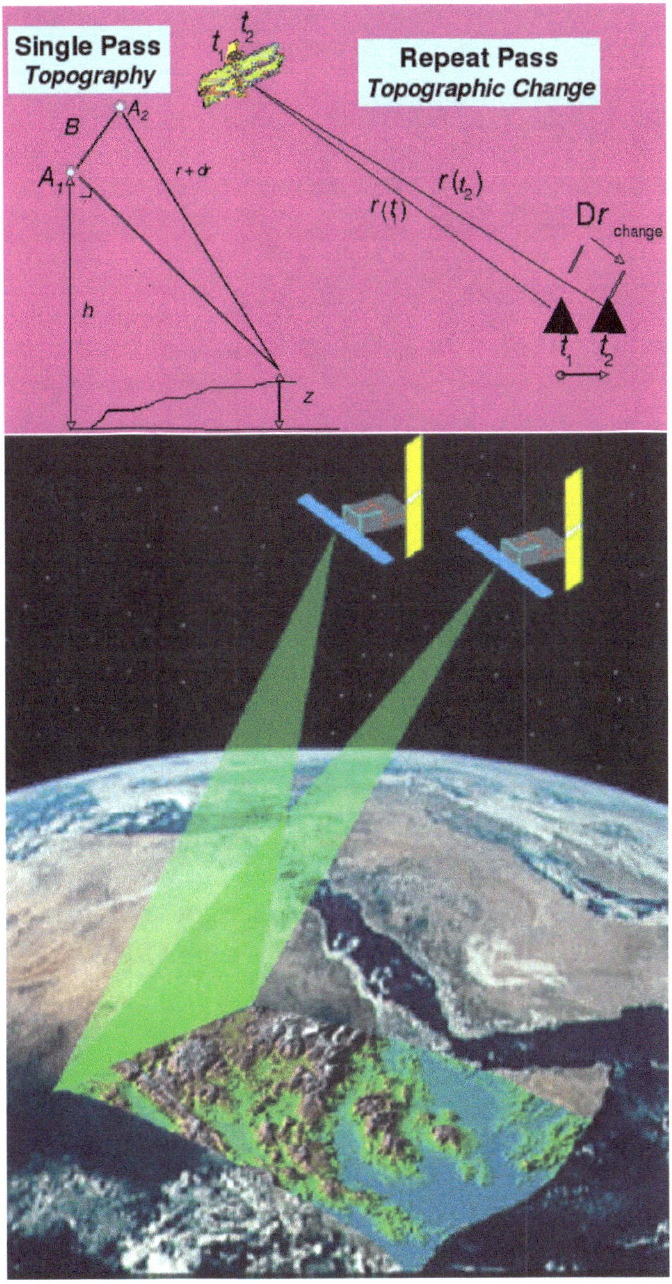

Fig. 2.26. Principle of InSAR. Two antenna separated in space can be used to determine topography (top), while an exact repeat pass of a radar instrument can be used to determine topographic or surface change (bottom).

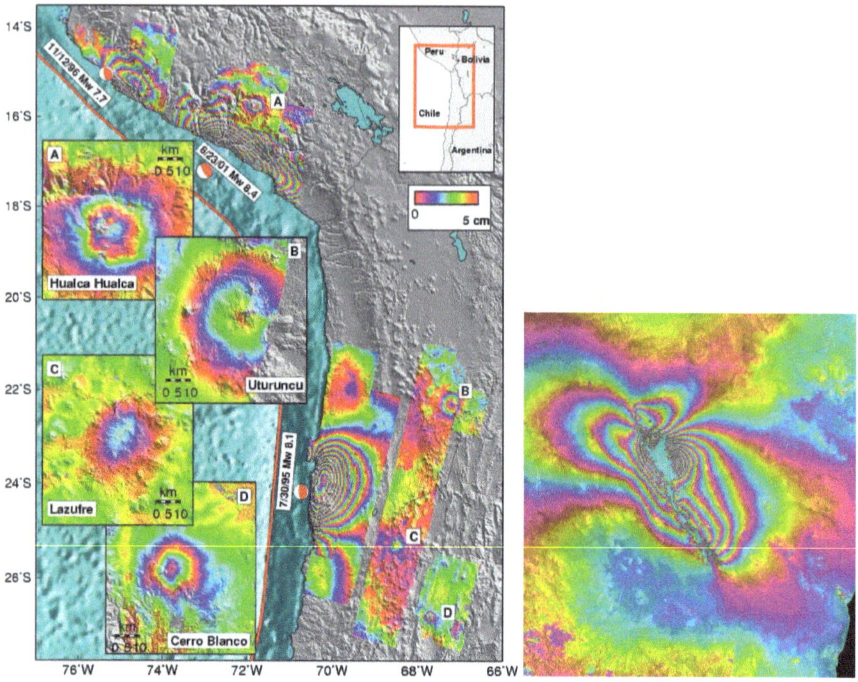

Fig. 2.27. Interferograms from ERS showing deformation. Each color cycle corresponds to 5 cm of deformation. Left image: InSAR has revealed that four Andean volcanoes (named on the small interferograms on the left of the image), thought to be inactive, are now known to be rapidly deforming. The top three volcanoes are inflating and Robledo is deflating (Pritchard & Simons, 2002; reprinted with permission from Macmillan Publishers Ltd). Right image: Hector Mine earthquake observed from ERS (courtesy G. Peltzer, UCLA).

Mission (SRTM) is an example of a radar mission that mapped 80% of the Earth's topography using this technique. In order to measure surface change, a single radar is used, measuring the surface at two times from an exactly repeated pass. A change in the line-of-site distance to the satellite results in a phase change that can be used to infer surface change (Figure 2.26, bottom panel).

Several radar missions have used interferometric techniques for topography and surface change. SRTM mapped 80% of the Earth's topography in a 10-day mission in 2000. The European ERS-1 and ERS-2 missions, the Japanese JERS-1 and ALOS missions, and the Canadian Radarsat missions have provided important data sets for measuring surface change. The European and Canadian missions are C-band instruments, and the short wavelength signal decorrelates over vegetated regions. A recently released report of the U.S. National Research Council (National Research Council, 2007) recommends an L-band InSAR mission with 8-day repeat to provide global coverage of Earth's deforming regions. The report recommends a launch in the 2010-2013 time frame, essentially the earliest possible juncture.

Successes from radar interferometry include the SRTM topographic map, discovery of actively inflating volcanoes that were thought to be dormant (Figure 2.27), measurement of interseismic, coseismic, and postseismic deformation related to earthquakes that have truly influenced physical models of Earth's crust, observation of incipient landslides, and subsidence due to water and oil withdrawal. Long-term systematic measurements will also provide insight into time dependent behavior of earthquake, volcanic, and other solid Earth and cryosphere systems.

The above documents emphasize the increasing importance of image geodesy. However, a major challenge is still the integration of point and image geodesy (Plag et al., 2007b; DESDynI Writing Committee, 2007). Solid Earth science and many applications require observations of Earth's surface displacements at the sub-cm level. Solid Earth processes exhibit temporal scales from seconds (e.g., coseismic displacements) to secular with respect to the lifetime of a mission (e.g., isostatic adjustments), and spatial scales from local (e.g., local subsidence, volcanoes) to global (e.g., great earthquakes, glacial isostatic adjustment). This wide range of temporal and spatial scales poses a major challenge for the extraction of unbiased surface displacements from InSAR observations.

The determination of surface displacements from InSAR requires at a minimum a high-resolution Digital Elevation Model (DEM) and information on tropospheric water vapor content. Additional data of ionospheric Total Electron Content (TEC), for example, from GPS/GNSS is likely to improve the correction of ionospheric path-delay based on InSAR observations alone. If *a priori* deformation models are available, tropospheric water vapor content can be estimated directly. However, the strategies for an optimal combination of *a priori* information on DEM, water vapor, surface deformation, and ionospheric TEC are still the object of research. Particular emphasis should be on consistent treatment of errors in the *a priori* information.

The "Decadal Survey" (National Research Council, 2007) states that a stable global geodetic reference frame is indispensable for all satellite missions, and this is also true for geodetic imaging missions. For most Earth science applications, the surface displacements need to be given relative to such a stable, global geodetic reference frame. For example, for local sea level studies, coastal subsidence or uplift need to be given in a reference frame well tied to the CM. Glacial isostatic adjustment is important for the conversion of ice surface displacements into ice volume and mass changes. The deformation of the solid Earth surface due to ice loads has large spatial scales and need to be referred to the same reference frame as that of the ice surface displacements. Large earthquakes have displacement fields exceeding by far the size of several adjacent images. Likewise, postseismic deformation, which is a key quantity for earthquake process studies, can have spatial scales of the order of 1000 km. For all these phenomena it is crucial to relate the displacements from different interferograms to the same unique reference frame in order to capture the large-scale displacement pattern. However, as discussed in Chapter 8 (see also DESDynI Writing Committee, 2007), the present approach to the realization of the ITRS has limitations that reduce the achievable accuracy and necessitate conceptual improvements.

In particular for early warning and disaster damage assessments, high temporal resolution and low latency are key requirements. Typical InSAR missions have repeat periods of several days of longer. Hazardous volcanoes and unstable slopes can be monitored with such repeat period, but in critical phases, early warning may need much shorter repeat periods. In these cases, supporting measurements with airborne LIDAR (see below) and InSAR can be used to achieve improved temporal resolution. Ground-based GPS/GNSS can also provide a higher temporal resolution, especially if the repeat time increases. In cases of earthquakes, landslides, and volcanic eruptions, emergency response requires rapid information on the extent of damage. Surface displacements are indicative of damage. In order to reduce the latency, again airborne LIDAR and InSAR can support the mapping. In all these cases, the appropriate algorithms for the combination of the spaceborne, airborne, and *in situ* observations need to be developed.

LIDAR: Another imaging technique to be mentioned here is LIDAR. Based on the same principle as RADAR the LIDAR instrument transmits light out to a target (Kavaya, 1999). The transmitted light interacts with and is changed by the target. Some of this light is reflected and/or scattered back to the instrument where it is analyzed. The change in the properties of the light enables some property of the target to be determined. The time for the light to travel out to the target and back to the LIDAR is used to determine the range to the target.

There are three basic generic types of LIDAR:

- Range finders: These are the simplest LIDARs. They are used to measure the distance from the LIDAR instrument to a solid or hard target.
- DIfferential Absorption LIDAR (DIAL): These LIDARs are used to measure chemical concentrations (such as ozone, water vapor, pollutants) in the atmosphere. A DIAL uses two different laser wavelengths which are selected so that one of the wavelengths is absorbed by the molecule of interest whilst the other wavelength is not. The difference in intensity of the two return signals can be used to deduce the concentration of the molecule being investigated.
- Doppler LIDARs: These are used to measure the velocity of a target. When the light transmitted from the LIDAR hits a target moving towards or away from the LIDAR, the wavelength of the light reflected/scattered off the target will be changed slightly. This is known as a Doppler shift - hence Doppler LIDAR. If the target is moving away from the LIDAR, the return light will have a longer wavelength (sometimes referred to as a red shift), if moving towards the LIDAR the return light will be at a shorter wavelength (blue shifted). The target can be either a hard target or an atmospheric target - the atmosphere contains many microscopic dust and aerosol particles which are carried by the wind. These are the targets of interest to us as they are small and light enough to move at the true wind velocity and thus enable a remote measurement of the wind velocity to be made.

2.5 Observing Earth's rotation

Most Earth rotation observations today originate from the geometric space-geodetic techniques described in the previous Section. In the following, focus is therefore only on the specific aspects related to rotation.

2.5.1 Space-geodetic techniques

VLBI: As described in Section 2.4 VLBI observes radio signals emitted by quasars. These fixed points constitute the ICRF (see Section 2.2), and variations in the orientation of the Earth are measured with respect to the ICRF. This technique is sensitive to processes that change the relative position of the radio telescopes with respect to the source, such as a change in the orientation of the Earth in space or a change in the position of the telescopes due to, for example, tidal displacements or tectonic motions. If just two telescopes are observing the same source, then only two components of the Earth's orientation can be determined. A rotation of the Earth about an axis parallel to the baseline connecting the two radio telescopes does not change the relative position of the telescopes with respect to the source, and hence this component of the Earth's orientation is not determinable from VLBI observations taken on that single baseline. Multibaseline VLBI observations with satisfactory geometry can determine all of the components of the Earth's orientation including their time rates-of-change. In fact, the motion of the axis of rotation of the Earth in space (precession and nutation) and the rotation angle around the axis of rotation are uniquely monitored by VLBI through its direct connection to the ICRF.

GNSS: GNSS signals observed by a network of ground stations can be used to determine the orientation of the network of receivers as a whole. In practice, in order to achieve higher accuracy, more sophisticated analysis techniques are employed to determine the EOPs and other quantities such as orbital parameters of the satellites, positions of the stations, and atmospheric parameters such as the zenith path delay (Bock & Leppard, 1990; Blewitt, 1993; Beutler et al., 1996; Hofmann-Wellenhof et al., 1997; Leick, 2003). Only polar motion and its time rate-of-change can be independently determined from GNSS measurements. UT1 cannot be separated from the orbital elements of the satellites and hence cannot be determined from GNSS data. The time rate-of-change of UT1, which is related to the length of the day, can be determined from GNSS measurements. But because of the corrupting influence of orbit error, VLBI measurements are usually used to constrain the GNSS-derived Length of Day (LOD) estimates.

SLR and LLR: Although a number of satellites carry retro-reflectors for tracking and navigation purposes (see Section 2.4.2), the LAGEOS I and II satellites were specifically designed and launched to study geodetic properties of the Earth including its rotation and are the satellites most commonly used to determine EOPs. Including range measurements to the Etalon I and II satellites has been found to strengthen the solution for the EOPs, so these satellites are now often included in the process. The EOPs are recovered from the basic range measurements in the course of determining the satellite's orbit and station coordinates. However, because variations in UT1 cannot be separated from variations in the orbital node of the satellite, which are caused by the effects of unmodeled forces acting on the satellite, it is not possible to independently determine UT1 from SLR measurements. Independent estimates of the time rate-of-change of UT1, or equivalently, of LOD, can be determined from SLR measurements, as can polar motion and its time rate-of-change.

In the case of LLR, the EOPs are typically determined from observations by analyzing the residuals at each station after the lunar orbit and other parameters such as station and reflector locations have been fit to the range measurements (Stolz et al., 1976; Langley et al., 1981; Dickey et al., 1985). From this single station technique, two linear combinations of UT1 and the polar motion parameters can be determined, namely, UT0 and the variation of latitude at that station. A rotation of the Earth about an axis connecting the station with the origin of the terrestrial reference frame does not change the distance between the station and the Moon, and hence this component of the Earth's orientation cannot be determined from single station LLR observations.

DORIS: Processing DORIS observations (see Section 2.4.2) allows the orbit of the satellite to be determined along with other quantities such as station positions and EOPs. As with other satellite techniques, UT1 cannot be determined from DORIS measurements, but its time rate-of-change can be determined, as can polar motion and its rate-of-change (Willis et al., 2006).

2.5.2 Ring laser gyroscopes

Ring laser gyroscopes are a promising emerging technology for determining Earth rotation (Figure 2.28). Ring lasers are active Sagnac interferometers: two monomode laser beams propagate in opposite directions around a polygon (ring) circumscribing an enclosed area. Since the ring laser gyroscope is rotating with the Earth, the effective path length of the beam that is co-rotating with the Earth is slightly longer than the path that is counter-rotating with it. Because the effective path lengths of the two beams differ, their frequencies differ, so they interfere with each other to produce a beat pattern. The beat frequency is strictly proportional to the rate of rotation experienced by the entire apparatus. Therefore, ring lasers are very sensitive to rotational, but entirely insensitive to translational motion. In fact, the beat frequency is proportional to that component of the instantaneous angular velocity $\omega(t)$ of the Earth that is parallel to the normal of the plane of the ring. Ring laser gyroscopes measure the absolute rotation of the Earth in the sense that, in principle, just a single measurement is required to determine the Earth's instantaneous rotation. All of the other techniques discussed above are relative sensors because they infer the Earth's rotation from the change in the orientation of the Earth that takes place between at least two measurements that are separated in time.

Fig. 2.28. Ring laser gyroscope for Earth rotation monitoring. The picture shows the G ring laser at Wettzell.

The sensitivity of the ring laser depends on the area enclosed. Ring lasers with an enclosed area between 1 and 833 m^2 have been built and they achieved sensor sensitivities reaching from $5 \cdot 10^{-10}$ to $5 \cdot 10^{-12}$ rad/s/$\sqrt{\mathrm{Hz}}$. However, sensitivity is only one of the important parameters. It is also critical to reduce the instrumentally induced drift.

The most stable ring lasers experience a non-negligible drift of $2 \cdot 10^{-6}$ degrees per hour, several orders of magnitude smaller than the best known commercial laser gyros. Therefore, these sensors capture the effect of diurnal polar motion and tilt effects from solid Earth tides. Earth rotation variations are resolved to approximately 1% at integration times of about 1 day. Recent progress in reducing the aging of the laser gain medium substantially reduced the drift by approximately 2 orders of magnitude.

Compared to other space-geodetic techniques such as VLBI and GPS, currently, ring lasers have a resolution about one order of magnitude worse. However, because ring lasers are local sensors, they are already revealing interesting crustal deformation effects from a region several hundred kilometers in diameter around the observatory. Furthermore they are operated continuously. Their main advantage over other techniques is the very high temporal resolution. Within the next decade, a substantial improvement in sensor stability as well as a much higher sensor resolution is expected. Apart from Earth rotation research, ring lasers are the first sensors that have shown their sensitivity for measuring rotations from seismic and tele-seismic events at high resolution. It is expected that this application will expand the global network of ring lasers considerably; a development beneficial for Earth rotation monitoring with ring lasers.

2.6 Observing Earth's gravity field

The gravity field of the Earth is observed with *in situ* airborne and spaceborne sensors. Relative gravimetry surveys gravity mainly in order to improve the geoid locally at short wave-lengths but also for exploration purposes. Superconducting and absolute gravimeters measure temporal variations of gravity locally and stationary at sites at the Earth surface (Sections 2.6.1 and 2.6.2, respectively). Modern gravimetry also supports studies of land motion (Section 2.6.3). Gravimeters on ships and airplane measure profiles along the track of the vessel (Section 2.6.4). Satellite orbits are affected by the gravity field at the satellite, and orbit perturbations can be integrated to determine a static gravity field model with low spatial resolution. Recently, dedicated satellite gravity mission have been designed and placed in orbit. One in particular (GRACE, see Section 2.6.5) not only gives the static field with increasing spatial resolution and accuracy but also the temporal variations of the gravity field with low temporal resolution.

2.6.1 Superconducting gravimetry

With the advent of the cryogenic, or Superconducting Gravimeter (SG), in the mid 1980s, the time resolution of the gravity field routinely increased from sampling intervals of minutes to 1 hour to sampling intervals of 1 to 10 seconds. SGs now overlap with seismometers in the recording of high frequency ground motions caused for example by earthquakes in the 1 to 1000 seconds range. Gravimeters measure acceleration, whereas seismometers are velocity recording devices. This difference determines the transfer function of the instruments and impacts the conversion of the observations to ground displacement. The accuracy of the SG in the time domain is on the level of 1 nms^{-2} (= 10^{-9} ms^{-2} = 0.1 microgal) or better, which translates into a frequency domain accuracy at high frequencies (< 1 d^{-1}) at the level of 0.01 nms^{-2} (= 10^{-11} ms^{-2} = 1 nanogal). SGs are known to have a small instrumental drift (a few microgal per year) that can be established by co-located measurements with an absolute gravimeter, and their calibration is very stable in time and determined to better than 0.1%.

The high temporal resolution of SGs is particularly useful in the high frequency domain for recording the long period normal mode spectrum (Figure 2.29), although a sampling interval of 1 s is insufficient for body wave seismology. In the time domain, the high temporal resolution allows for precise determination of effects

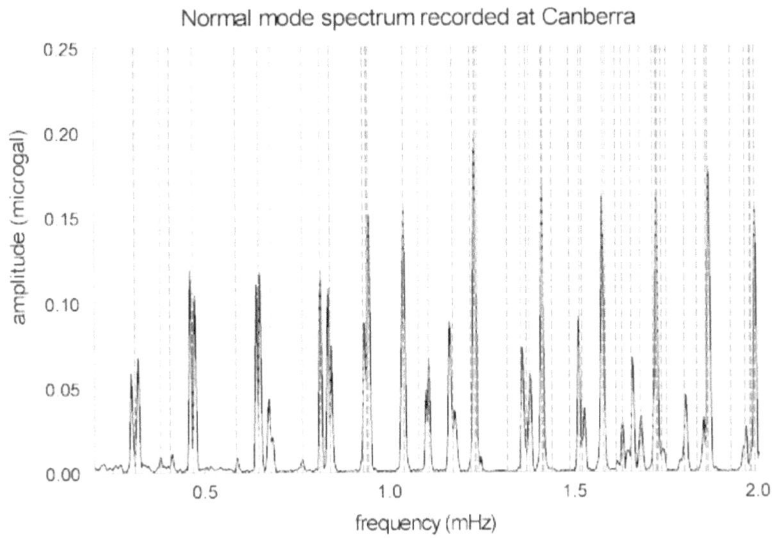

Fig. 2.29. Long period normal modes from the Mw = 9.1 Sumatra-Andamen earthquake (2004/12/26) recorded by the SG at Canberra. The vertical lines are the theoretical multiplet peaks. The high signal-to-noise ratio is generally high.

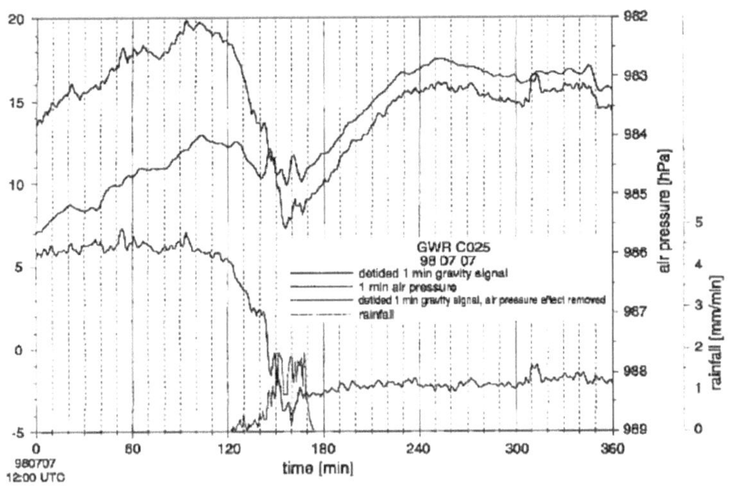

Fig. 2.30. Example of atmospheric mass transport during heavy rain. The signal at the top of the figure (at the start of the record) is gravity with tides removed, and the curve beside it is the pressure. After correcting with a frequency dependent admittance, the residual gravity is the lower curve (left). Note this residual gravity begins to decrease sharply just before the onset of the rain (lowest curve) due to a mass increase above the station that is not seen in the surface pressure.

such as coseismic mass changes associated with of earthquakes, offsets due to rapid atmospheric changes (Figure 2.30), and at periods of minutes the changes in gravity due to hydrological effects such as extreme rainfall. The traditional goal of high accuracy relative gravimetry has been the recording of Earth tides from ter-diurnal to annual periods, mainly for studies of solid Earth an ocean loading tides. Today,

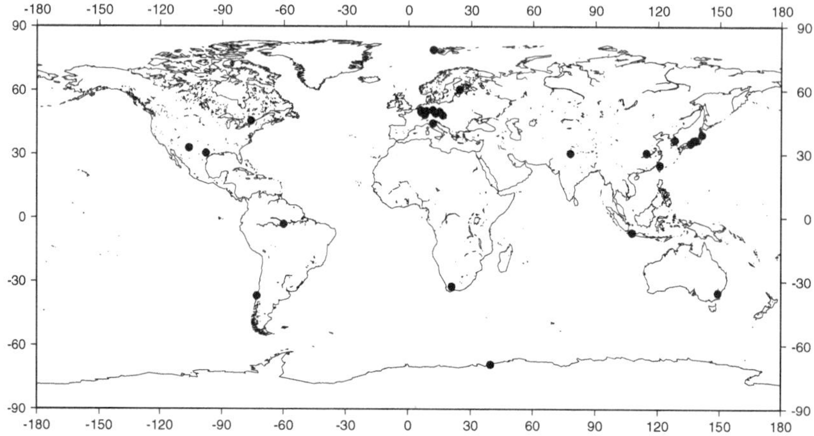

Fig. 2.31. Global network of SG stations contributing to GGP. The stations shown are either operating or planned to start operation in 2007 or 2008.

the solid-Earth tidal component is considered to be a known phenomenon that can be predicted theoretically at the 1 nanogal level. Current interest is in the discrimination between models of ocean tidal loading, which amounts to a few percent of the total tidal signal. With in the frame of the Global Geodynamics Project (GGP), some 30 SG are currently operated or planned in a global network (Figure 2.31, Crossley et al., 1999; Hinderer & Crossley, 2004). .

2.6.2 *Absolute gravimetry*

The low-frequency variations in gravity at a site are usually determined by episodic observations with an Absolute gravimeter (AG). Today, AG are almost invariably of the free-fall type (FG5) manufactured by "microG" (now LaCoste-Romberg). This instrument is an absolute measurement device that registers the value of gravity over the period of time it takes the mass (a small corner cube) to traverse approximately one meter in free fall. Typical measurement sessions take a few days. Single drops carried out every 20 seconds or so have a high scatter, but an accuracy of 1-3 microgal is achieved from the mean of a large number of drops that are done over typical campaign durations of several hours to days. In order to extract the secular signal from these observations, high-frequency variations caused by solid-Earth and ocean loading tides, polar motion and atmospheric loading have to be modeled and corrected for. Hydrological loading is usually not included in these corrections.

In order to check the instrument stability and calibration, intercomparisons between AGs are done every few years. These intercomparisons have established agreement at the level of a few microgal between the best instruments.

AG measurements have been very successful in measuring long-term gravity changes such as the post-glacial rebound in regions such as Fennoscandia and North

America. For example, Figure 2.32 shows the secular trend cause mainly by post-glacial rebound combined with an interesting long-term saw-tooth signal due to episodic slip on the Cascadia subduction zone. In this example the use of a continuous recording SG would enable interpolation between the AG values and thus give the time history of each slip event. At many sites it has become common to perform intercomparisons between the SG and AG instruments, both from the point of view of calibrating the SGs and to monitor the continuous gravity changes during, and in between, the AG observations.

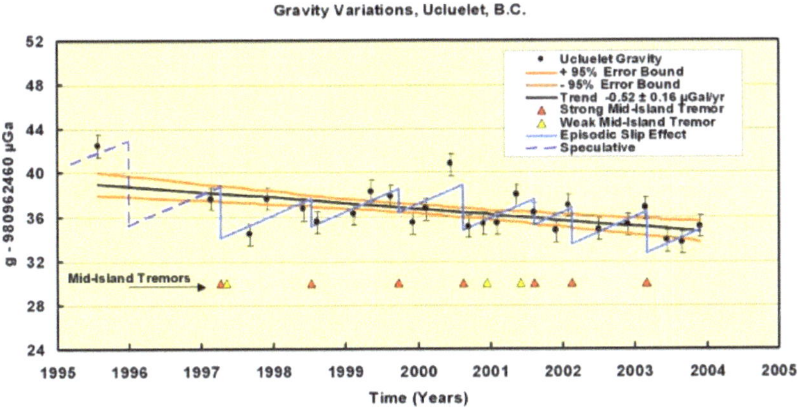

Fig. 2.32. Variations in absolute gravity at Ucluelet (western coast of Vancouver Island) showing some concordance with the episodic slip and seismic tremor activity above the Cascadia subduction zone (figure courtesy of T. Lambert.). The downward trend is due to postglacial rebound.

Gravity changes at a point on the Earth's surface are generally associated with displacements of the Earth surface or some other processes. The gravity anomaly measured by a gravimeter is therefore the sum of the effect due to the vertical motion of the gravimeter through the unperturbed gravity field and the contribution from mass changes in the vicinity of the gravimeter. In order to separate these two effects, gravimeters need to be co-located with geometric instruments such as a GNSS receiver. Wahr et al. (1995) discussed combined gravity and geometric observations, which, in principle, can be used to detect mass changes, for example, in ice sheets, while Plag et al. (2009) showed that spatially distributed observations of secular trends in gravity and vertical displacements constrain the tie between the RFO and the CM, thus supporting SLR in this function.

2.6.3 Land movements and terrestrial gravimetry

Among the terrestrial observation techniques used for estimating vertical land movements, gravimetry is a completely independent method with respect to space geodetic techniques. The task of gravimetry is the measurement of gravity, which is the magnitude of the acceleration due to the force of gravity, and of the gravity gradient at the surface of the Earth, or near to it. Time-dependant gravity variations are important in the study and comprehension of phenomena leading to crustal deformation. The study of crustal deformation plays a key role in the determination of mean sea-level changes. A crustal deformation process implies a variation of the position (coordinates) and a variation of the gravity field. This last because the gravity field is directly affected by the variation of the position of the measuring point (mainly of the vertical component) and because crustal deformation is associated with changes in the density field in the Earth's interior (due to viscoelastic deformation, pre-seismic dilatancy, dislocation or transfer of internal masses). Therefore, the combination of gravity and position changes allows the computation of changes of the potential and can provide important information on the dynamics of the phenomena (Marson, 2000).

Over the last five decades, gravimetry has made impressive progress. The precision of both absolute and relative measurements has improved by almost three orders of magnitude to presently 10^{-9}. The instrumental accuracy of the absolute gravimeter FG5 is about 10-20 nms^{-2} at good stations for a 24 hours observation period (Niebauer et al., 1995). Continuous measurements are not feasible because of the wear and tear of the mechanical system. Van Camp et al. (2005) demonstrated that gravity trends with uncertainties of 1 $nms^{-2}yr^{-1}$ can be achieved over a time span of 7 years with annual observations. A technology to measure the temporal variations of the gravity field continuously at a given site by means of superconducting gravimeters (SG) exists. The SGs are relative instruments but very stable in time. Absolute gravimeter observations taken at the location of a SG allows the identification of outliers and the correction for long-period, mostly environmental signals. In this way the accuracy mentioned above can be achieved in a much shorter time span (Zerbini et al., 2002; Richter et al., 2004). Continuous monitoring of height and gravity changes allows the separation of the gravity potential signal due to mass redistribution from the geometric signal due to height changes and the sound interpretation of crustal deformation processes (Zerbini et al., 2006).

2.6.4 Airborne gravimetry

Airborne gravimetry is an effective way to cover the medium-range wavelengths (10-1000 km) of the Earths' gravity field, supplementing the satellite gravity field missions, which at best gives gravity field information for wavelengths longer than some 400 km (corresponding to 200 km resolution on the surface). The high-resolution gravity field information is essential for determining the geoid with sufficient accuracy, especially relevant for unifying height systems and geometric information around core GGOS sites.

The development of airborne gravity has been made possible by the use of the kinematic GPS technique as well as improvement in airborne gravity acceleration sensor systems (Figure 2.33). Current accuracies are routinely in the 1 to $5 \cdot 10^{-5}$

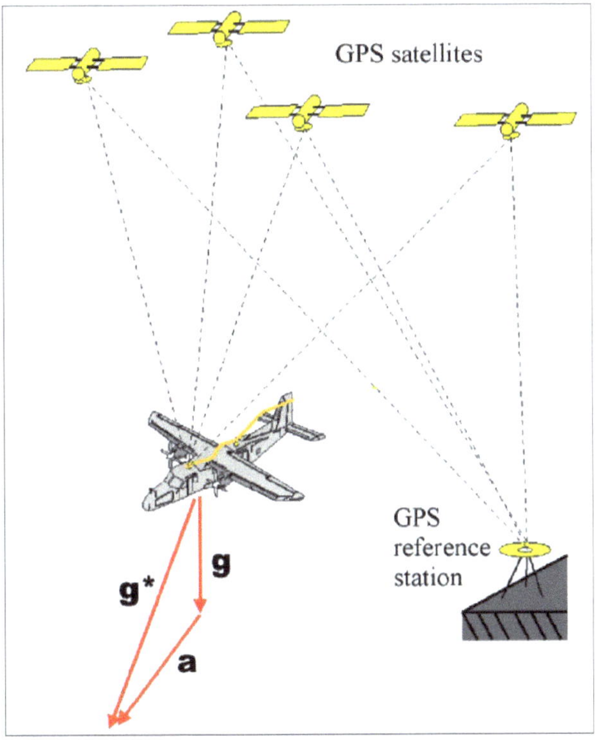

Fig. 2.33. Principle of airborne gravimetry.

ms^{-2} r.m.s. domain, with relatively large differences between different sensor systems and implementations. Major commercial airborne activities are ongoing in connection with geophysical exploration for oil and gas; for mining airborne gradiometry systems at accuracies of 1 E or better have been developed in recent years. Commercial gravity and gradiometry survey projects are generally restricted to relatively small areas, and data are usually not available for more widespread geodetic use. Long-range airborne gravity surveys for geodetic gravity field applications (geoid and spherical harmonic reference models) have been operational since the early 1990s, and many regions of the Earth has been covered, including major parts of the Arctic, and major countries such as Malaysia, Mongolia, Afghanistan and Ethiopia. Currently US, European, Russian and Chinese groups are active in carrying out such surveys.

Albeit many airborne surveys are currently classified or proprietary, experience has shown that many such surveys may fully or in part be included in future high-resolution spherical harmonic reference models. Such reference models, like the new EGM2008, complete to degree and order 2159, would be the major static gravity field product of GGOS. To improve the quality of such models, generally there is a need for continued surveys in many inaccessible areas of the globe, especially the Amazon, mountainous areas, large parts of Africa, coastal regions (high accuracy geoid across the coastal zones) and especially Antarctica, which is the largest continental void of gravity on the globe. Coordinated global surveys should be accompanied by effort to secure release of terrestrial gravity data, still unavailable for large parts of the Earth.

2.6.5 Satellite missions

Artificial satellites have played a dominating role in determining the gravity field of the Earth since the early sixties (e.g., Kaula, 1966). The observations of non-Keplerian variations in orbital motion using either terrestrial (radio or laser) tracking or space-based GPS have long been analyzed to extract the long-wavelength components of the gravity fields. Earth gravity models such as the EGM96 (Lemoine et al., 1998) used decades of tracking data to Earth orbiters to derive the mean long-wavelength gravity models. Determinations of the time-variability were limited to the hemispheric scales, however. The significance of time-variable gravity to climate sciences was well established from the study of three-decade long time series of the Earth's oblateness (J_s), determined from satellite laser ranging to LAGEOS satellites, and showing clear signals from Post-Glacial Rebound (PGR), atmospheric and hydrological mass redistribution, and ice-mass changes (e.g., Cheng & Tapley, 2004).

Data from a large number of space missions have contributed to the determination of the Earth's gravity field in the past. Some recent examples are given in Table 2.2. In addition, terrestrial and space-based tracking to nearly twenty satellites, some dedicated to geodesy and other missions of opportunity, has contributed to the determination of the Earth's gravity field in the past.

A significant step forward in the determination of the gravity field from satellites with respect to resolution and precision has been provided by the satellite missions CHAMP (e.g., Reigber et al., 1999), and GRACE (e.g., Tapley et al., 2004b,a), in orbit since 2000 and 2002, respectively. GRACE (Figure 2.34) has enabled the improvements in our knowledge of the static gravity field to centimeter level accuracy in the geoid determination to spherical harmonic degree 70, with further improvements forthcoming as longer data spans are analyzed. The European GOCE mission (e.g., Le Provost et al., 1999; Drinkwater et al., 2003; Ilk et al., 2005) will complement the results achieved so far with an extremely high precision and resolution of the static part of the gravity field.

Gravity field determination using space missions has also contributed tremendously to advances in geodesy. Improvements in gravity field models obtained over the last three decades have gone hand-in-hand with improvements in the reference frames and Earth orientation from the LAGEOS and other low-orbiting satellite laser-ranging targets. The innovative sensor technologies used in these gravity field missions have already contributed to a substantial improvement of the Earth static gravity field recovery (e.g., Reigber et al., 2003; Tapley et al., 2004b). Figure 2.35 shows the dramatic improvement of the gravity field during the last decade. Gravity field models from GRACE have benefited the space geodetic analysis of the DORIS tracking data (Willis & Heflin, 2004). They have been used to improve the

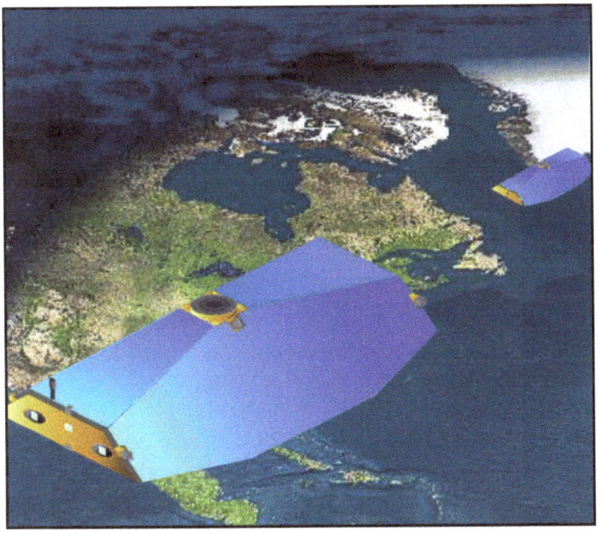

Fig. 2.34. The GRACE satellites.

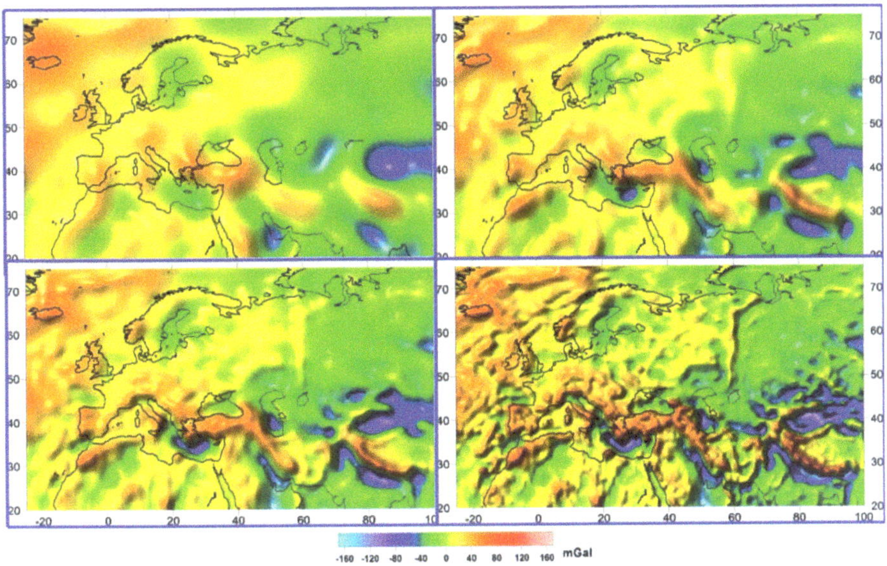

Fig. 2.35. Improvement of the Earth's gravity field models. The models are (from top left to bottom right) GRIM-5S1: Best gravity field model before CHAMP and GRACE computed from SLR data only; EIGEN-CHAMP03S: Gravity field from CHAMP; EIGEN-GRACE03S: Gravity field from GRACE; EIGEN-CG03C: Gravity field from GRACE combined with terrestrial data. Reprinted from Reigber et al. (2005), Copyright 2005, with permission from Elsevier.

knowledge of the orbits of ocean radar altimetry satellites (Haines et al., 2004), and for laser altimeters, thereby enhancing the geodetic contributions from other space missions. Gravity missions are also of central importance for altimetry, because the precise geoids are required to refer the sea surface topography to the geoid. The

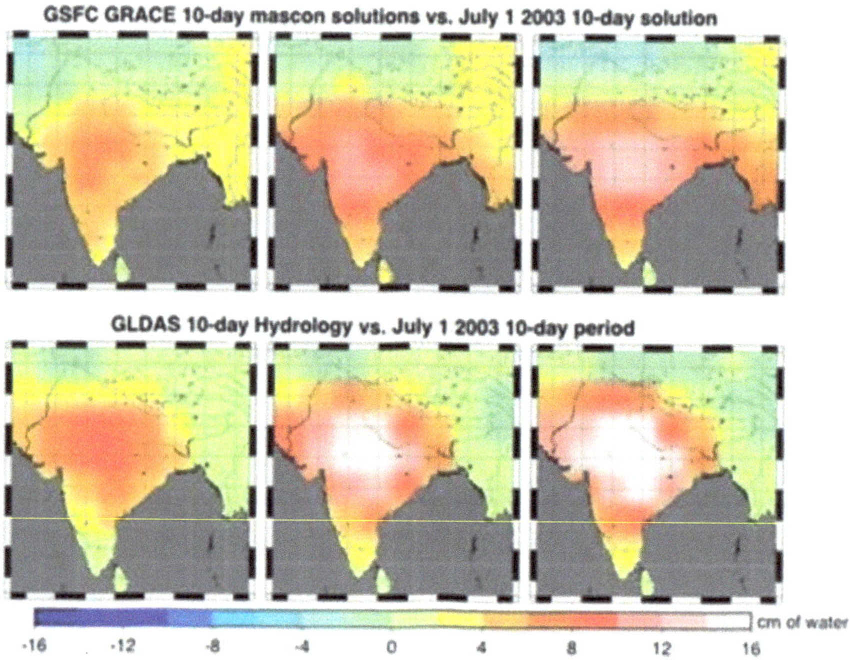

Fig. 2.36. GRACE-determined variations in water storage on land. Upper row: Ten-day estimates of the mass change with respect to a multi-year averaged gravity model in a $4^\circ \times 4^\circ$grid. The values shown are the mass change mapped into an equivalent change in a surface layer of water in units of cm. To estimate these values, the effects of atmospheric pressure changes and solid Earth and ocean tides have been removed based on model predictions. Lower row: Numerically modeled soil moisture and snow mass fields from the Global Land Data Assimilation System (GLDAS) by Rodell et al. (2004). From http://grace.sgt-inc.com/.

integration of all the satellite missions with the existing space-geodetic techniques for the determination of the Earth's shape creates new opportunities to determine and study the mass transport in the Earth system in a globally consistent way (e.g., Kusche & Schrama, 2005; Wu et al., 2006; Gross, 2006) or to derive information on changes in part of the water cycle (e.g., the large ice sheets, see Velicogna & Wahr, 2005, 2006).

The GRACE mission in particular is providing unprecedented insight in the water cycle on regional scales and on intraannual to submonthly time scales. This mission is designed to monitor local, regional, and global changes in the Earth's gravity field. The changes observed in the gravity field are mainly caused by mass transport in the hydrology cycle, in particular the oceans, atmosphere, and on land. Analysis of the data delivered by GRACE using an approach based on Stokes coefficients yields a direct measure of mass flux with high spatial resolution on the Earth's surface with a temporal resolution of one month and a spatial resolution of $\sim$500 km (e.g., Wahr et al., 2004; Davis et al., 2004; Tapley et al., 2004b,a; Crowley et al., 2006). Recent developments using a mass concentration (mascon) approach have been successful in recovering submonthly mass flux at a high spatial resolution over certain regions of interest. The mascon gravity representation largely mitigates the spatial and temporal aliasing problems encountered with monthly GRACE solutions using Stokes coefficients (Luthcke et al., 2006).

Figure 2.36 shows a time series of discrete ten-day estimates of the mass change with respect to a multi-year averaged gravity model in a $4^\circ \times 4^\circ$grid for the Indian sub-continent and adjacent land areas together with the predictions of the GLDAS. GLDAS ingests satellite- and ground-based observational data products and uses advanced land surface modeling and data assimilation techniques, in order to generate optimal fields of land surface's hydrological state and its fluxes (Rodell et al., 2004). Agreement of the GRACE-derived and model predicted changes in water mass are on the few centimeter level.

2.7 Observing time

2.7.1 Relativity: proper and coordinate time; realized time scales

Relativity distinguishes locally measurable (proper) quantities from coordinate quantities which are, by definition, dependent on conventions. Therefore one should distinguish proper time, which is the output of an ideal clock, from coordinate time, which is one of the coordinates chosen to represent the four-dimensional space time. In its Resolution A4, the IAU in 1991 explicitly introduced general relativity as the theoretical background for space-time reference frames. For the geocentric system, it defined two time coordinates the Geocentric Coordinate Time (TCG) and Terrestrial Time (TT), which differs from TCG by a constant rate so that the scale unit of TT agrees with the International System of Units (SI) second on the geoid. To account for upcoming improvements in accuracy, the IAU refined the relations between these relativistic coordinate times in its Resolution B1 in 2000. International Atomic Time (TAI), established by the Bureau International des Poids et Mesures (BIPM), is a realization of TT. TAI has stability well below $1 \cdot 10^{-15}$ for averaging times between 5 days and 6 months and can be accessed with an uncertainty of about 1 ns with modern time transfer techniques (see Section 2.9.4). UTC differs from TAI by an integer number of seconds. UTC has therefore the same metrological characteristics as TAI and is universally used to date events.

2.7.2 Geodetic measurements and geodetic coordinates

Time enters geodesy in (at least) two ways. First, present-day geodetic measurements (VLBI, GNSS, Doppler, Laser ranging, Radar) are all based on local measurements of proper time or frequency. These raw measurements are subsequently processed to obtain geodetic coordinates. Second, a reference coordinate time scale is required to date all measurements and results. Because the magnitude of relativistic effects in the vicinity of the Earth is close to 10^{-9} in relative value, a complete relativistic treatment is mandatory for all techniques. As a result, (geodetic) coordinates must be understood in a fully relativistic sense and have no direct relationship with a measurable quantity (meter stick). However, coordinate differences, for example, between results from different techniques or the variation of coordinates with time, are small enough to be directly interpreted as physical quantities, provided that the different sets of coordinates have been determined in a consistent manner. Note that the IUGG in 1991 adopted the IAU relativistic framework to define its CTRS. However, as two time coordinates are possible (TCG and TT), geodetic coordinates may differ in scale by $7 \cdot 10^{-10}$ depending on the time coordinate used.

2.7.3 Clocks and geodesy: future trends

The performances of clocks, counters, and other time/frequency devices seem, at least in principle, sufficient to cover the present and foreseeable needs of geodetic measurements. However progress is needed on the one hand in calibration techniques, in order to obtain unbiased measurements. On the other hand, the requirements posed by geodesy to a reference coordinate time scale seem to be fulfilled. For example, a 1-year integration of the motion of a satellite with 1 mm accuracy requires about $1 \cdot 10^{-15}$ accuracy in the reference time scale.

Nevertheless improving clocks and timescales should provide several improvements related to geodesy, in two domains. First, some progress is possible in the geodetic techniques: for example, GNSS will benefit from more stable clocks on board satellites by allowing less frequent updates of clock parameters and yielding a better modeling, i.e. a better determination, of the transmitted clock parameters. VLBI could also benefit from more stable clocks at the stations, however this would necessitate that the entire hardware chain has stability characteristics similar to those of the clock itself. Second, the development of a new domain, that of relativistic geodesy, can be envisaged. Because the relativistic frequency shift experienced by a clock is about 10^{-16} per meter of altitude at the surface of the Earth, clocks accurate to $1 \cdot 10^{-17}$ or $1 \cdot 10^{-18}$ can sense geopotential with 10 cm or 1 cm accuracy, respectively, with respect to some reference. This reference would be free of the limitations inherent to any geophysical realization like the geoid. Ultimately, the fundamental time/frequency reference would be provided by accurate clocks in space, where the relativistic frequency shift can be modeled with $1 \cdot 10^{-18}$ accuracy, while accurate clocks on Earth would be measuring the geopotential. It would also be necessary to reconsider the definition and procedure of realization of TAI in order to benefit from such improvements, in the accuracy range $1 \cdot 10^{-17}$ and below.

Important progress has been accomplished in recent years, bringing new horizons to terrestrial time scales and promising the future development of new ultra-stable and ultra-accurate clocks. Two main directions are being explored for these clocks: laser cooling of atoms and ion traps. In the first direction, several Cs fountains have been in routine operation since the early 2000s, and they realize the SI second with uncertainties that, since 2006, reach a few parts in 10^{16}. It is expected that an accuracy of $1 \cdot 10^{-16}$ may be reached with such a fountain and that a fountain using rubidium atoms may be even more stable. Based on a slow beam of cold atoms, similar devices operating in space in zero gravity may reach an accuracy of $1 \cdot 10^{-17}$. A first step towards operating such clocks in space will be PHARAO/ACES which should fly on board the ISS in 2013. In the second direction, clocks based on optical transitions promise to achieve still better performance in stability and in accuracy, thanks to a transition frequency several orders of magnitude larger. Already in 2006, a clock based on a transition in Hg+ has demonstrated that all systematic effects could be modeled at the level of $7 \cdot 10^{-17}$. The prospects of relativistic geodesy look bright, even though the technical challenge is formidable.

2.8 Ensuring consistency of the observations of geometry, gravity field, and rotation

The "observations" that GGOS will eventually disseminate are really the products of the various supporting IAG Services, i.e., results of the analysis and reduction of the raw observations gathered by various ground and space-based systems. Consistency across these products can only be assured if the raw data are collected using consistent standards and practices, and if their analysis and reduction follows again consistent standards and conventions across all three pillars. Of similar importance is the integration of the various techniques on the observation level, that is through co-location of techniques at the same location and with known local ties between the respective reference points. In the following, we first summarize the situation concerning co-location and then describe the main issues related to common standards and practices across the techniques.

2.8.1 Consistency through co-location

Co-location of techniques at the same location is not only a means to ensure consistency across techniques but it allows full exploitation of the different strength of the individual techniques and mitigation of their weaknesses. Core geodetic sites are those site with three or more space-geodetic techniques co-located and connected through well monitored (on the level of 1 mm) local ties between the techniques. In most cases, a core site will include at least three out of SLR, VLBI, GNSS, and DORIS and also be co-located with absolute and relative (superconducting) gravimeters. However, the number of core sites with three or more of the space-geodetic techniques co-located is only of the order of fifteen sites (see Table 2.6 for

Table 2.6. Co-location sites. Listed are those stations that currently have three or more space-geodetic (geometric) techniques co-located.

Site Name	Latitude	Longitude	GNSS	SLR	VLBI	DORIS	Gravimeter (1)	
							Cryogenic	Absolute
Arequipa	-16.47	-71.49	X	X	-	X	-	-
Concepcion	-36.84	-73.03	X	X	X	-	X	X
Greenbelt	39.02	-76.83	X	X	X	X	-	-
Hartebeesthoek	-25.89	27.69	X	X	X	X	X(2)	X
Kokee Park	22.13	-159.66	X	-	X	X	-	-
Matera	40.65	16.7	X	X	X	-	-	-
McDonald/Fort Davis	30.68	-104.01	X	X	X	-	-	X
Metsahovi	60.22	24.7	X	-	X	X	X	X
Monument Peak	32.89	-116.42	X	X	-	X	-	-
Mount Stromlo	-35.32	149.01	X	X	-	X	X	X
Ny Alesund	78.93	11.87	X	-	X	X	X	X
Shanghai	31.10	121.20	X	X	X	-	-	-
Simeiz	44.41	33.99	X	X	X	-	-	-
Syowa	-69.01	39.58	X	-	X	X	X	X
Tahiti	-17.58	-149.61	X	X	-	X	-	-
Wettzell	49.14	12.88	X	X	X	-	X	X
Yarragadee	-29.05	115.35	X	X	X(3)	X	-	-

NOTES:
(1) Where there is a SCG operating it is assumed that there will also be ABSOLUTE measurements done, since they are part of the SCG's calibration process.
(2) Located in Sutherland
(3) Future VLBI occupation

the current network of core sites) and in fact decreasing over time. In the late 1990s, more than 20 core sites existed, as demonstrated by the larger number of core sites used for the determination of ITRF2000 (Figure 2.37, upper diagram).

The international space geodesy network has recently suffered several debilitating closures and reductions in the last several years due to budgetary cutbacks. NASA support for the SLR stations at GSFC and Texas has been reduced to single shift. The Arequipa and Maui stations have recently reopened after a 2 year hiatus. The budgetary situation has also been a factor in the delayed completion of the Next Generation SLR (NGSLR). Despite clear international recommendations to avoid a "weekend effect" on space-geodetic products, budget cuts in Italy forced weekend operations at the Matera station to be discontinued. There, also lack of funds for necessary maintenance and analysis have hampered the station operation and data processing. In 2005 and 2006, the VLBI network lost both the Algonquin and Yellowknife stations as a result of Canadian government budget cuts and the Gilmore Creek/Fairbanks station in Alaska due to NASA funding reductions. Several stations have been threatened with closure which was averted in part through strong international support.

Table 2.6 also indicates the gravimetric observations carried out at the core sites. Only about half of these sites have co-located superconducting gravimeters. On the other hand, the GGP currently operates or has plans to start operation for about 30

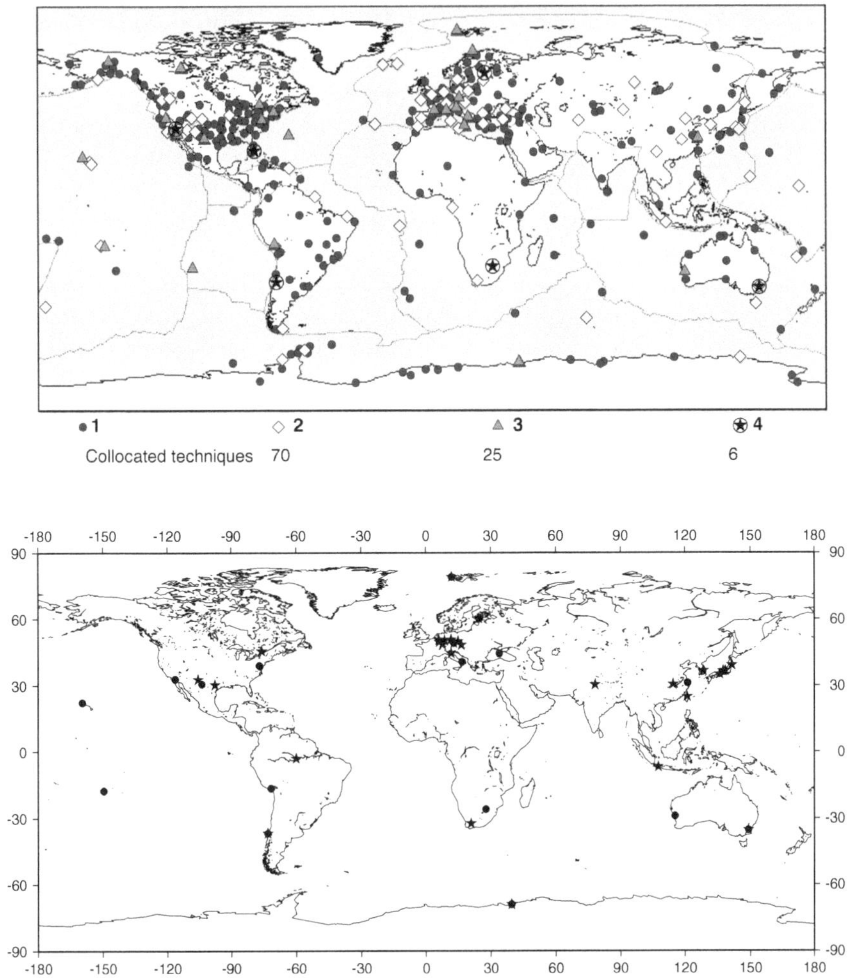

Fig. 2.37. Network of core geodetic sites and the temporal evolution. Upper diagram: Sites used in the determination of ITRF2000, which included 25 sites with three co-located techniques. Lower diagram: Current network of core sites with three or more space-geodetic stations co-located. Circles are core sites, stars indicate the GGP stations (see Section 2.6.1).

superconducting gravimeter sites (see Section 2.6.1). Figure 2.37 illustrates these two networks, and shows that enhanced coordination of the GGP station selection with the choice of core sites has the potential to significantly increase the tie between the geometric and gravimetric techniques.

2.8.2 Consistency of data collection and processing: conventions

Since the very early days, international geodesy has always adhered to some form of standards and conventions, the best known of which being the Geodetic Reference System (GRS), revised appropriately on decadal scales, the last version being GRS80. GRS consistently covered geometry, gravity and rotation, albeit at the very top level of required constants and the most basic formulae, with an eye towards classical techniques and approaches, which at the time were still the main source of geodetic products. At that time however, a new project was conceived and successfully executed with international participation at all levels, including design, execution and evaluation; a project that would eventually lead geodesy from the classical era to that of the space age. The project Monitoring Earth Rotation and Inter-comparison of Techniques (MERIT) (see e.g., Mueller et al., 1982), acted as the pilot for what was later to become the IERS. Along with it came an expanded compilation of constants and standard formulas, mostly associated with the reference frame and Earth rotation, to be used by the project participants. These came to be known as the MERIT standards and with the establishment of the IERS, they became the basis for the development of the IERS Conventions as we know them and use them today (for the last version, see McCarthy & Petit, 2004).

While, at the beginning, the Conventions mainly served as a guideline for the purpose of data analyses and reduction for Earth orientation monitoring only, they gradually developed as "the" reference for geometry and reference frame work as well, including all aspects of the required techniques, from geometric modeling of the observables to all of the required geometric and dynamic corrections in order to achieve the accuracy that IERS expected for these products. To achieve this, the Conventions slowly expanded to encompass models and constants that were well beyond the observations for geometry and rotation, including the gravity field and all of its temporal variations (tides and secular changes as well as loading effects from the oceans and atmosphere), relativistic corrections and environmental corrections (e.g. atmospheric delays). The area where these Conventions are focused is that of the space geodetic observations, leaving out most of the constants and practices for ground-based geodesy. This is perhaps due to the fact that the products that concern IERS are of global nature and none of the ground-based geodetic techniques can contribute significantly or compete with the satellite-borne or space-based techniques. Looking at it from a spectral view, they cover the long-wavelength part of the spectrum of products. Geodesy however can deliver significant information at the high-frequency end of the spectrum, albeit in some areas only. One of these areas, the most important one, is that of the gravitational field of Earth. Ground and airborne surveys provide very high quality and high-resolution local information that is used along with the long-wavelength information obtained from spaceborne instruments (CHAMP, GRACE, GOCE), to develop extremely high resolution global Earth gravity models that will never be derived from space data alone (see, e.g., Reigber et al., 2005, , and the new EGM2008 complete to degree and order 2159). This is the area that the Conventions need to cover in more detail, both, in the description of the required constants and the standard formulas and practices in reducing such data. Once this is accomplished, the foundations of all three pillars will be ably supported by the same, unique set of Conventions and Standards.

While the expansion and enrichment of the existing Conventions and Standards is a rather simple task, the actual enforcement in practice is by far a more challenging task. While most institutions seek to be part of the appropriate IAG Service in order for their products be granted the seal of approval from that Service, it is usually very difficult to force the required changes in the software and the procedures followed by that institution to make it conform with the IERS rules. As most Services discovered, it took years for the various Analysis Centers within a technique to achieve this harmonization. It will take quite an effort to ensure that this harmonization exists also across techniques, since the geodetic products are for the most part a combination of inputs from several if not all of the Services.

An even more difficult and taxing effort will be required in making sure that not only the same constants, theoretical or empirical models, and reduction procedures are consistent, but also all of the background information used in forward-modeling geophysical processes are also consistently derived and applied in the various analyses and reductions of geodetic observations.

When all the above are accomplished, there is still going to be an issue concerning the parameterization of the same effects across techniques. Recognizing that not all techniques are equally sensitive (or sensitive at all) to all of the "geodetic products", we will need to identify what parameters each technique should deliver and at what frequency, in order to ensure that this information can be easily and readily combined with inputs from other techniques. This issue has been given enough attention for the set of parameters that cover the geometric and rotational group, with only minor attention given to some very long-wavelength gravity information.

To some extent this approach has been reasonable since the very short wavelength gravitational information is well below the sensitivity of any space technique at this point, and for many years to come. There are other areas though where part of such information can be applied in a different form, as a constraint to the results obtained from the global space techniques. For example, incorporating some absolute gravity measurements at a few points on Earth in the development of a precise orbit from some type of tracking data is practically meaningless. On the other hand, imposing a constraint on the height change of a tracking station based on repeated absolute gravity measurements at that site is a very useful piece of information independent of the primary source of data determining the position and motion of that site. A global network of combined absolute gravity and space-geodetic stations can constrain the tie between the RFO and CM (Plag et al., 2009).

Such synergistic use of various inputs with a common, single output can only be done if the information from all sources adheres to one set of conventions.

2.9 Essential additional observations and applications

2.9.1 Atmospheric sounding

Besides the variables of direct geodetic interest, the space-geodetic infrastructure enables soundings of the atmosphere and ionosphere by electromagnetic waves of the GNSS. Properly equipped GNSS receivers on the ground can for example observe the integrated precipitable water vapor content in the atmosphere and the total ionospheric electron content in the ionosphere, respectively.

A number of studies conducted in the 1990s have shown that the amount of precipitable water contained in the neutral atmosphere can, in fact, be retrieved using ground-based GPS receivers (Figure 2.38). Assimilation of this information from ground-based GNSS networks into numerical weather forecasting models may improve particularly the prediction of extreme events (e.g., Elgered et al., 2005). Practically, zenith total delay observations collected by European ground-based GPS stations are assimilated operationally in numerical weather prediction since 2006 (Poli et al., 2007).

In addition, it has been demonstrated (Kursinski et al., 1995) that a GPS receiver aboard a microsatellite in a low Earth orbit, supported by a ground-based network of receivers, can be used to collect observations of atmospheric refraction as a function of altitude during the event of satellite occultation by the Earth's atmosphere and ionosphere (Figure 2.39). Thus, the availability of remote sensing observations from GPS radio occultation sensors provides a unique opportunity to improve the quality of ionospheric and meteorological analyses, particularly over the traditionally under-sampled regions, as well as promise higher vertical and temporal resolutions, if a sufficient number of sensors is launched and supported by an adequate ground-based tracking network.

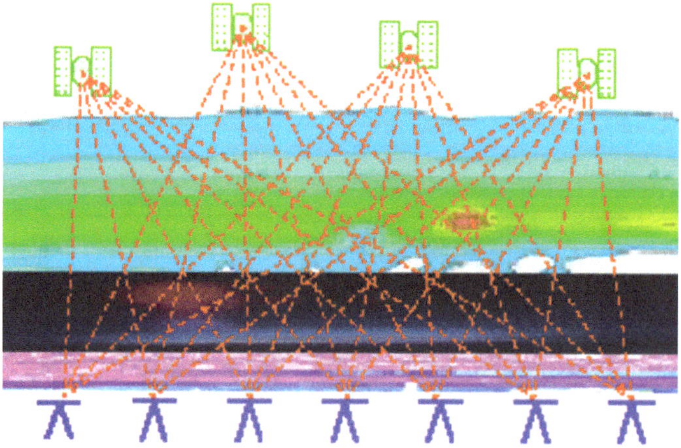

Fig. 2.38. Atmospheric sensing with ground-based GPS receivers.

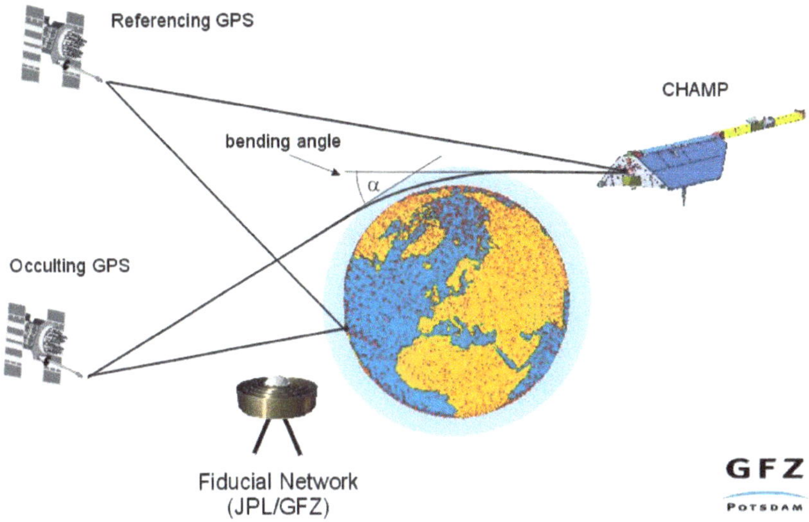

Fig. 2.39. Geometry of GPS occultation illustrated here with the CHAMP satellite.

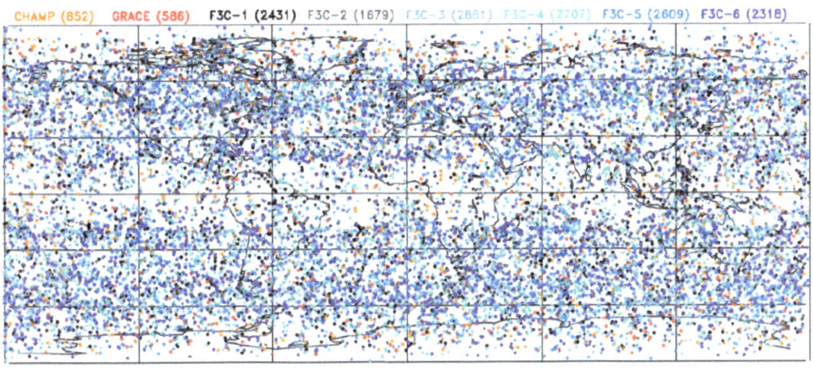

Fig. 2.40. Global coverage of GPS radio occultations. Shown are geographic locations of the soundings collected by CHAMP, GRACE, and FORMOSAT-3/COSMIC (F3C), 1–7 March 2007, and as received in near-real time by national numerical weather prediction centers. The number of GPS radio occultations collected by each GPS receiver is shown in parentheses.

As of 2008, there are twelve satellites in orbit carrying GNSS occultation-capable receivers: FORMOSAT-3/COSMIC (F3C) 1 to 6, METOP, CHAMP, GRACE-A and B, TERRASAR-X, SAC-C. Only the first nine of these produce near-real time observations of GNSS occultations. Such GNSS occultations are particularly promising for meteorological applications and are already today providing routinely information to operational weather services. Figure 2.40 shows the spatial coverage achieved by the radio occultation experiments CHAMP, GRACE, and the six-satellite F3C (Anthes et al., 2008). Each point on the map corresponds to a radio

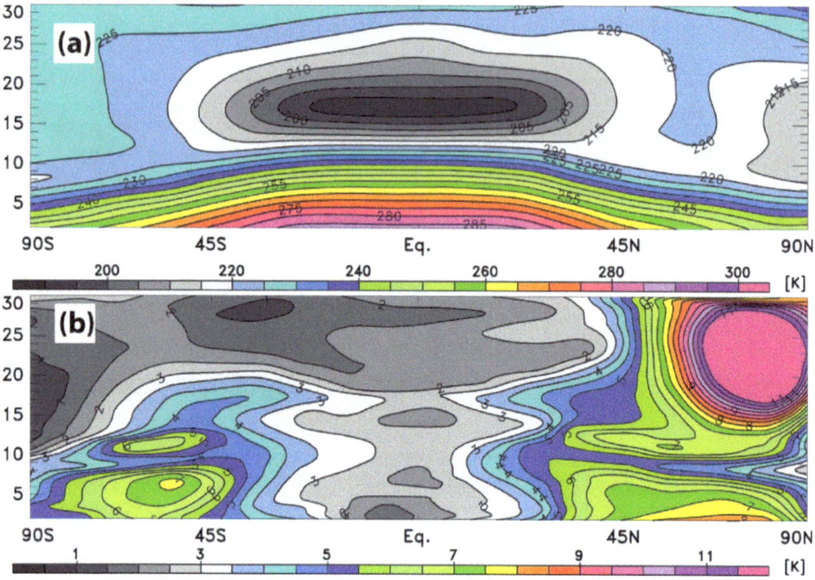

Fig. 2.41. Atmospheric temperature retrievals from GPS radio occultations. (a) Zonal average of one week of FORMOSAT-3/COSMIC retrievals (March 1-7,2007), binned to 1 km vertical and 10 degrees latitude resolutions. (b) Standard deviation of the temperature retrievals within each bin.

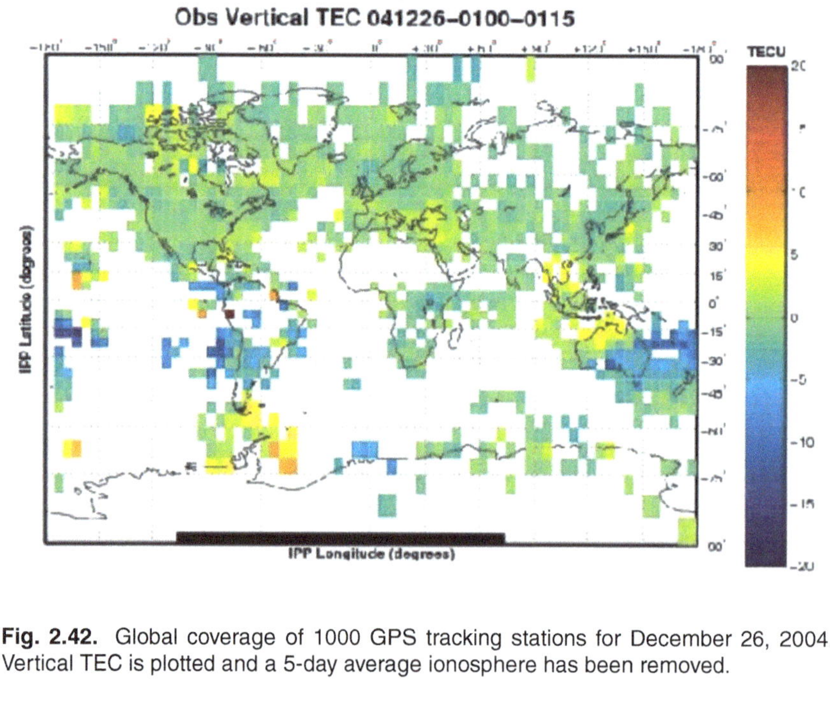

Fig. 2.42. Global coverage of 1000 GPS tracking stations for December 26, 2004. Vertical TEC is plotted and a 5-day average ionosphere has been removed.

occultation event probing the neutral atmosphere from the near-surface up to the upper stratosphere (about 40 km altitude) at 200 meter vertical resolution. The data for the points shown were received by national weather prediction centers between March 1-7, 2007. Figure 2.41a shows the zonal mean temperature retrieved from the F3C GPS refraction measurements. Note that in the lower troposphere (below about 7 km altitude in the tropics, about 2 km altitude in the mid-latitudes), the retrieval of temperature information from GPS radio occultations requires the use of *a priori* information as constraints and hence this information cannot be considered as completely independent measurements. The zonal temperature structure observed by the sole F3C retrievals is consistent with known climatology (for example, tropopause around 15 km altitude in the tropics, double structure around 60°N latitude). As expected, the Tropics present a smaller variability than the mid-latitudes (Figure 2.41b). A region of strong variability can be observed in the stratospheric Northern polar vortex as the Arctic region emerges from the winter polar night. Because of the multitude of receivers, these results can be generated with only one week of GPS radio occultation data with high vertical resolution. In the future, more GNSS receivers in space could decrease the time needed to get such a global picture of the atmosphere. The temperatures retrieved from the GPS radio occultation technique as shown here are invaluable in the sense that they provide atmospheric physicists with a fairly new and now near-complete coverage of the Earth's atmospheric mass field in the upper troposphere and stratosphere, complementing passive measurements from existing infra-red and micro-wave sounders.

Another emerging technique for atmospheric sounding is LIDAR, which, in principle, can be used to measure atmospheric CO_2 (see Section 2.4.5). One currently developed approach is a ground based zenith viewing LIDAR to measure CO_2profiles as function of time (roughly hourly) with an altitude range of a few km, that is essentially to the top of the boundary layer (Burris et al., 2006). The other alternative is a down-looking CO_2 sounder to measure CO_2 content in the column below an aircraft or, eventually, from space (Abshire et al., 2007). The implication of these new developments are further discussed in Section 5.7.1.

2.9.2 Ionospheric remote sensing: one person's signal is another person's noise

The signals from the GNSS satellites must travel through the earth's ionosphere on their way to receivers on or near the earth's surface. To achieve the highest possible positioning accuracies for geodetic and surveying applications, one must correct for the propagation delays imposed on the signals by the ionosphere. Whereas these effects may be considered a nuisance by most GNSS users, they will provide the researchers with an opportunity to use GNSS satellites as a tool to better understand the plasma surrounding the Earth. The dispersive nature of the ionosphere makes it possible to measure its TEC using dual-frequency e.g., GPS observations collected by ground and spaceborne receivers.

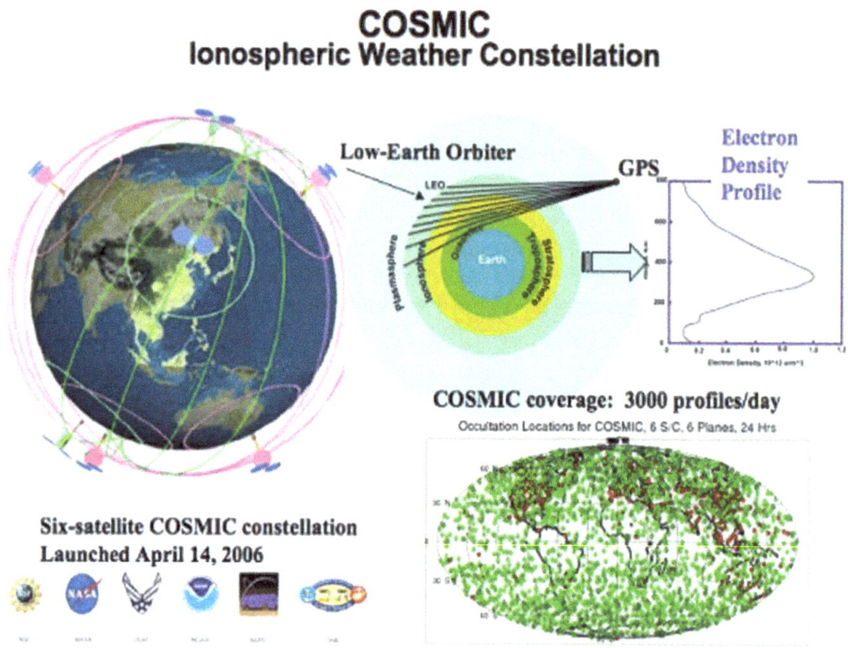

Fig. 2.43. Schematic view of COSMIC ionospheric occultations and the expected 3000 daily profiles.

There are a number of techniques available to mitigate the ionospheric effect including global empirical and physics-based ionospheric models. For geodetic applications, the most effective technique has been to use dual-frequency GPS observations to estimate TEC. Between 1997 and 2007, the number of GPS ground receivers has increased approximately by an order of magnitude. Currently, there are more than 1500 globally-distributed dual-frequency, ground-based GPS receivers available using publicly accessible networks including, for example, the IGS and Continuously Operating GPS Stations (CORS). To take advantage of the vast amount of GPS data worldwide, researchers use a number of techniques to estimate parameters e.g., satellite and receiver inter-frequency biases, directly affecting the GPS TEC measurements of the ionosphere. Most techniques estimate vertical ionospheric structure and, simultaneously, hardware-related biases treated as nuisance parameters (e.g., Mannucci et al., 1998, 1999; Schaer et al., 1998). Other approaches take advantage of all available GPS receivers and calibrate the biases using processing algorithms based on Global Ionospheric Mapping (GIM) techniques developed at various research centers (for illustration using about 1000 GPS stations, see Figure 2.42). These techniques are designed to estimate receiver biases for all stations in the global network and solve for the instrumental biases by modeling the ionospheric delay and removing it from the observation equation (Komjathy et al., 2005).

We seem to be in the midst of a revolution in ionospheric remote sensing driven by not only the abundance of ground but also the space-based GPS receivers, new UV remote sensing satellites, and the advent of data assimilation techniques for space weather. The GLONASS constellation is nearing its completion and GALILEO satellites are expected to contribute significantly to ionospheric data coverage starting in the early next decade. As for spaceborne data coverage in particular, the COSMIC 6-satellite constellation was launched in April 2006 (see Figure 2.43). COSMIC now provides unprecedented global coverage of GPS occultations measurements (1700 per day as of May 2007), each of which yields electron density information with unprecedented $\sim$1 km vertical resolution. Calibrated measurements of ionospheric delay suitable for input into assimilation models is currently made available in near-real time (NRT) from COSMIC with a latency of 30 to 120 minutes. Similarly, NRT TEC data are available from two worldwide NRT networks of ground GPS receivers ($\sim$75 5-minute sites and $\sim$125 additional hourly sites, operated by NASA JPL and others). The combined NRT ground and space-based GPS data sets provide a new opportunity to more accurately specify the 3-dimensional ionospheric density with a time lag of only 15 to 120 minutes. With the addition of the vertically-resolved NRT occultation data, the possibility exists of retrieving the hour-to-hour ionospheric "weather" much more accurately than previously possible.

New Global Assimilative Ionospheric Model (GAIM) techniques are used to monitor space weather, study storm effects, and provide ionospheric calibration for various users including NASA flight projects. GAIM is a physics-based 3D data assimilation model that uses both 4DVAR and Kalman filter techniques to solve for the ion and electron density state and key drivers such as equatorial electrodynamics, neutral winds, and production terms (e.g., Mandrake et al., 2005; Schierless et al., 2004; Spencer et al., 2004). Daily GAIM runs typically accept as input ground GPS TEC data from more than 1200 sites, occultation links from CHAMP, SAC-C, and the COSMIC constellation, UV limb and nadir scans from the TIMED and DMSP satellites, and in situ data from a variety of satellites (DMSP and C/NOFS). Real-Time GAIM (RTGAIM) ingests multiple data sources in real time, updates the 3D electron density grid every 5 minutes, and solves for improved drivers every 1-2 hours.

The abundance of ground and space-based GPS ionospheric observations is expected to help create new and exciting applications including e.g., space weather monitoring during ionospheric and geomagnetic storms (e.g., Fedrizzi et al., 2005) and developing a tsunami early warning system using GPS-derived ionospheric signals. Researchers have shown considerable progress in understanding the geophysics of tsunami-atmosphere coupling and determine the feasibility of using GNSS technology as part of an improved future tsunami warning system complementing more traditional methods of tsunami detection.

2.9.3 Tide gauges

Sea level measured by tide gauges is an important parameter for geodesy for several reasons. For example, geodetic datums in most countries have been defined historically in terms of sea level measured at their coasts. A second example concerns the linkage of GGOS to other components of global observing, notably the Global Climate Observing System (GCOS) and the Global Ocean Observing System (GOOS).

Historical tide gauge records are mainly derived from float and stilling well devices. Tide gauges based on mechanical float devices have lasted for more than 150 years. Still in 1983, a survey conducted by the International Oceanographic Commission (IOC) of United Nations Educational, Scientific and Cultural Organization (UNESCO) showed that 94% of the tide gauges were mechanical. The situation has considerably changed since then. The floating gauges are progressively replaced by new technologies. Modern types of gauges are mainly based either on the measurement of the subsurface pressure, or on the measurement of the time of flight of a pulse, acoustic or radar. It is worth pointing out here that, whatever the technique is employed, the basic quantity provided by tide gauges is an instantaneous height difference between the level of the sea surface and the level of a fixed point on the adjacent land. Hence, tide gauges not only record ocean tides but also a large variety of sea-level signals that can be caused by variations in atmospheric pressure, density, currents, continental ice melt, as well as vertical motions of the land upon which the measurement instrument is located. The recorded processes have characteristic time scales from several minutes to centuries. Many other scientific applications than tidal research and modelling may therefore benefit from tide gauge records (Woppelmann et al., 2006).

Sea-level data from tide staffs or tide gauges have been used for more than a century to establish vertical reference systems on land and on sea in order to define the height and depth datums. The main elements in a height-system definition are an origin, a vertical reference surface of zero level, and a "type" of height, for example dynamic heights. The geoid, defined as that equipotential surface of the Earth's gravity field that most closely coincides with the mean sea level, was originally selected as reference surface because it was believed that the average level of the sea was constant over long periods of time, which we now know it is not true. In general, each Country chose one tide gauge station for the computation of the "mean sea level" over a certain arbitrary period of time. However, whatever the choice of the site, the mean sea level varies from place to place and at one specific place over time. Therefore different height datums may refer to different equipotential surfaces, resulting in constant offsets between them. Space geodesy provides the mean to evaluate these offsets in a well-defined geocentric reference system (Woppelmann et al., 2006).

At present, vertical crustal motions at tide gauges can be measured to high accuracy independently of the sea-level reference surface by means of space techniques, therefore it will be possible to separate the crustal motions from the absolute sea-level variations. Tide gauge measurements are difficult to compare because tide gauges are referred to local reference systems and they have not yet been connected on a common datum. However, it should be pointed out that several international efforts are underway both at global (IOC, 1997; International GPS Service, 2001) and regional scales (Zerbini et al., 1996; Becker et al., 2002) which aim to overcome this problem.

Nowadays, gauges are available based on many technologies (digital float, radar, acoustic, pressure) which can provide low latency, accurate sea level information optimized for the particular installation. For tide gauge details see IOC (2006). Traditionally, tide gauges measure local sea level with respect to a nearby benchmark on land. Modern geodetic techniques provide the means to position the tide gauges in a global geodetic reference frame (see Figure 2.44). Today, permanent GNSS stations (and in some cases DORIS stations) at tide gauges are used to determine the land motion at these sites in a global reference frame, and to position the tide gauge in the same frame as the satellite altimeters (see also Section 3.7 and below). Measurement of vertical land movements at gauge sites allows the determination of sea surface height changes in the same reference frame as the altimeter data. Absolute gravity measurements provide an independent control of the vertical land motion rates determined from the GNSS observations, and help to eliminate a bias of these rates due to a potential secular motion of the reference frame origin with respect to the CM. At some sites, an additional GNSS station is used in a dual-CGPS approach (Plag et al., 2000a) to control the stability of the tide gauge monument with respect to the adjacent land, thus replacing or augmenting episodic leveling.

Internationally, tide gauge sea level measurements are coordinated through the Global Sea Level Observing System (GLOSS) of the IOC (Woodworth et al., 2003). GLOSS defines a worldwide Core Network of approximately 300 stations (see Figure 2.45 on page 83), which is densified by means of inclusion of regional and national networks. The use of GPS at gauge sites is the topic of the current IGS TIde GAuge Pilot Project (TIGA).

GLOSS does not dictate to tide gauge operators which technology is preferable; GLOSS standards simply require measurements to better than 1 cm accuracy in all weather conditions. However, especially since the Sumatra tsunami of December 2004, one would expect that any new GLOSS installation would consist of dual gauges (e.g. a "sea level" gauge based on radar, and a "tsunami" gauge based on pressures) and dual telemetry. Data flow would be both near-real time (especially so for tsunami and storm surge applications) and delayed-mode for scientific applications.

Geodetic techniques have extended the number of ways by which local sea level can be measured. Techniques which have been developed in the last few years include the use of GPS on buoys (in effect extending coastal tide gauge measurements off-shore), and the use of GNSS scatterometry and reflectometry (see Section 2.4.4). The emerging use of GNSS receivers for earthquake magnitude determination (Blewitt et al., 2006b), with tsunamis being the potential sea level consequence, indicates another role for space geodetic techniques in a sea level observing system.

However, even in the established methods, geodesy has resulted in major improvements. Positioning of sensors (such as tide gauges and ocean buoys) in a global reference frame has already been mentioned above. The provision of precise timing

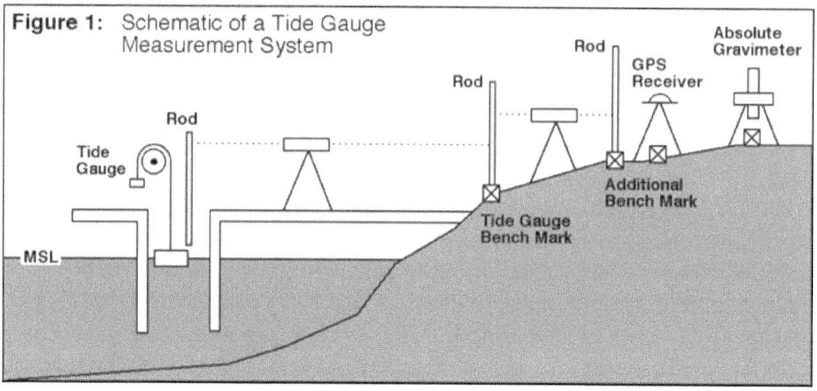

Fig. 2.44. Principle of tide gauge measurements. From http://sealevel.colorado.edu/tidegauges.html. See also IOC (2006).

through GNSS (see Section 2.9.4) to the equipment (for example, the clock of a tide gauge) is another example. Before GNSS, positioning and timing were accomplished by almost as many methods as research groups. The result of the new techniques and a more standardized approach is more precise data and meta-data with consequent improvement in our knowledge about sea level.

Geodesy has in effect terminated some traditional areas of work. An example concerns the replacement of chart datum as the height reference on nautical charts, hitherto based on interpolations of information on lowest astronomical tide at tide gauges, with the use of geo-located tide gauge data and off-shore mean sea surface information from altimetry, together with the availability of GPS positioning to mariners. Another example includes the replacement of long distance leveling by GNSS-minus-geoid, thanks to the availability and accuracy of GPS and regional geoid information, with most recent geoid improvements following GRACE operations and further ones anticipated from GOCE (see Section 2.6.5).

Today, the largest database of monthly mean tide gauge data is provided by the Permanent Service for Mean Sea Level (PSMSL). Since 1933, PSMSL has been responsible for the collection, publication, analysis and interpretation of sea level data from the global network of tide gauges. The PSMSL is a member of the Federation of Astronomical and Geophysical Data Analysis Services (FAGS) established by the International Council for Science (ICSU), and it is based in Liverpool at the Proudman Oceanographic Laboratory (POL).

The database of the PSMSL contains monthly and annual mean values of sea level from almost 2000 tide gauge stations around the world (Figure 2.45, upper diagram) received from almost 200 national authorities. On average, approximately 2000 station-years of data are entered into the database each year, and in December 2006, the database contained over 55000 station-years.

The data are provided in two data sets, namely the METRIC data set containing basically all data, and the Revised Local Reference (RLE) data set containing records for which the history of the local reference is known so that time series anal-

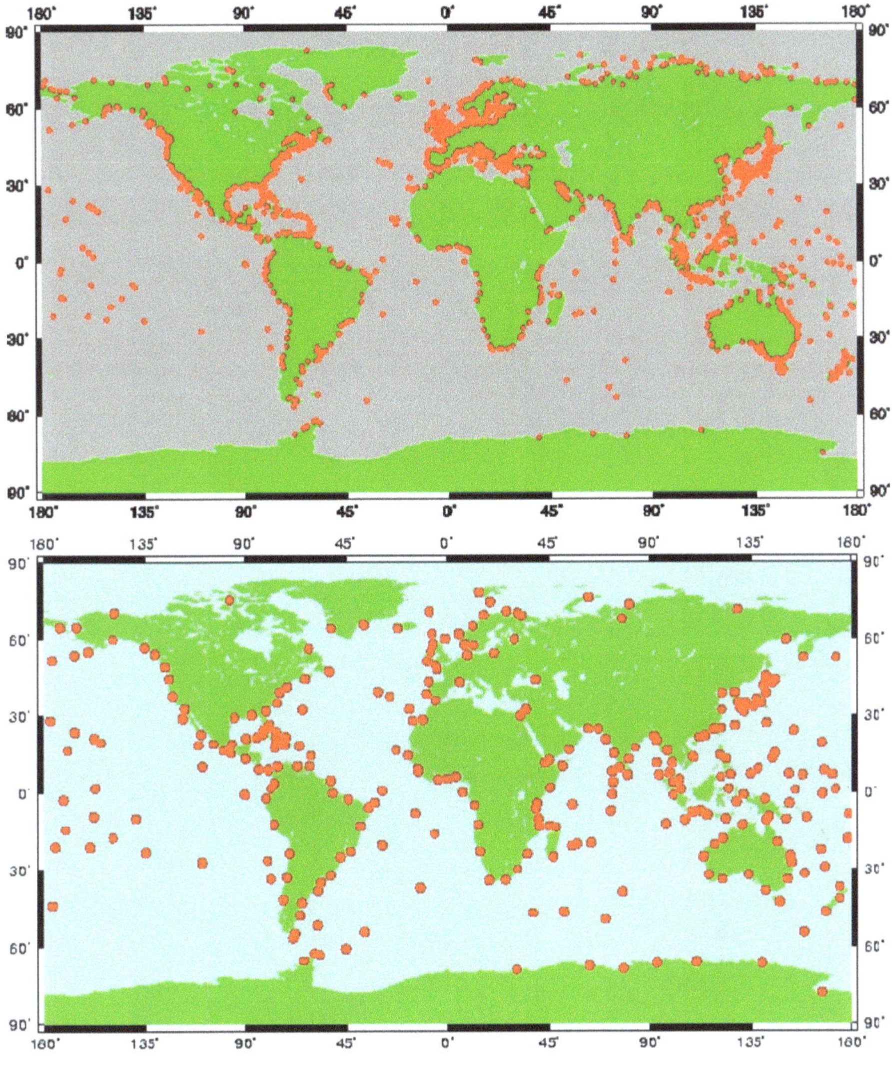

Fig. 2.45. Upper diagram: location of the roughly 2000 tide gauges for which data are stored in the PSMSL data base. Lower diagram: locations of tide gauges in the GLOSS core network.

ysis of secular sea level changes can be performed (Woodworth & Player, 2003). Long records from this data set have been the basis of most analyses of secular changes in global sea level during the last century. The geographical distribution of longer RLR records contains significant geographical bias towards the northern hemisphere, a situation which is being rectified by the establishment of the GLOSS global sea level network (Figure 2.45, lower diagram). A major conclusion from the

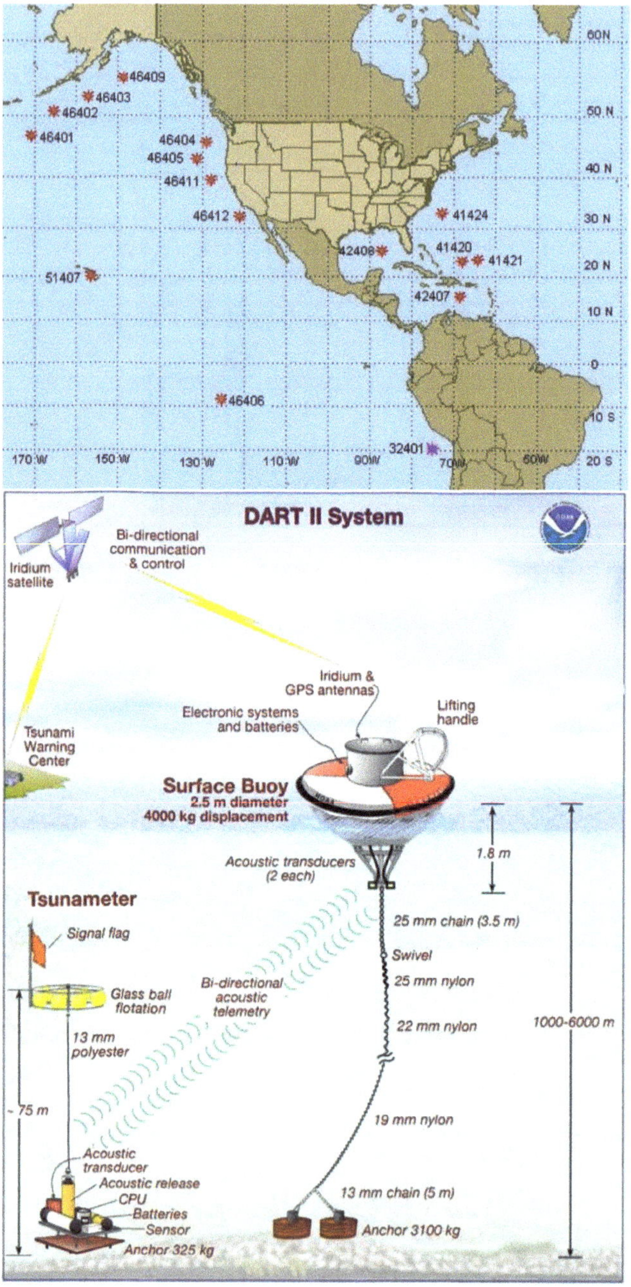

Fig. 2.46. NOAA's DART stations. Top: Location of NOAA's DART stations. Bottom: Schematic illustration of the DART system. For explanation, see text. Figures taken from NOAA's DART system page at http://www.ndbc.noaa.gov/Dart/dart.shtml.

global tide gauge data has been that global sea level has indeed risen by approximately 10 to 20 cm during the past century (Church et al., 2001).

BottomBottom Pressure Recorders (BPRs) use similar pressure sensors to those in coastal pressure tide gauges with two main differences. One is that the sensors obviously have to be capable of operating at greater depths (often down to 5000 m) and as a consequence are more expensive. The other is that they have to be 'absolute' sensors, recording total pressure at the sea bed, which includes the pressure due to the water plus atmospheric pressure. In coastal pressure gauges, it is more normal to use a 'differential' sensor, which is compensated for atmospheric pressure, although absolute sensors employed in combination with conventional barometers are also available and are preferred by some operators.

Data from deep ocean bottom pressure recorders are particularly relevant for comparison to temporal space gravity data from missions such as GRACE. However, only a few BPRs have been deployed so far explicitly for such comparison purposes; the POL BPRs in the South-West Atlantic being one example (Hughes et al., 2007).

BPRs have a long history in oceanography, but were developed most intensively in the 1970-1980s for tidal research (Spencer & Vassie, 1997). Instruments were placed on the sea bed for typically a year and recovered during a second visit by a research ship. This provided a one-year record which was adequate for a tidal analysis. More recently, BPRs have been deployed for longer periods (up to 5 years for the POL Multi Year Return Time Level Equipment (MYRTLE)) for non-tidal studies, such as monitoring the variability of ocean currents. Data retrieval remains a major issue, and recovery by means of acoustic release of the whole BPR by a ship is still the main method. MYRTLE additionally contains a number of 'data podules' which are released by a timing mechanism at regular intervals (e.g., once a year) with data transmitted from the podule to a satellite when on the surface. The podule itself may be recovered if a ship happens to be nearby but can otherwise be considered disposable.

However, this technology can never provide real-time information required for tsunami warning systems. For that, one requires undersea cables or the use of acoustic transmission from a BPR to a surface buoy. The most advanced systems currently in use are the Deep-ocean Assessment and Reporting of Tsunamis (DART) stations deployed by NOAA mainly in the Pacific (Figure 2.46a). DART systems consist of an anchored seafloor bottom pressure recorder (BPR) and a companion moored surface buoy for real-time communications (Gonzalez et al., 1998, see Figure 2.46b). An acoustic link transmits data from the BPR on the seafloor to the surface buoy. The data are then relayed via a GOES satellite link to ground stations (Milburn et al., 1996), which demodulate the signals for immediate dissemination to NOAA's Tsunami Warning Centers.

A major source of uncertainty in understanding sea-level variations from tide gauges is the accurate knowledge of vertical crustal movements which are embodied in the sea-level measurements. In fact, tide gauges measure sea-level changes as the difference between the height of a geodetic benchmark attached to the Earth's crust and the height of the sea surface. Vertical land movements need to be accounted for if tide gauge records are to be compared to satellite altimetry measurements of sea surface height changes. At global scale, post glacial rebound, a vertical crustal motion due to the isostatic readjustment of the Earth's crust to the last deglaciation, is the only coherent geological contribution to the long-term sea-level change for which a thorough understanding of the physical process has been achieved (Mitrovica et al., 1994; Peltier, 2004). Isostatic adjustment is the process by which the Earth attains gravitational balance with respect to superimposed forces. If a gravitational instability occurs, the crust rises or sinks to compensate this instability. Modeling the post-glacial rebound effects, however, still leaves in the vertical crustal rates different regional and local isostatic components as well as tectonic effects which are difficult to model.

At present, vertical crustal motions at tide gauges can be measured to high accuracy by means of space techniques such as, for example, the GNSS DORIS (Soudarin et al., 1999). Continuous GPS, however, has shown to be the technique of use in this particular application due to the ease of use, high precision, and its direct connection to the ITRF through the products of the IGS. On the other hand, by means of simultaneous GPS measurements performed at tide gauges and at fiducial reference stations of the global reference system, tide gauge benchmarks can be tied in a global well-defined reference system (Becker et al., 2002; Zerbini et al., 1996). The possibility to refer the tide gauge data to the same high precision global reference system allows the comparison between the different tide gauge data sets to be made. This was not the case until about 15 years ago when tide gauge benchmark coordinates were mostly available in the different national height systems.

The long-term sea-level trends at tide gauge stations is measured to about 0.3-0.5 mm/yr (Zerbini et al., 1996), provided that the time series are long enough (20-50 years). The accuracy required by GPS shall be in the same range; tide gauge positions must be monitored at the level of 10 mm absolute position error so that a long-term trend with a realistic error of 0.3 mm/yr can be obtained over 20 years or so (Becker et al., 2002). The current accuracy of GNSS products provided by International GNSS Service (IGS) is 3, 3, 6 mm for weekly mean values of the north, east and up coordinates respectively and 2, 2, 3 mm/yr for the associated linear velocities (see for instance Altamimi et al., 2002, 2007), with a significant contribution of the error in the velocities originating from long-term stability of ITRF (Blewitt et al., 2006a). The height determination using GPS data is a delicate task because of several reasons, among them, the atmospheric refraction in the troposphere and the geometric weaknesses in the height component of the GPS in general, and the complicated interactions of the GPS receiver and antenna hardware imperfections (like antenna phase-center variations and multipath). Moreover, with the exception of areas with natural or anthropogenic subsidence, active tectonics and strong seismic events, vertical rates are smaller by an order of magnitude as compared to the horizontal crustal motions, i.e. they are in the mm/yr range (Woppelmann et al., 2006).

2.9.4 Geodetic time and frequency transfer

High-accuracy geodetic methods using dual-frequency GPS observables are now routinely employed to produce positioning repeatabilities globally at the centimeter level for one-day integrations, as demonstrated for example in products of the IGS. Similarly, the same methods have been shown to produce equivalent global time and frequency comparisons with precisions approaching about 100 picoseconds at each analysis epoch, but whose accuracies are limited to roughly the 1-ns level because of instrumental calibration uncertainties, particularly for the GPS antennas. Current techniques yield calibration uncertainties of about 3 ns for standard GPS antennas (Plumb et al., 2005).

The essential ingredients for the geodetic GNSS method involve the availability of dual-frequency carrier phase as well as pseudorange (code) observables, recorded typically at 30 s intervals, together with an analysis modeling of one-way signal propagation accurate to the millimeter level. Standard errors for GPS phase and code data are about 1 cm and 1 m respectively with multipath believed to be the dominant source of error for both. The code data are needed to separate the otherwise indistinguishable clock offset and phase cycle ambiguity. The effect of utilizing both observables in this way is that the noisier code data are effectively smoothed by the more precise carrier data and that the overall accuracy of the time transfer is determined from the code data; the precision within a continuous analysis arc (typically 1 day) is determined from the quieter carrier data. Formal errors for the geodetic clock estimates are typically about 120 ps (3.6 cm), but have been shown to be highly optimistic in many cases. A more realistic measure of the accuracy may be determined by performing a classic repeatability test, comparing the agreement at successive analysis arc boundaries. Such a test is only feasible if the underlying clock stability is sufficient, which effectively restricts its use to GPS receivers equipped with an external H-maser frequency standard. A detailed analysis of day-boundary clock estimate discontinuities was performed for a subset of stations contributing to the IGS Combined Clock Products (Ray & Senior, 2003, 2005). The analysis showed that performance is highly site-specific, varies widely among the stations studied, and is independent of the choice of receiver or antenna model used. In many cases, poor performance or abrupt changes in performance was traced to changes in equipment or installation problems such as loose cable connections or poor external frequency distribution. Some stations showed distinct seasonal variations in the level of discontinuities which can not be fully explained by thermal effects. However, in the best cases sites have day-boundary discontinuities (rms) that are commensurate with the formal errors. The stability floor for the current state of the art geodetic time transfer technique has been inferred to be about $2 \cdot 10^{-13}\tau^{-1/2}$ for τ intervals up to 1 day, consistent with a random walk process. Deducing the limit of the method beyond 1 day will require comparisons using more precise frequency standards such as cold atom clocks.

As evidenced in the above performance measure, the limit of geodetic timing is determined from the quality of the pseudorange data. Therefore, in order to achieve the highest quality time and frequency comparisons, there are some special considerations for monumentation and instrumentation which should be made to minimize multipath and signal reflections. Receivers vary widely with respect to their sensitivity to thermal effects and so thermal control of the receivers is generally necessary. Also, phase stable cabling with low thermal sensitivity should also be employed with cable runs having minimal length and environmental exposure. Thermal control of the antennas is not required (Ray & Senior, 2001, 2003; Rieck et al., 2003), however the antenna siting should strive to minimize code and phase multipath. Some recent work has also indicated the possibility that long-wavelength multipath from below may also be an issue (Ray & Senior, 2005; Elósegui et al., 1995). In the near future, the largest gains in performance will likely come from new GNSS broadcast signal modulations whose multipath characteristics are likely to be greatly improved over those of the current GPS system.

Chapter 3
Understanding a dynamic planet: Earth science requirements for geodesy

R. Rummel, G. Beutler, V. Dehant, R. Gross, K. H. Ilk, H.-P. Plag, P. Poli, M. Rothacher, S. Stein, R. Thomas, P.L. Woodworth, S. Zerbini and V. Zlotnicki

3.1 Introduction

The complexity of the Earth system has been discussed in many fundamental documents in recent years. Trying to understand the Earth system and improve our forecast capability step by step are the great challenges of Earth system science. The opening sentence of the recent NRC report "Earth Science and Applications from Space: Urgent Needs and Opportunities to Serve the Nation" (2005) nicely expresses this current situation: "*Understanding the complex, changing planet on which we live, how it supports life, and how human activities affect its ability to do so in the future is one of the greatest intellectual challenges facing humanity. It is also one of the greatest challenges for society as it seeks to achieve prosperity, health and sustainability.*" Chapter 1 discussed what the role of GGOS will be in this challenge. It is evident that only a joint effort of many scientific disciplines will make it possible to shed more light into the complexity of the Earth system. Despite enormous progress in recent years the uncertainties of predictions are still rather high. A recent analysis of the deeper reason for the limited quality of climate forecasts led to the unanimous conclusion that "data are still poor" (see Hogan, 2005, and the references therein). Data from space will and must play a central role in Earth system research. Only satellites are capable of providing data globally, of uniform quality, and with acceptable repetition rate. Moreover, complementary sensor systems can be combined and data delivered in near real time, (compare, e.g., Battrick, 2006). Observations are also the crucial element of a second central element of Earth system research, up-scaling and down-scaling. This is the process of establishing representative and reliable connection from local data to global processes and vice versa. Establishment of this connection will unify the wide range of temporal and spatial scales in Earth system models.

Geodesy is a "measurement discipline" much like astronomy. Its modern development and success is intimately connected with progress in space science and engineering. The successes in the scientific applications of satellite and lunar laser

H.-P. Plag, M. Pearlman (eds.), *Global Geodetic Observing System*,
DOI 10.1007/978-3-642-02687-4_3, © Springer-Verlag Berlin Heidelberg 2009

ranging, microwave satellite ranging, radar altimetry, VLBI and many more result from this development.

Geodetic space techniques have reached a level of maturity and precision that make them an important tool for Earth system sciences. Important added-value and new areas of application will result from a combination of the fundamental three types of geodetic parameters: surface geometry, Earth rotation and gravity. This is what GGOS intends to provide. Examples of this modern development are detection and monitoring of tectonic, ice and ocean motion, the determination of mass anomalies and implicitly density anomalies, observation and quantification of mass transport processes in the hydrosphere and in the oceans, estimation of global and regional mass changes in the Earth components, separation of the thermal and mass components of sea level change, ionospheric and tropospheric sounding.

This chapter deals with the science prospects resulting from the GGOS and with the science requirements connected to this task. There are two dimensions to this theme. One is the analysis of the challenges geodesy faces in the realization of a global observing system at a precision level of 1 ppb relative to Earth dimension and with decadal stability in space and time. This part will be discussed in Section 3.2. The second dimension is the analysis of the benefits for Earth system science and application that will result from GGOS. One can view this from two opposite directions. The first is as Earth scientist; to elaborate on possible benefits of GGOS for solid Earth geophysics, glaciology, oceanography and climatology. This will be done in Section 3.3 for geophysics, Section 3.4 for glaciology, Section 3.5 for oceanography, Section 3.6 for weather and climate, Section 3.7 for sea level studies, and Section 3.8 for hydrology. Alternatively the benefits of GGOS for understanding the Earth system can be considered from the geodesist's view: the expected added-value for Earth system sciences resulting from a combination of the three data types surface geometry, Earth rotation and gravity. Taking this perspective Section 3.9 will deal with mass transport and mass anomalies in Earth system, Section 3.10 will describe the link between Earth rotation and geophysical fluids, and Section 3.11 explores what Earth rotation tells us about core and mantle processes.

In these discussions it is worth noting two uses of the terms "model" and "modeling". Geodetic models, e.g. gravity or Earth rotation models, essentially condense large numbers of observations (satellite, VLBI and terrestrial data) into a meaningful set of parameters. In contrast, we often use the term "models" to describe the mathematical representation of some geophysical processes such as climate, atmosphere, ocean, ice or solid Earth. These two types of modeling also appear in the scheme of Figure 3.1 where the geodetic modeling is termed "observation modeling" while "influence/modeling" refers to modeling in geosciences.

3.2 The scientific and technological challenges for GGOS

Geodesy is a measuring science. Loosely speaking it provides data in space, time and gravity at a level of 1 ppb relative to the Earth's dimension. One of the novel and most important features of GGOS results from the integration of the traditional three fundamental types of geodetic parameters: surface geometry, i.e., the determination and monitoring of the surfaces of land, ice, and ocean, Earth rotation comprising nutation, polar motion and variation in length of day, and gravity with the Earth's gravity field and geoid. This integration will permit the separation of the thermal and mass contribution to sea level change, and the study of mass anomalies, mass transport, and mass exchange in the Earth system.

Global change phenomena are very small and therefore difficult to measure. Often changes cannot be measured directly but are inferred from data derived by complementary sensor and observation systems and by comparison with numerical models (see Figure 3.1). A good example is dynamic ocean topography. It is derived from accurate radar altimetric measurements of the ocean surface along satellite tracks in combination with a geoid surface based on global gravity models derived from the data of gravimetric satellite missions. Both altimetry and the geoid model have to refer the same global coordinate system, free of any distortion. Processing of altimetry (a data model in Figure 3.1) requires elimination of effects such as ionosphere, troposphere, tides and sea state biases and has to be consistent with a similar processing chain that leads to the geoid model. Hence a variety of sensor systems, mission characteristics, tracking systems, and sampling patterns have to lead — with high precision — to a unified and consistent model of dynamic ocean topography. In a second step ocean transport estimates are derived from the assimilation of dynamic ocean topography into global or regional numerical circulation models.

The resulting research challenges are:

1. The various geodetic satellite systems, comprising both their instrumentation and observatories that establish the link from ground stations to the orbiters, have to operate as a global entity in a global reference frame. Thus, space techniques (SLR, VLBI, GPS, DORIS, (differential) INSAR, ocean altimetry, ice altimetry), gravimetric space techniques (orbit perturbation analysis, high-low and low-low satellite-to-satellite tracking, satellite accelerometry and gradiometry), relevant astrometric techniques and missions, and geodetic techniques of atmospheric sounding from GNSS satellites to low Earth orbiters or to ground stations have to be unified and integrated at the 1 ppb level. Moreover, the global network of observatories and receivers must operate in one Earth fixed coordinate system at this precision level. Specifically, its 3D positions and geopotential heights must be known and monitored with such precision and with the same long term stability over time. For an overview see Table 2.1.
2. The space segment has to be complemented by terrestrial and airborne/shipborne techniques and campaigns. This implies the combination of measured data of very different density and resolution. It may be referred to as the geodetic "up and down-scaling problem," (Section 3.1). Terrestrial and airborne measurements serve a threefold purpose. They provide calibration and validation to the space segment, lead to a regional densification in terms of spatial and temporal resolution and accuracy and are essential in the attempt to separate individual geophysical effects which can be observed from space only in their superposition.

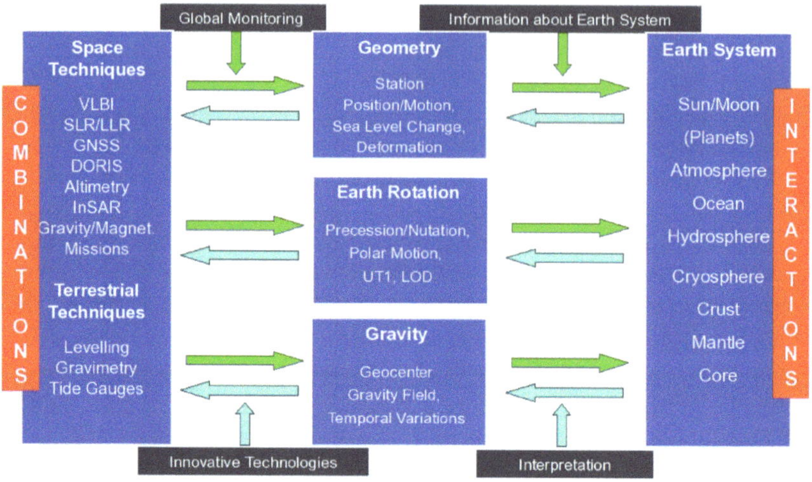

Fig. 3.1. Measuring and modeling the Earth system.

The land, ice and ocean surfaces are to be monitored with high spatial resolution. The required resolution depends on the surface type (land, ice, or ocean) and on the region (tectonically active zones, major ocean currents etc.). The latter implies the necessity of establishment and integration of regional projects in areas of particular geophysical relevance. These projects have to comply with the overall GGOS standards. The development of new terrestrial and airborne/shipborne techniques needs attention and encouragement.

3. A link has to be established between the global time series of geodetic parameters delivered by GGOS and relevant geophysical process models. This is a demanding and highly interdisciplinary task that requires a close cooperation with geophysicists/geologists, glaciologists, oceanographers, hydrographers and atmospheric physicists. Thus, geodesists will need to be involved into numerical Earth modeling. The ultimate goal is the development of a comprehensive Earth model constrained by geodetic time series of global surface processes, rotation parameters, gravity models and mass transport and exchange data. They should contribute to our understanding of solid Earth processes such as Glacial Isostatic Adjustment (GIA), tectonic motion, volcanic activity or earthquakes, ice mass dynamics and balance and the dynamics of sea ice, the structure and changes of ocean circulation, and mass and heat transport in the oceans, to the various components of sea level change and to their separation and quantification, to the global water cycle, and to atmospheric dynamics.
4. The measured temporal variations of Earth rotation and gravity and geoid represent the total integrated effect of all mass changes in the Earth system. Thus, strategies have to be developed for their separation into individual contributions (although interaction between the various processes at or above the accuracy level may not allow for separation and rather require an integrated modeling approach). The complementarity of satellite techniques, sampling strategies, for-

mation flights of several satellites, terrestrial calibration sites, permanent recordings and campaigns and geophysical models will prove important for this task. Of similar character is the problem of the limited resolution in time and space of any space configuration, which leads to aliasing problems that need careful analysis.

The rationale of this program is summarized in Figure 3.1. The left box lists the most important geodetic measurement techniques. They have to be processed and combined in a consistent manner to form a unified and global observatory. They yield geodetic parameter sets and time series for surface geometry, Earth rotation and gravity. Their contribution to the study of the Earth system results from their introduction, assimilation and/or integration into models of Earth processes. In addition to forward modeling, there results a reverse feedback. Improved Earth process modeling can also improve analysis of the geodetic parameter sets and lead to a more consistent data processing.

These goals are at the cutting edge of what is possible today and in the near future. They require a joint effort in theory, numerical methods, data handling, measurements and campaigns, instrument development, organization and management. The defined goals for precision, uniformity, consistency and stability result in a series of geodetic requirements for all components and stages of geodetic data processing. Several, that are complex and not yet fully understood, have to be regarded as research topics. Agreement has to be reached in terms of:

- fundamental constants;
- geodetic world datum in space, time and geopotential at 1 ppm;
- standards;
- geophysical background models (solid Earth and ocean tides, atmosphere, oceans, ice, loading, ionosphere, troposphere) used for data reduction, remove-restore steps and separation of effects;
- combination and comparison of global and regional/local data;
- determination of field quantities (gravity field) from finite sampling in space and time (aliasing, leakage, truncation, filtering, regularization, etc.);
- complementary measurements from other disciplines.

In the past the use of separate geodetic and geophysical models for data reduction and analysis was acceptable. However, when generating time series of the very small geodetic "global change" parameters, resulting from a combination of the three parameter types surface geometry, Earth rotation and gravity, the consistency of these models is mandatory. Developing a unified geodetic Earth system model that is applicable to all geodetic observation techniques is challenging but certainly worthwhile considering.

A central contribution of geodesy to Earth science is the provision of a stable reference frame and tools to assign coordinates in this reference frame to any point with high accuracy, and spatial and temporal resolution. This enables scientists to reliably monitor processes on land, on ice, and on the ocean with high short-term accuracy and long-term stability. It has allowed accurately positioning sensors in motion (e.g., on air planes, ships, satellites), and thus facilitated the development of observation techniques with high spatial coverage. Because geodetic observations reflect to the mechanical processes in the Earth system, they constitute a fundamental data set for Earth sciences.

3.3 Solid Earth physics

Space geodesy has revolutionized the study of solid Earth processes through its ability to measure the deformation of the Earth's surface and the Earth's gravity field with extraordinary accuracy. These measurements provide our best data about the motion of the great plates of the Earth's lithosphere, crucial insight into the resulting hazards posed by earthquakes and volcanoes, and powerful constraints on the forces within the Earth that drive them.

Nonetheless, despite significant progress over the last few centuries, many questions remain about fundamental processes in the solid Earth (e.g., NASA, 1991a; Solomon & the Solid Earth Science Working Group, 2002; Board on Earth Sciences and Resources, 2003; Ilk et al., 2005; Space Studies Board, 2005; Battrick, 2006). These reflect the complex nature of the Earth system in which chemical, physical, and biological factors jointly yield a highly non-linear system in partial homeostasis (Lovelock, 1979; Schellnhuber & Wenzel, 1998). Understanding the complex Earth system requires integrated sets of observations on global to regional spatial scales and with high spatial and temporal resolution. In the pre-space era, few parameters were observed with global coverage or sufficient spatial and temporal resolution. In the space era, this has changed for many parameters, but not for all.

These limitations of data are being addressed by regional and global collaborative programs. The NASA Crustal Dynamics Project (CDP) (Smith & Turcotte, 1993a,b,c) and the Working group of European Geoscientists for the Establishment of Networks for Earth-science Research (WEGENER) (e.g., Plag et al., 1998a) were established in the beginning of the 1980s as interdisciplinary programs applying space geodetic and other techniques to the study of geodynamics and crustal dynamics. A recent example is the U.S. EarthScope Program, a major national undertaking applying modern observational technologies and analytical methods to the study of the structure and evolution of the North American continent and the underlying physical processes that cause associated phenomena such as earthquakes and volcanic eruptions (Carlson & 42 others, 2002). EarthScope is developing new facilities for seismology, geodesy, and borehole geophysics, to provide a foundation for fundamental and applied research throughout the United States. This network of geodetic and geophysical instruments is significantly expanding capabilities to observe the structure and ongoing deformation of the North American continent. EarthScope seeks to promote multidisciplinary research addressing some of the grand scientific challenges in Earth science.

Geodetic observations play a major role in these programs because they are fundamental for the understanding and modeling of Earth system processes. Changes in the Earth's shape, its gravitational field, and its rotation are caused by external forces acting on the Earth system and internal processes involving mass transfer and exchange of angular and linear momentum. Thus, variations in these geodetic quantities reflect and constrain mechanical and thermo-dynamic processes in the Earth system.

Understanding these processes transcends purely scientific goals because these processes have significant societal impact (Solomon & the Solid Earth Science Working Group, 2002). Thus, understanding of these processes and their interactions is important for sustainable development and has important consequences for natural hazard mitigation.

This Section examines some of the scientific problems in solid Earth physics that would benefit from improved geodetic observations. In particular, it elaborates on the added-value of the combination of time-dependent positioning (geometry), Earth rotation and gravity/geoid.

This analysis is designed to motivate the design of the future GGOS in terms of requirements for accuracy and resolution of geodetic observations. Typically, scientific requirements are simply 'as good as possible'. Nevertheless, we attempt to identify quantitative threshold and target requirements that are likely to help distinguish between models and hypotheses and thus improve our knowledge significantly.

We also attempt to anticipate the impact of likely technological developments. One is the maturing of geodesy on the ocean bottom. Because the ocean covers two-third of the Earth's surface, seafloor geodesy could lead to significantly better understanding of geodynamic processes. Spiess (1990) predicted that *by the year 2000 we will be discussing at least a few real multi-year data sets and using them to constrain our models of the structure and dynamics of the crust beneath the sea - its genesis, its evolution as it moves away from the mid-ocean ridges, its destruction in the trenches, and the effects of its interaction with continents and islands*. However, progress has been slower than anticipated in that statement. The program documents of the International Year Of Planet Earth (IYPE) again emphasize the need to extend geodesy to the ocean floor (Chen et al., 2005).

Another example are time measurements, where accuracy and daily stability of 10^{-16} should be possible with Atomic Clock Ensemble in Space (ACES) clocks. Time is the geodetic observable from which geodetic parameters are derived. The anticipated progress in time measurements should significantly improve geodetic measurements, and thus make them even more valuable.

Improved geodetic data should advance our understanding of many open questions related to mass movements in the Earth system, deformation of the Earth's surface, and dynamics of the solid Earth:

- **Convection:** are the anomalies in seismic velocities detected by seismic tomography in the Earth's mantle due to chemical anomalies or temperature anomalies? This is crucial for the question of whether convection extends throughout the whole mantle or is layered, which has major consequences for the thermal, chemical, and mechanical evolution of the Earth.
- **Plate tectonics:** the location of some plate boundaries and the processes that occur at plate boundaries still pose many questions. Large uncertainties exist par-

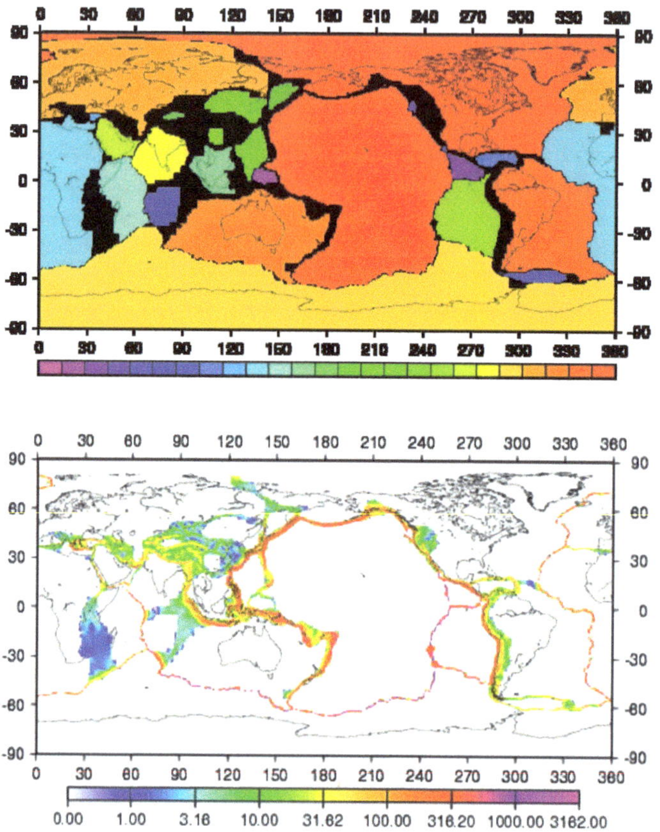

Fig. 3.2. Model of tectonic plates. The model of the stable parts of the tectonic plates (top) is derived from the strain field (bottom) (data from Kreemer et al., 2003). The upper diagram shows the individual stable parts of the plates by different colors. For large areas, the surface is deforming and the exact location of the plate boundaries is often uncertain.

ticularly for the ocean bottom, which covers more than two-third of the Earth's surface. Likewise, the extent of deformation zones is uncertain in many regions of the Earth's surface (see Figure 3.2). Strain rates, which vary in space and time, are known only for parts of the Earth's surface, as illustrated by the problems in defining the "stable parts" of the continents (e.g., Nocquet et al., 2001) and assessing seismic hazards there. The strain field of the ocean floor is mostly unknown, which is a severe limitation assessing earthquake and tsunami hazards.

- **Ice sheets/glaciers and sea level:** there are large uncertainties with respect to the ice load history, in particular, for Antarctica. Even the sign of the present-day changes in ice sheets are still uncertain for parts of the ice sheets. Consequently, their contribution to sea level changes are highly uncertain.
- **Rheology:** despite its fundamental importance, the appropriate rheology (linear or non-linear) of the Earth's mantle and its dependence on time scales (transient

versus steady-state) is not well understood. Moreover, understanding the effect of lateral heterogeneities in mantle and crust (including plate boundaries) is still in an early stage (Plag et al., 1998b) and models incorporating these effects are just emerging (e.g., Latychev et al., 2005).

- **Core-mantle dynamics:** much remains to be learned about processes at the core-mantle boundary, the dynamics of the core, and its coupling with the mantle.
- **Hydrological cycle:** better quantification of the fluxes between the different reservoirs is required. How large are groundwater movements? What are the variations in continental water storage?
- **Solid Earth response to loading:** much remains to be learned about the mass loads on the Earth's surface, in particular continental water storage, non-tidal ocean loading, and ice loads, and the resultant deformations of the Earth (including gravity changes), which depend on crust and mantle composition and rheology.
- **Rotational dynamics:** despite recent progress, issues remain in the areas of the coupling of angular and linear momentum equations, free modes of the ocean on a rotating Earth, and the forcing of rotational perturbations.
- **Tides:** indextidesvalidation of ocean tide models and an improved knowledge of tidal friction is both important on its own and for analysis of other geodetic data.
- **Earthquakes:** Plate tectonics provides a general kinematic framework for relating individual earthquakes to geological deformations . However, understanding earthquakes as a physical process, as a source of societal hazard, and as manifestations of Earth deformation remains a major challenge. Among the fundamental questions remaining are how earthquakes result from the cycle of strain accumulation at faults, how rupture begins, and whether earthquakes can be predicted. It is unclear why in some areas crustal strains localize on major faults, whereas others show more continuous deformation. It is not clear how brittle crustal deformation couples to the ductile motion of the convecting solid mantle. Such questions connect the study of earthquakes to many basic aspects of solid-Earth research (from Board on Earth Sciences and Resources, 2003).
- **Earth structure:** Despite great recent progress, much remains unknown about the structure and composition of the deep Earth, and its relation to lithospheric, asthenospheric, and mantle dynamics.

3.3.1 Plate motion

Space geodesy is revolutionizing tectonics by providing our first measurements of plate motions over time scales of years — rather than millions of years — and our first clear insight into the motions within the deforming zones at plate boundaries that cover about 15% of the Earth's surface and are homes to about 40% of the human population.

Using space-based measurements to measure plate motions was suggested by Alfred Wegener when he proposed the theory of continental drift in 1915. Wegener realized that proving continents moved apart was a formidable challenge. Unfortunately, surveying methods available at that time offered no hope of measuring these slow motions. Wegener thus decided to measure the distance between continents using astronomical observations. However, because measuring continental drift called for measurement accuracies far greater than ever before to show small changes in positions over a few years, Wegener's attempts failed, and the idea of continental drift was largely rejected.

By the 1970s the story was very different. Geologists accepted continental drift, in large part because paleomagnetic measurements showed that continents had in fact moved over millions of years. It thus seemed natural to see if modern space-based technology could accomplish Wegener's dream of measuring continental motions over a few years. Three basic approaches were attempted. Each faced formidable technical challenges - and all succeeded. Hence plate motions can now be measured to a precision of a few mm/yr or better using a few years of data from systems including VLBI, SLR, and GPS.

Space geodesy measures both the rate and azimuth of the motions between sites, and can thus be used to compute relative plate motions. One of the most important results of space geodesy is that plate motions have remained generally steady over the past few million years. This is shown by the striking agreement between motions measured over a few years by space geodesy and the predictions of global plate motion models that average over the past several million years. The general agreement is consistent with the idea that although motion at plate boundaries can be episodic, as in large earthquakes, the viscous asthenosphere damps out the transient motions and causes steady motion between plate interiors. This steadiness implies that plate motion models can be used for comparison with earthquake data.

Space geodesy also gives detailed views of the spatial distribution of deformation within broad plate boundary zones like western North America, the Mediterranean, and the Himalayas. This surmounts a major difficulty faced by geologic plate motion models, which predict only the net motion across the boundary. Hence they give only partial insight into earthquakes, volcanism, and other deformation that occur in boundary zones. Understanding this deformation is a major geological problem, which also has social relevance because of the resulting geologic hazards to populated areas. Thus space geodetic data play a rapidly growing role in assessing hazards and developing strategies to mitigate them.

Building on these successes, space geodesy is poised for significant advances in several areas in years to come. As geodetic data improve due to additional sites, longer time series, and higher precision analyses, we are likely to see major progress on topics including:

- Resolving plate motions in complex boundary zones: Even after the giant 2004 Sumatra earthquake, we know little about the plate geometry and motions that caused it, and hence how often such earthquakes recur. The earthquake resulted from subduction of the Indian plate beneath the Burma microplate, a sliver plate between India and the complex deforming zone in southeast Asia that can be regarded as a single Sunda plate or a series of microplates. The southern extent of rupture may be where the subducting plate changes from India to Australia. None of the plates' motions are well constrained, because of the limited GPS data and because many of the sites are near trenches and thus influenced by interseismic strain accumulation. Similarly, the plate geometry and motions are very poorly known in areas like Northeast Eurasia (from Siberia to northern Japan) or the East African Rift. New GPS and InSAR data, ideally complemented by seafloor geodesy, will be of great value.

- Time variability of plate motions: Space geodesy has reached the point where discrepancies between plate motions inferred from space geodesy and from geologic models are no longer regarded as necessarily due to errors in one or both techniques. Increasingly, they appear to indicate real changes in motions over the past few million years. For example, GPS data show that as the Andes mountains grew, the subduction of the Nazca plate beneath South America slowed. Similarly, we appear to be seeing a slowing of the opening of the South Atlantic, and changes in the plate motions around the Adriatic. As these data improve due to additional sites, longer time series, and higher precision analyses, they will give new insight into the physical processes changing plate motions.
- Relation between earthquakes and deformation in plate boundary zones: By mapping the present strain fields in boundary zones like the eastern Mediterranean, western North America, or the India-Eurasia convergent zone, space geodesy is giving crucial insight into the dynamics of these areas. A major research thrust is understanding how the geodetic deformation seen today relates to the historic record of earthquakes and geologic record of faulting. It is becoming increasingly clear that loci of seismicity and faulting have migrated with time within some boundary zones, and in some places are doing so today. Improved geodetic data are crucial for characterizing these effects, modeling their dynamics, and using the results to better understand earthquake hazards.
- Intracontinental earthquakes: Space geodesy is giving new insight into the mysterious, relatively rare, but sometimes large, earthquakes within plates. It shows that the deformation causing these earthquakes is surprisingly slow. The geodetic data are being combined with earthquake locations, focal mechanisms, and other geological and geophysical data to investigate the motions and stresses within plates and how they give rise to earthquakes. A growing body of evidence finds that continental intraplate earthquakes often occur in temporal clusters on faults that remain active for some time, and then have long quiescent periods during which seismicity migrates to other faults. Space geodetic data are letting us observe these processes as they occur today, with important prospects for hazard estimation.

3.3.2 Earthquake and volcano physics

Space geodesy observes the deformation that occurs during the long intervals between earthquakes and volcanic eruptions, and thus provides important insight into the physical processes that control them and for strategies to mitigate their hazards.

Traditional earthquake and volcanic eruption studies focus on what occurs during these dramatic events and the results, with a goal of understanding the underlying processes. Space geodesy can now expand this view by observing the deformation field between events and spanning a broad region around a fault or volcano. GPS measurements are being complemented by InSAR from satellites, which does not require monuments on the ground.

The advent of space-based methods like GPS and InSAR, which make collecting geodetic data faster and easier, have made geodesy and seismic wave studies complementary approaches to earthquake studies. Hence although seismology and earthquake geodesy had been viewed as distinct, owing to the different instrumentation, earthquake geodesy is increasingly viewed as very-low-frequency seismology (or earthquake seismology as high-frequency geodesy).

In years to come, we expect these advances to continue. In particular we anticipate:

- Continued advances in observing and understanding the physics of faulting: Space geodetic measurements are providing important information about earthquake mechanics. Data from many regions show that significant afterslip, not detectable seismologically, is a common phenomenon. These observations offer insight into the effects of the stress changes during earthquake, and provide constraints on the rheology of the lithosphere and asthenosphere. Such data are of special importance given the emerging view that stress transfer between faults may contribute to earthquake triggering. GPS and InSAR data are also showing transient deformation between earthquakes in some regions, notably at subduction zones, and thus providing new insight into the processes by which strain accumulates and hence where and how it can be released. These data are increasingly being used in earthquake and tsunami hazard assessment. In addition, GPS and InSAR data permit comparison of geodetic, seismological, and geologic estimates of the rates and directions of deformation within active regions. Initial data from around the world suggest that these rates can be quite different, and should lead to an improved understanding of the partitioning between seismic and aseismic deformation. The issue of this partitioning is crucial for seismic hazard assessment.
- Major advances in volcano monitoring: Although less appreciated by the public, volcanoes can pose dangers just as severe as the weather. Eruptions directly threaten over 200 million people worldwide, can potentially knock jet aircraft out of the sky by choking their engines with ash, and can disrupt global commerce by suddenly blanketing key regions with thick layers of debris. In contrast to earthquakes, which commonly strike without warning, volcanoes typically show telltale signs of unrest. Nonetheless, our ability to forecast the timing, magnitude and impact of future eruptions is frustratingly imprecise. For every major predictive success, like at the Philippines' Mt. Pinatubo in 1991, which saved over $ 1 billion of equipment and tens of thousands of lives, there are tragic failures, like at Colombia's Nevada del Ruiz in 1985, where mudflows triggered by a small eruption killed over 25,000 people. Most hazardous volcanoes are not presently monitored, largely due to the costs involved. Space geodetic data are

proving a powerful tool for volcano monitoring. They provide a less expensive, rapid and remote (hence safe) way of measuring surface deformation associated with volcanic processes, which will both, significantly improve our fundamental understanding of volcanic processes and will aid in eruption forecasting.

3.3.3 Deep Earth dynamics

Observations of motions at the surface, together with gravity data that constrain mass distribution at depth, provide crucial constraints on mantle dynamics and thus the Earth's thermal and mechanical structure and evolution.

Plate tectonics is the primary surface manifestation of the heat engine whose nature and history govern the planet's thermal, mechanical, and chemical evolution. Because the lithosphere is the cold outer boundary layer of the mantle's convection system, our most important constraint on this system comes from the rates and directions of plate motion. Space geodetic data are crucial for this purpose, especially via their ability to resolve plate motions on time scales of years for comparison with those observed over geologic time. Our ability to observe these changes provides important constraints for understanding how they result from plate driving forces. Similarly, understanding the deformation pattern in boundary zones like mountain belts is prompting new models of the processes at work, some which involve complex interactions between tectonics and climate.

Space geodetic data also provide other constraints. GPS and gravity data from the GRACE mission are giving new insight into PGR or GIA, the response of the solid Earth to the changing surface load brought about by the waxing and waning of ice sheets and glaciers. In the past 20,000 years GIA has caused up to several hundred meters of relative sea-level fall and over 100 m of sea-level rise in different parts of North America and Europe. Until recently, present-day observations of GIA were limited in two important ways. First, horizontal motions could not be observed. Second, vertical motions were measured only along coasts via sea and lake level changes, which requires climatic and hydrographic corrections. The advent of space-based geodesy, which can measure crustal velocities of less than a few mm/yr and provide detailed images of the changing gravity field, has changed this situation.

This is important because GIA is the subject of active research for three major reasons. First, the delayed response to deglaciation is one of the few ways of constraining the viscosity structure of the mantle, which is crucial for understanding the mantle convection process. Second, GIA can provide a powerful constraint on the distribution and thickness of ice since the last glacial maximum, about 21,000 years ago. Although the general pattern is known from glacial geomorphology, significant questions remain on which GIA can provide important information. Third, GIA has been suspected to be a major cause of deformation within continental plates interiors.

The gravity data are also important for mapping the Earth's deep interior. Although we seek to understand the composition, mineralogy, and temperature at depth, our measurements are sensitive instead to parameters like density and seismic velocity that can result from various combinations of composition, mineralogy, and temperature at a various depths. As a result, our ideas reflect inferences from combining geodetic and seismological data with results from geology, geomagnetism, cosmochemistry, and the physics and chemistry of materials at high temperature and pressure. These give a snapshot of the present stage of the Earth's thermal and chemical evolution, which is our best constraint on the Earth's evolution and crucial in developing our ideas about the other terrestrial planets.

In years to come, we anticipate that the increased density and accuracy of space geodetic data will dramatically improve models of both the rheology of the Earth's interior and of the history of glacial loading. This potential is shown by the fact that although the vertical motions are generally consistent with the predictions of GIA models, the horizontal data illustrate the need and opportunity to improve the models via more accurate descriptions of the ice load and laterally-variable mantle viscosity.

3.3.4 Surface loading

Mass relocation on the solid Earth's surface and in the fluid envelope constantly loads the solid Earth and induces deformations and changes in the gravity field. On time scales of minutes to years, most of the loading is due to mass re-location in atmosphere, ocean, terrestrial hydrosphere and cryosphere, i.e., the global water cycle. On decadal to century time scales, slow climatological changes in land water storage, glaciers and ice sheets are major sources of loading. The cycle of ice ages with the associated large changes in ice sheets has typical time scales of 10^3 to 10^4 years. Sediment loading induces significant load on even longer time scales of up to several million years.

While the theory for the elastic response to loading is well developed (see, e.g., Farrell, 1972), major uncertainties exist in the atmospheric, oceanic and hydrological loads (see, e.g., Van Dam et al., 2003). For ocean tidal loading, the accuracy of the ocean tidal models has increased considerably over the last years, and the ocean tidal loading signal in surface displacement and gravity changes can be predicted with high accuracy, particularly for locations not directly at the coast. For non-tidal ocean loading, the ocean bottom pressure field is still a major source of uncertainty. Uncertainties may be reduced with the help of gravity satellite missions. For atmospheric loading, the surface pressure field is a major uncertainty (Plag et al., 2007a), which can be reduced by improved processing. Models for changes in land water storage show still large inter-model differences, indicating large uncertainties. Here, too, satellite gravity missions may be a viable source for improvements.

For longer time scale, the rheology of the Earth's mantle and simplifications of the Earth model still constitutes significant contributions to the uncertainties (see, e.g. Plag et al., 1998b). For PGR, these uncertainties also impact the ice history derived from relevant observations. Ice histories determined on the basis of spherical symmetric Earth models and model predictions of the present-day PGR signal in surface displacement, gravity field, and rotation show a wide range of variations, which in turn hampers the interpretation of geodetic observations in terms of present-day changes in, for example, ice sheets and global sea level.

Recent attempts to include lateral heterogeneities in the Earth model (e.g., Latychev et al., 2005) are necessary steps but these models are just a beginning. 3-D Earth models need to be developed. However, results from seismic tomography also show a wide range of possible Earth models and research will have to focus on these issues in order to improve the interpretation of the geodetic observations in terms of mass re-locations.

3.4 The cryosphere

Ice sheets, glaciers, and sea ice are intricately linked to the Earth's climate system. They store a record of past climate; they strongly affect surface energy budget, global water cycle, and sea-level change; and they are sensitive indicators of climate change.

Geodesy is crucial for these studies because of its ability to measure the motions of ice masses and changes in their volumes. Since the mid 1990s, new geodetic observation techniques have shown rapid changes: Arctic sea ice is shrinking, both in extent and thickness (e.g., Stroeve et al., 2008); low-latitude glaciers and ice caps are losing mass at rapidly accelerating rates (e.g., Meier et al., 2007); and even parts of the vast ice sheets in Greenland (Zwally et al., 2002; Velicogna & Wahr, 2005; Tedesco, 2007; Khan et al., 2007, e.g.,) and Antarctica (e.g., Thomas et al., 2004; Zwally et al., 2005; Velicogna & Wahr, 2006) are shrinking.

Various geodetic techniques, largely involving measurements from aircraft and, increasingly, from satellites have led to the recent improvement in our knowledge of what is happening to ice on Earth. Detailed imagery, at wavelengths ranging from visible to microwave, precise altimetry, and measurements of temporal changes in gravity have yielded the largest returns: highly-accurate measurements of ice velocity from repeated GPS surveys; spatially detailed measurements of glacier motion from interferometric SAR (InSAR); ice thickness measurements over glaciers and ice sheets from low-frequency radar; ice-sheet thickening/thinning rates over vast areas from laser and radar altimeters; estimates of rates of change of the mass of entire ice sheets from satellite measurements of temporal changes in Earth's gravity field; time series of sea-ice extent and motion from passive-microwave and SAR images; and routine mapping over almost entire ice sheets of characteristics, such as temperatures, iciness, and wetness, of surface snow. Although some of these data have been available for many years, recent advances allow much more data to be acquired and allow more accurate quantitative estimates. These advances involve both measurement systems and improved geodetic reference frames. In particular, many techniques that are now the "bread and butter" of glaciological research became possible only after GPS data became easily available and widely applied.

The importance of geodesy to ice research is highlighted by the rapid changes in the ice sheets that have been revealed only recently. Their detection, and investigation of their causes, would not have been possible without accurate geodesy. A key question posed by these changes was how much could be explained by natural variability in snowfall and melt rates, and how much required longer-lasting changes in ice-sheet dynamics. Ice velocities measured by interferometric SAR data from satellites showed that, in many regions, the observed rapid thinning was accompanied by local glacier acceleration, indicating substantial changes in glacier dynamics.

Calculation of highly accurate orbits and aircraft trajectories requires a reliable reference frame and continued tracking of key satellites and Earth crustal motion. This has direct benefits for glaciological research by improving the accuracy of measured parameters, such as ice-surface elevation, detection of temporal changes in ice cover and ice-surface velocities, and indirect benefits, such as improved knowledge of rates of sea-level change and post-glacial uplift.

Until recently, the most reliable information on the mass balance (net rates of mass change) of glaciers and ice sheets was provided by our knowledge of sea-level change. This provided bounds on how rapidly the mass of land ice could be changing, that were more reliable than the results of decades of glaciological measurements. Recent advances in our ability to measure ice behavior have changed this situation. Ten years ago, it was not possible to determine even whether total mass was increasing or decreasing. Now, measurements show that, over those ten years, ice has contributed an average of about 1 mm/yr to sea-level rise. Setting this information in context requires accurate measurement of total sea-level change, which also depends on geodesy, both to provide a reliable reference frame to tie together tide gauges from all over the globe, and to provide accurate orbits for altimetric satellites used to map sea-level over entire oceans. Results show that recent changes in ice mass are responsible for $\sim$30% of a total sea-level increase of $>$ 3 mm/yr.

There are three ways to measure the mass balance of the large ice sheets in Greenland and Antarctica: comparison of total snowfall with total losses; measurement of volume changes, using altimetry of the ice surface; and measurement of temporal changes in gravity that are indicative of mass changes. All but the first require correction for changes in the elevation of rock beneath the ice. This is particularly so for interpretation of gravity changes, because rock is much denser than ice. By necessity, the required estimates of crustal motion beneath ice sheets come from models, which become progressively more reliable as more information becomes available on actual vertical motion. This in turn depends heavily on highly accurate geodetic measurements.

In addition to these approaches for measuring ice-sheet mass balance, changes in length of day and in the direction of the Earth's rotation axis also reveal mass redistribution. These techniques are still under development, but also require precise geodetic measurements.

3.5 Ocean processes and their climatological implications

Geodetic instrumentation and techniques are at the heart of present day ocean studies. First, geodesy provides precise positions for shipboard samples or those from floating instruments. Second, geodesy provides direct observations. Satellite altimetry has become an essential input for numerical ocean models. Gravimetric satellites provide the horizontal for altimetric determinations of surface pressure gradient. Time varying gravity data from gravimetric satellites, the latest entry into this field, yields information on time varying ocean heat content and bottom currents, as well as ice mass variation, a sensitive indicator of climate change.

Geodesy will play an increasing role in ocean circulation and climate studies in years to come: It will provide techniques to monitor the oceans and cryosphere for seasonal to interannual changes, which is of immediate societal relevance. This ability to provide long term climate-quality time series to see our planet change on decadal time scales must be a fundamental goal of GGOS. Hence the technology should be designed to increase coverage and resolution and monitor new observables.

3.5.1 Providing the reference frame and the means for precise positioning

The accurate determination of position at sea is much more demanding in terms of reference frame than on land, where landmarks provide information about position. Marine observations such as a sample from the sea floor, the temperature and salinity of seawater, or wind at the surface, require accurate measurement of the position of the sensors. Tide gauges are peculiar instruments, always affixed to land, but measuring an ocean property (sea level). Their position in a reference frame is essential to determine whether the sea surface went up, or the land surface went down. GPS, and more general GNSS, has become the standard for all these measurements, although positioning by Argos transmitters (http://www.cls.fr/html/argos/welcome_en.html) is still popular with moving platforms (such as for tracking animals), due to the low mass and power consumption of the transmitter. GLONASS, and the imminent GALILEO constellation will greatly expand the positioning service, and thus minimize the length of time a float needs to be at the surface, while increasing the accuracy of its position, and thus of the derived water velocity. This crucial application requires accurate reference frames to which highly accurate satellite positions can be referenced. Although these reference frames have little visibility outside the geodetic community and often are taken for granted, they are crucial to the success of present day ocean studies.

3.5.2 Altimetry and ocean circulation

Satellite radar altimetry has existed for over 30 years, since Skylab carried the first radar altimeter. The subsequent TOPEX/Poseidon mission, launched in 1992, and its successor JASON-1, produced a large improvement in accuracy to 4 cm or better overall, which opened the door to a variety of new discoveries in ocean physics. The largest source of that improvement was the precision orbit determination for the altimeters. Since tides are the largest ocean signal and must be removed before other signals can be studied, renewed interest in modeling ocean tides was spurred. This not only produced the most accurate global ocean tidal models to date (Andersen et al., 1995), and better estimates of their energy dissipation, but also led to a new understanding of the possible energy source for the meridional overturning circulation, the slow (order of 100 years) predominantly vertical transport of waters of different densities. Another fundamental observation was that Rossby waves, the slow, westward propagating waves crucial in carrying information from one location in the ocean to another, had a velocity versus latitude function that differed significantly from that in standard theory. This led to a revision of theories to describe these waves.

These advances have practical applications. Satellite altimetry lets us 'see' an El Nino/Southern Oscillation mode of interannual variability in its early stages of development, allowing accurate forecasts of its consequences. Radar altimetry lets us map the ocean's heat content at low latitudes, a quantity termed 'hurricane potential' for its ability to feed these storms, thus helping forecast their path and intensity. Moreover, global satellite altimetry data are being assimilated into numerical models of the ocean circulation data to determine the state of the oceans at any one time which is consistent with ocean physics and the available data. This capability has become the basis for operational oceanography.

An interesting indirect application of satellite altimetry to ocean circulation studies comes from the close relationship between the shorter scales (order 100 km) of the gravity field over the oceans and bathymetry. Thus gravity data are used to derive maps of bathymetry (Smith & Sandwell, 2004) which provide information on the ocean bottom roughness that is a source of mixing rate variation (Kunze & Smith, 2004), with direct effect on climate.

The next generation of instruments should permit further advances. To date radar altimetry uses downward (nadir) pointing instruments on satellites usually flying alone, or at best from two or three satellites not intended for joint operation. This has prevented extending detailed mapping of coastal waters, where the length scales are much shorter than in the deep ocean, and thus shorter than the 150 or 300 km between tracks of nadir altimeters. We envision several advances: a) an imaging instrument, essentially an interferometric SAR that draws a wide swath along the ocean, rather than a narrow thin one; b) a constellation of relatively inexpensive nadir altimeters; these would avoid many of the characteristics that made TOPEX/Poseidon expensive, such as an onboard radiometer, etc, paired with one TOPEX/Poseidon class high accuracy altimeter for crossover adjustments, and c) a constellation of satellites of opportunity equipped with antennae and electronics capable of detecting reflections from the signals transmitted by the GPS satellites.

3.5.3 Satellite gravity, ocean circulation and climate

A basic application of satellite gravimetry in oceanography is to provide the horizontal at every point on the ocean surface, because the slope of the sea surface measured by radar altimetry relative to this horizontal is the pressure gradient associated with the geostrophic component of surface currents (and the other component, called the Ekman component, is associated with winds and has no signal in sea surface height). For this calculation, time changes in the gravity field are much smaller than uncertainties in the mean sea surface. This calculation was dramatically improved by the CHAMP and GRACE gravity missions. Further refinements will come from the gravity gradiometric satellite mission GOCE launched in March 2009. The surface component of velocity can also be derived from *in situ* data, with the latter providing the shorter scales and satellite altimetry minus the geoid providing the longer scales.

Precise measurements of the time variations in the global gravity field from GRACE offer a powerful new method to study the oceans and the cryosphere. Measurements of the temporal variations in gravity averaged over the global oceans show the exchange of water between the atmosphere, land, cryosphere, and ocean reservoirs. Such measurements show changes in deep pressure gradients, indicative of temporal changes in deep currents or in vertically-averaged currents.

When combined with sea surface heights inferred by radar altimetry, the gravity data provide information on the geographic distribution of monthly or submonthly changes in ocean heat content. Altimeter record changes in local water column height, whether produced by expansion due to heat (and, to a lesser extent, salt), or by mass addition, whereas gravity is only sensitive to the latter. Hence the difference reveals changes associated with heat and salt, dominated by the former.

The gravity data also are valuable for cryospheric and climatic studies. One of the most dramatic results from GRACE is detection of ice mass losses from Antarctica and Greenland (Velicogna & Wahr, 2005, 2006).

The next section will discuss further applications of time-averaged and time-varying satellite gravity measurements, in combination with other data. Given what has been accomplished in the 6 and 4 years of the CHAMP and GRACE missions (see Section 2.6.5), both of which were the first of a kind and produced new data scientists were unaccustomed to and are still learning to exploit, further important discoveries are likely ahead of us.

Because of the long lead times needed to get new satellite missions approved, and the need to demonstrate results from first-of-a-kind missions such as CHAMP and GRACE, it is essential that further missions to measure the gravity field from space, especially its time-varying component, are planned and budgeted. GOCE, which will provide much shorter scales of the global gravity field than CHAMP or GRACE, is an excellent first step.

3.5.4 Synergistic combination of measurements

The greatest power of the measurements discussed above comes when two or more technologies measure essentially the same quantity. For example, surface ocean velocity is measured both from space and by surface drifters (although the positioning relies on GNSS). The velocity includes both Ekman and geostrophic components, the latter of which is directly linked to much deeper velocities. After modeling and removing the Ekman component, in principle one is left with the geostrophic component. Thus, current efforts are blending the two data types to resolve their different scales and error characteristics.

Similarly, time varying ocean heat content distributions can be obtained from satellite gravity and altimetry, or from sea surface (Argo) floats. However, owing to limited coverage, data from the Argo floats need to be combined with GRACE and altimetry time-varying data.

A third example of synergistic application of data is estimating the flux of moisture from the ocean into the land. One approach combines data from atmospheric sounders that yield atmospheric water vapor with surface ocean vector wind measurements from scatterometers. For comparison, river runoff from the land can be obtained from poorly constrained climatological estimates of evaporation minus precipitation over land. However, by mass conservation, the balance of the previous terms is the storage of water over the land, and so can be estimated from GRACE data. Ultimately, all the data are assimilated into numerical models to constrain a physically plausible description of the coupled system including the oceans, cryosphere, land, and atmosphere. Thus, the satellite data are combined with observations at or below the sea surface and assimilated into numerical models of the oceans. The satellite data improve this process significantly.

3.5.5 Future needs

Longer-term altimeter observations from multiple missions are clearly needed in the future and with sufficient overlap. Since gravitational field observations, such as those from GRACE, are sensitive to processes that change the Earth's mass distribution, they can be used to investigate sea level rise and ice sheet volume changes. Moreover, since the Earth's gravitational field is not sensitive to the thermal expansion of sea water, observations of the gravitational field can be used in concert with sea level change observations to separate the change due to thermal expansion or contraction from that due to oceanic mass changes, thereby helping to quantify the extent to which greenhouse warming is sequestered in the oceans. Satellite altimetry provides measurements of the time-varying sea level which, when assimilated into oceanic general circulation models along with other remotely sensed and in situ measurements, provide improved estimates of the three-dimensional oceanic temperature, salinity, and velocity fields. The Earth's gravitational field will change as the ocean-bottom pressure changes, and, under the principle of the conservation of angular momentum, the Earth's rotation will change as the oceanic angular momentum varies due to fluctuations in the ocean-bottom pressure and velocity fields. These three data sets (satellite altimetry, gravity, and Earth rotation) therefore provide a powerful means of investigating the causes and consequences of sea level change (Blewitt et al., 2006a).

3.6 Studies of weather and climate processes

The contribution of geodesy to the study of the weather is four-fold. Geodesy can help observation and prediction of the weather by (1) geo-referencing meteorological observing systems, (2) providing atmospheric weather models with space- and time-varying gravity fields, (3) collecting observations of the stratospheric mass and lower tropospheric water vapor fields, and (4) contributing to tracking atmospheric global change.

3.6.1 Geo-referencing of all meteorological observations

Since the early days of meteorology, the need for time synchronization of atmospheric observations has been recognized in order to be able to compare simultaneous measurements and establish a dynamically consistent picture of the weather. Observations must also be referred spatially, both in horizontal and vertical position. As measurements have moved away from the ground to be made aloft, from aircraft or balloons, pressure was chosen as the independent coordinate to reference measurements in the vertical. Horizontal positioning was obtained by various systems and now more commonly by GPS. These *in situ* measurements are now complemented by thousands of remote measurements from satellites. Many of these are passive emissions for which vertical positioning does not apply. However, more and more such measurements are (and will be) LIDAR returns and limb sounding radio occultation refraction (or delay) data, both with an unprecedented vertical accuracy. In order to exploit these measurements and use them in combination with *in situ* measurements, considering all atmospheric measurements in a single reference frame is crucial .

The World Meteorological Organisation (WMO) recently (November 2006) adopted a world geodetic system and global geoid model for positioning all weather observing stations. Meteorological measurements reported to WMO are now located in the horizontal (latitude, longitude) using WGS 84, and in the vertical (altitude) with respect to Earth Gravity Model (EGM)96. Hence the continued improvement of these reference frames is a powerful tool for meteorologists. The close alignment of WGS84 to ITRF ensures the accuracy of WGS84.

3.6.2 Providing atmospheric weather models with space- and time-varying gravity fields

Weather observations today are assimilated into numerical models to predict the weather and issue forecasts. Most of these model atmospheric circulation assuming a perfectly spherical Earth, a constant gravity acceleration, and a thin layer of atmosphere. However, the use of observations from the real world forces atmospheric models to recreate surface pressure fields consistent with observations and in line with the actual shape of the Earth. It is currently only in that process where the model assumptions of spherical symmetry and constant gravity are starting to be accounted for. Future atmospheric models could use geodetic data to refining some of these assumptions.

Geodesy could provide weather models with three-dimensional gravity fields at regular time intervals. These fields would be used to map the observations' altitudes onto geopotential heights and constrain newer models whose dynamics have been upgraded to account for non-constant gravity fields.

3.6.3 Collecting observations of the upper-atmospheric mass and lower tropospheric water vapor fields

Extreme weather events are often associated with significant rainfall and latent heat release. The prediction of the timing and intensity of these events is critical to ensure proper warning and preparation of the affected populations. Although atmospheric water vapor is currently observed globally, using passive sounders and *in situ* measurements, the time- and space-resolution is too poor to support efficient warning for extreme events.

This situation is changing due to space-based geodesy. Instruments using GPS and VLBI technology rely on radio signals that traverse the atmosphere between platforms on the ground and space-based instruments or distant radio sources outside the atmosphere. These signals are modified by the atmosphere, which introduces a noise from the geodesist's point of view that is now routinely accounted for to yield atmospheric-corrected geodetic observations. The converse of that process is that geodesy produces atmospheric observations from the complex web of radio links that constitute the core of GGOS.

Weather prediction already benefits from atmospheric observations made from the ground with GPS receivers (e.g. Elgered et al., 2005). For example, Meteo France now uses observations of zenith radio path delays as part of its global and European weather forecast systems after trials showed that these data helped improve the forecasts.

Similar observations made with the radio occultation technique using GPS receivers on LEO such as CHAMP have been shown to help improve weather forecasts in the stratosphere at several national weather centers (see also Section 2.9). Radio occultation observations of (refraction) bending angles provide indirect information on the stratospheric mass field and hence information on the stratospheric temperature (assuming hydrostatic equilibrium).

Geodesy already helps atmospheric science by providing high-temporal resolution observations of the lower troposphere via atmospheric delays that contain information on the atmospheric water vapor field. Densification and improved timeliness of such observations could benefit the prediction of intense rainfall events.

Radio occultation observations are now collected in near-real time by the recent COSMIC mission, with an array of 6 receivers (see Section 2.9). Current missions such as CHAMP and GRACE and future missions are expected to deliver similar observations. Ensuring the processing and delivery of these to weather prediction centers would help them observe and forecast temperatures in the stratosphere.

3.6.4 Tracking global change in the atmosphere

The attempts of atmospheric physicists to monitor long-term atmospheric trends have been limited by the issue of instrument calibration. Some of the instruments that make up the GGOS hold the key for auto-calibrated, long-term stable atmospheric measurements. Because GPS satellites rely on accurate (atomic) clocks, the phase of the signals they transmit is calibrated from atomic clocks on board, which are regularly updated with clocks on the ground. In contrast, no other observation of Earth's atmosphere relies on such a recurrent, atomic calibration procedure. For example, measurements of passive infrared radiation need to be calibrated with respect to a blackbody of known temperature to compensate for optics and detector aging in the instrument. However, monitoring the decay of that blackbody over a long period of time proves difficult (if not possible), unless it is brought back to known experimental conditions in a laboratory. Because this is impossible to do with in-orbit radiometers, their time drift is thus difficult to estimate. GGOS could develop a record of atmospheric delays from a set of ground stations whose position is monitored over time using other means such as VLBI. A similar record based on atmospheric-induced frequency shifts could also be collected by radio occultation missions. Thus, GGOS could provide data on the climate change effects anticipated in: temperatures, altitudes of constant pressure levels, and atmospheric mass transport.

The metric most often used for assessing climate change is the rate of change of atmospheric temperature near the surface, because it is related to human activities and controls our environment. To monitor that temperature, instruments and methods must be designed to collect measurements with an accuracy within the suggested climate trends of 0.1 K per decade. While it is important to plan and realize such measurements with the desired accuracy, it is equally important to measure the consequence of such trends on the static and the dynamic structures of our atmosphere.

For example a magnifying effect of temperature change is on air density and hence atmospheric layer thicknesses. Assuming for example a 0.1 K homogeneous warming throughout any given atmospheric layer whose boundaries are defined by fixed pressure levels, that atmospheric slab would expand by about 0.04% of its original thickness. In real terms and with a tropospheric average temperature of 250 K, this would raise the mid-latitude near-tropopause level of 200 hPa by 5 meters. With GGOS system capable of ensuring a reference frame with a decimeter accuracy over a decade to locate upper-air atmospheric pressure *in situ* sensors, and accounting also for possible changes in height at the Earth surface in the same time frame, this trend could be identified.

Temperature change would also affect patterns of atmospheric circulation. As horizontal temperature gradients change, the cells that make up the general circulation are affected in their strength and shape (including extent and position). (Vecchi et al., 2006) have found evidence of a weakening in the tropical Pacific Walker circulation between 1861 and 1992 based on sea level pressure observations in that region. Using climate models to elaborate on the origin of that decline, they found that anthropogenic changes in the atmosphere could explain the observed decrease in sea level pressure gradient. Similarly, Intergovernmental Panel on Climate Change (IPCC) simulations made at Meteo France, predict a weakening of the Hadley cell. The total mass of air flowing through the upper branch of that cell at the latitude 15°N and between 200 and 100 hPa pressure levels is currently about 50 Megatons per second in January. A shift in the location of that upper branch of the Hadley cell predicts that by January 2030 this flow would decrease by up to 5 Megatons per second. GGOS could help in measuring such changes by looking at mass displacements, which would complement atmospheric measurements of air density and wind flow measurements in the example mentioned here.

3.7 Sea level change

From a geophysical point of view, sea level has a twofold interest both because the ocean itself is one of the major components of the Earth system and because sea level is the boundary between the ocean and the atmosphere. Recently, due to the increasing attention and evidences for anthropogenic changes induced in the Earth system, one aspect of sea-level has come into focus: a rise in global sea level is considered as one of the more severe consequences of the predicted global warming. However, besides this impact aspect, sea level may also prove to be an important indicator of global warming, especially if an acceleration of the sea-level rise can be detected.

Observations of sea-level variations on different time scales are highly relevant. On the one hand, they allow for the analysis and description of sea-level variability which is a prerequisite to a better understanding of the causes behind these variations. On the other hand, they constitute crucial constraints for models related to sea level, such as hydrodynamical models or coupled atmosphere-ocean circulation models (Plag et al., 2000b).

The contributions of geodesy to sea level science can be divided into two main groups. Geodesy provides the essential data sets needed to monitor sea level change and vertical land movements. It can also supply a range of information by which the reasons for sea level change can be understood, leading eventually to more precise sea level change prediction.

3.7.1 Geo-location of sea and land levels and their changes

Information on historical sea level changes takes the form of measurements of the height of the ocean surface relative to that of the nearby land (see Section 2.9.3). This is true of all "relative" sea level measurements, whether geological, archaeological or tide gauge. The long term changes in sea levels estimated by such methods are necessarily affected by vertical movements of the land itself with respect to the CM which for some locations and timescales can exceed the changes of the sea surface with respect to the CM. Carefully conducted combined GNSS and absolute gravity measurements can provide an independent monitor of vertical land movement, and thereby facilitate decoupling of sea surface and land level changes in tide gauge records. Such a decoupling supports an improved understanding of the various processes responsible for change.

The use of GNSS at tide gauge sites is far from straightforward, requiring suitable benchmark monumentation and precise leveling in port areas which are often extremely busy and subject to frequent redevelopment. In addition, operations at the coast can potentially lead to larger tropospheric variability in GPS data sets than those from inland.

The spatial variability of vertical land movements can be studied with networks of GNSS receivers and from space by INSAR. The latter is of special interest in providing insight into the spatial variations in land movement in the local areas near to tide gauges, and thereby into how well the movements measured by GPS at the gauge apply also at some distance from it (e.g., whether very localized subsidence exists). With the use of data from various geodetic techniques, complementary global, regional and local models of vertical land movement can be constructed for application to tide gauge data for which no corresponding GNSS information might exist (e.g., in developing countries).

Satellite radar altimeters measure the sea surface positions in a geometric reference frame through precise orbit determination of satellites such as TOPEX/Poseidon and Jason-1 by SLR, GPS and DORIS (see Section 2.4.3). Such space data sets can be calibrated by means of tide gauge sea level measurements located in the same geometric reference frame with GNSS, thus enabling effective combination of measurements from the two sources.

While sea surface measurements are relatively straightforward to undertake from space, a major challenge is to make measurements relative to a geopotential surface (geoid) by means of which absolute ocean currents and fluxes can be determined, enabling direct comparison of transports measured *in situ* (e.g., by current meters) to those inferred from sea level data, and the constraint of such parameters in ocean and climate models. The GOCE mission (see Section 2.6.5) should provide such insight into the geoid for application to oceanographic research, including sea level science, and to an extended range of geophysics.

Temporal space gravity measurements such as those by GRACE (see Section 2.6.5) provide information on the variability of mass around the globe, which over the ocean can be studied as a parameter akin to ocean bottom pressure (essentially barotropic sea level). Such changes, when combined with data on sea level variability from altimetry and tide gauges, enable a decoupling of barotropic and baroclinic ocean variations. The latter in effect allow oceanographers to study ocean properties beneath the surface of a global ocean, something which can be performed less comprehensively only with hydrographic vessels or a network of sub-surface floats (Argo). Complementary sub-surface information derived from *in situ* measurements and from space can be assimilated into ocean models providing ongoing 3-D assessment of the state of the ocean.

3.7.2 Understanding sea level change

Changes in global sea level due to volumetric changes of the ocean water are thought to constitute the "climate signal" in sea level, which is composed of two parts: (1) melting of land-based ice adds water to the ocean, and (2) a change in the heat content of the water of the oceans leads to a volume change, for example, warming of the ocean increases the volume of the water. Therefore, knowledge of the global ocean water mass and volume as function of time constitutes a crucial constraint for the reconstruction of past climates, as well as the validation of models used to predict future changes in the global ocean mass and volume. However, extracting this climate signal from sea level records is a delicate task. One reason for this delicacy is the complicated mass budget of the ocean which is the major reservoir of the global hydrological cycle. Thus, a once established change in global sea level cannot directly be equated to the climate signal (Zerbini et al., 1996). Ignoring in this context changes in the volume of the ocean basins which are believed to be relatively small, the global ocean mass and volume may be affected by all changes in the other reservoirs of this cycle such as groundwater, soil moisture, humidity of the air, terrestrial surface, ice sheets, glaciers or perma-frost. At present, human interference such as deforestation, groundwater extraction, irrigation, river basin developments or reduced infiltration due to infrastructures and urban development are at a level where the global hydrological cycle is significantly affected. Various studies have been conducted on these anthropogenic influences; however, there are considerable differences in the results obtained, mostly because these studies were based on insufficient data.

A major contribution to the ocean mass changes is due to land ice which, if it were all to melt, would cause more than 60 m sea-level rise. Glaciers in most mountain regions are known to be retreating. In Greenland thinning of the ice sheets predominates at rates that are increasing with time. The picture is less clear for Antarctica (Pfeffer et al., 2008, e.g.,), but net loss appears probable, with dynamic losses also increasing with time (Steffen et al., 2006).

Monitoring the changes of the surfaces of the large ice sheets still poses considerable problems to the remote sensing methods currently in use including satellite radar altimetry and satellite and airborne laser altimetry. These problems will not be addressed here. The GRACE mission data allow deriving information on temporal changes in the mass distribution of the ice sheets and underlying rock. Because of the GRACE satellite altitude, mass balance estimates with spatial resolution of only several hundred km are possible; however, there is the advantage of covering entire ice sheets, which is extremely difficult using other techniques (Steffen et al., 2006).

Besides mass exchange with the cryosphere, the volume increase due to thermal expansion of the ocean is considered as a major contributor to sea-level change and variability. Significant progress has been made during the past 20 years in observing and understanding the decadal variability and, to a lesser extent, the multi-decadal trends in global ocean heat content and thermosteric sea level. As of the 1980's, the sign of global thermosteric sea-level change was unknown due to insufficient sampling. Among other reasons, advances in technology have contributed to change the nature of the problem dramatically. The World Ocean Circulation Experiment (WOCE) provided a global top-to-bottom survey of ocean temperature and salinity during the 1990's and it enabled the global Argo array whose implementation began in 2000. Future prediction of ocean thermal expansion and thus future sea-level rise (including its regional distribution) depends on coupled ocean atmosphere models (Roemmich et al., 2006).

Geodetic techniques have revolutionized our understanding of sea level variability through the provision of new data sets and more accurate versions of existing ones. Such understanding is reviewed and summarized at regular intervals in the assessments of the IPCC, with the 4th Assessment Report (Solomon et al., 2007) being the most recent one.

One might summarize several of the main reasons for sea level change and how geodesy contributes as follows:

- **Thermal expansion:** space gravity missions (e.g. GRACE) provide observations that allow the determination of effective bottom pressure; GNSS locates Argo floats.
- **Cryosphere changes:** GNSS allows the measurement of changes in the height and extent of glaciers and ice sheets and their flow rates and thereby provides insight into their dynamics (see Section 3.4); space gravity missions provide estimates of changes in mass balance of glaciers; laser and radar altimetry measure topographic changes (see Section 2.4.3).

- **Hydrosphere changes:** Space gravity missions provide observations related to ground water changes in large basins.
- **Geosphere changes:** GNSS, in particular if combined with absolute gravity measurements can aid improvement in models of postglacial rebound and tectonic processes.
- **Meteorological changes:** GNSS-meteorology contributes to improved atmospheric models and subsequent air pressure and wind fields; water vapor measurements are of direct importance to altimetric data accuracy.

The importance of geodesy to ocean and climate modeling can be mentioned in its own right. GPS on research vessels and advanced hydrographic techniques (acoustic depth gauges, Doppler current meters etc.), together with satellite altimetry and gravity data, provide information on shape and bathymetry of ocean basins and the currents in them which modelers require. Space gravity is a particularly exciting recent development: GOCE should enable estimation of absolute ocean transports down to short ($\sim$100 km) spatial scales for application throughout deep and coastal ocean and climate modeling, and temporal gravity (GRACE) data may help to select best choices of parameterization of physical processes in Atmosphere Ocean General Circulation Models. Such modeling results in improved modeling and understanding of the oceanic reasons for sea level change.

Sea level is intimately connected to the three "pillars of geodesy". Consequently, geodetic observations can characterize highly precise spatial and temporal changes of the Earth system that relate to sea level changes. The challenge for quantifying long-term change in sea and land levels imposes most stringent observation requirements, and can only be addressed within the context of a stable, global reference frame, such that measurements today can be meaningfully compared with measurements several decades later with millimeter accuracy. The reference frame becomes the foundation to connect observations in space and time and defines the framework in which global and regional observations of sea level change can be understood and properly interpreted.

However, long-term stability of the reference frame alone is not sufficient. Since the sea surface adjusts closely to an equipotential surface of the Earth's gravitational field, the RFO needs to be tied to the CM. A potential secular translation of the RFO with respect to the CM is expected to bias global sea level trend estimates on the order 0.2 to 0.3 mm/yr (see Section 2.2).

Consequently, improvements of ITRF are a crucial requirement for sea level studies. Moreover, observations of changes in Earth's shape with GNSS and other techniques (VLBI, SLR, DORIS), in LOD, and in polar motion are essential data sets for the understanding of the processes forcing sea level variations. Deformations on land as a result of load changes (ice, ground water, etc.) and geological processes, which can subsequently result in a sea level signal, can also be monitored by GNSS, complementing space measurements by GRACE. Therefore, a rich source of geodetic information needs to be maintained if the reasons for any future observed sea level change are to be properly identified and understood.

3.8 The hydrological cycle

Considering the importance of the hydrological cycle for the functioning of the biosphere as well as of the most near-surface processes of the Earth system, and its role as a major constituent of the climate engine, the need for innovation in techniques applicable to the monitoring of water-mass movements cannot be over-emphasized. There is a clear need for novel approaches in this field including, for example, the continuity of the new generation of gravity missions such as GRACE (Tapley et al., 2004b) designed to detect relatively small mass movements in the Earth system (Plag et al., 2000b; Ilk et al., 2005).

Many parameters characterizing the hydrological cycle such as ground moisture, effects of infiltration on groundwater renewal, percolation of groundwater, subsurface discharge of groundwater into the ocean are known with large uncertainty limits. The uncertainties are basically due to spatially and temporally insufficient observations of the transport of water within nearly all components of the hydrological cycle. To a large extent, this lack of observations is caused by the absence of technologies that allow sufficient monitoring of the relevant parameters within given economic constraints.

Understanding of the land component of the global water cycle is of profound importance for bio productivity, global water supply, and climate change. This is so, even though its total volume amounts to only 3.5% of the total water cycle. However the land component is by far the least understood part of the global water cycle. This is due to the fact that there exist (1) a severe scaling problem with the relevant scales reaching from molecular to global and from sub-daily to secular, (2) some quantities such as evaporation, soil moisture, bio productivity or photo synthesis are difficult to measure and (3) the residence time of water masses depend on vegetation soil, and rock storage compartments, hydro geological conditions and topography and are therefore highly uncertain.

The observability of the essential variables will improve in the near future. New satellite techniques are under development which are able to measure bio mass volume, photosynthesis (directly related to the actual bio activity), soil moisture, snow cover and precipitation. Thus, on a global scale it is an evolving area of research.

Geodesy and GGOS alone will not be able to solve the above problems. However GGOS is able to provide very important contributions to the study of the continental water cycle. Recent studies of the Amazon region, of the monsoon cycle and of smaller river catchments employing GPS, satellite altimetry and temporal gravity changes from GRACE are indications of the geodetic potential for hydrological research.

GGOS will be able to provide the following measurements relevant for hydrological studies, and they will be delivered in a unique global reference system:

- sub-seasonal, seasonal and secular movements of continental water masses as measured by satellite gravimetry such as GRACE (challenges: higher spatial resolution, separation of continental water mass changes from other time-varying gravimetric signals)

- the wet part of troposphere from atmospheric sounding using the global ground network of GNSS receivers and geodetic VLBI telescopes. (challenge: this technique is still evolving but it is used already routinely by several weather services)
- loading and un-loading of the land surface due to seasonal changes of groundwater, to be measured by the global network of GNSS stations (challenge: separation from other local station movements)
- local measurement of the integral variation of ground water from permanent gravimetric tidal stations (challenge: separation of the hydrological signal from signals such as temperature, pressure and environmental effects)
- measurement of water level of major lakes and rivers by satellite altimetry (challenge: locking algorithm of altimeters)
- improved digital terrain models, as basis for flux modeling of surface water and flood modeling (challenge: actual water flow is influenced by plant cover and soil characteristics)
- geographic information systems (GIS) for the establishment of comprehensive geo-references data modeling.

Nowadays uncertainty of the hydrological signal is a main uncertainty in modeling Earth rotation (compare Section 3.10). Its improvement will help Earth rotation modeling. In turn, one day, Earth rotation parameter may provide important global constraints on the climatological trends of the global water cycle.

3.9 Mass transport and mass anomalies in the Earth system

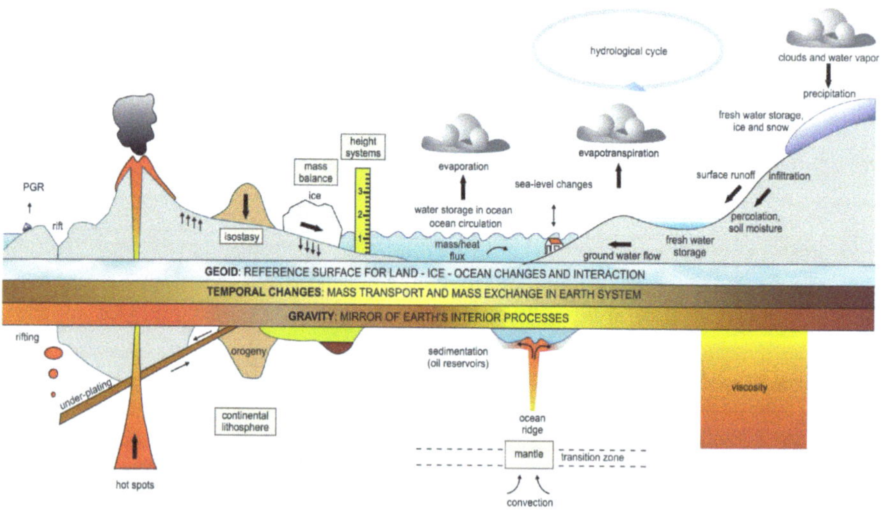

Fig. 3.3. The interrelation of gravity, gravity variations, mass transport and distribution. From ESA (1999).

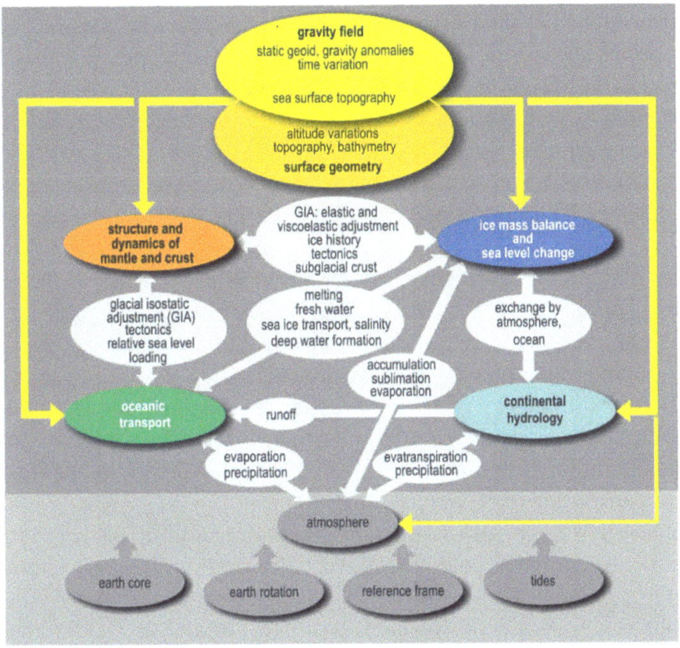

Fig. 3.4. Interconnections between processes and research themes related to mass transport and mass distribution. Arrows in the center of the figure indicate mass exchange and dynamic feedback mechanisms. Other arrows connect the gravimetric and geometric observations (on top of the figure) to the physical processes or indicate external influences and complementary fields (at the bottom of the figure). From Ilk et al. (2005).

3.9.1 Mass redistributions and geodesy

The gravity field and its variations - measured by satellites with unprecedented accuracy - are closely interrelated with mass transport and mass distribution. Figure 3.3 gives an overview of gravity-related phenomena, associated with anomalous signals in the geoid, in gravity or with temporal changes of geoid or gravity. The atmosphere, hydrosphere, ice covers, biosphere, land surface and solid Earth interact in various ways, ranging from sub-seasonal and interannual to decadal and secular variations on a global to regional spatial scale. This makes it difficult to develop realistic models that are capable to yield realistic predictions. Rather sophisticated partial models exist, for example, for weather predictions, the coupled atmospheric and ocean circulation, of local hydrological processes, or of glacial isostatic mass adjustment, but we are still far from a comprehensive description and understanding of the dynamics of Earth system. An important, and so far missing, segment of Earth system models is the determination of mass anomalies, mass transport and mass exchange between Earth system components and, ultimately, the establishment of global mass balance.

Quantification and understanding of mass transport and mass distribution requires a close cooperation of many Earth system research areas: oceanic transport, continental hydrology, ice mass balance and sea level, dynamics of mantle and crust, and geodetic signal analysis of the satellite missions. Such an interdisciplinary approach is necessary due to two reasons. The first reason is the importance of water mass exchange across the boundaries of the system components oceans, land, ice and atmosphere. The goal is a consistent modeling, where mass output from one model (e.g. for an ice sheet) is used as mass input for another model (e.g. for the neighboring ocean). The other reason is the integral character of the satellite observations. The satellite gravity data as well as surface geometry changes observed by satellite altimetry contain a complex superposition of various mass signals. For instance, in Antarctica gravity and height changes due to ice mass changes are superimposed by similar signals from mass change in the underlying crust and mantle due to glacial isostatic adjustment, from tectonics, and from mass changes in the adjacent oceans and in the atmosphere. To enable a reliable separation of such effects, an intensive exchange of results is required.

The quantities to be delivered from the combination of the three fundamental pillars geometry, gravity/geoid, and Earth rotation are small and therefore difficult to determine. In order to be useful for global change studies they have to be derived free of bias and consistently in space and time. In general they are derived from the combination of complementary sensor and observation systems. For example, dynamic ocean topography is to be derived from the accurate measurement of the ocean surface by radar altimetry in combination with a geoid surface provided by gravity satellite missions. It shows that a variety of sensor systems, mission characteristics, and tracking systems have to be combined with utmost precision. The interconnections between mass transport processes, and the relations between observable parameters of gravity and geometry and the different processes are sketched in Figure 3.4.

Earthquakes, volcano eruptions, tectonic deformations, land slides, glacial isostatic adjustment, deglaciation, sea level rise, ocean mass and heat transport, deep ocean circulation, the water cycle, atmospheric and ocean loading and many more are typical and well known phenomena of this kind. Mass anomalies, the transport and exchange of masses and mass balances are not measurable by any other means and add significantly to the understanding of global Earth dynamics. But it is clear that not all of these phenomena can be detected because of limitations in space and time. Figure 3.5 shows for the gravity field the requirements on the static gravity field in terms of spatial resolution and typical time periods for the temporal variations. In the following, mass transport phenomena are briefly listed which can be detected by a careful signal analysis and signal balancing. Subsequently, these phenomena are addressed in more detail in individual sections.

Ocean circulation and transport: The combination of geoid and altimetry allows for the first time the direct determination of the global dynamic ocean topography. The geostrophic balanced surface currents can be deduced from the inclinations of the dynamic topography. From these, complete profiles of the ocean circulation can

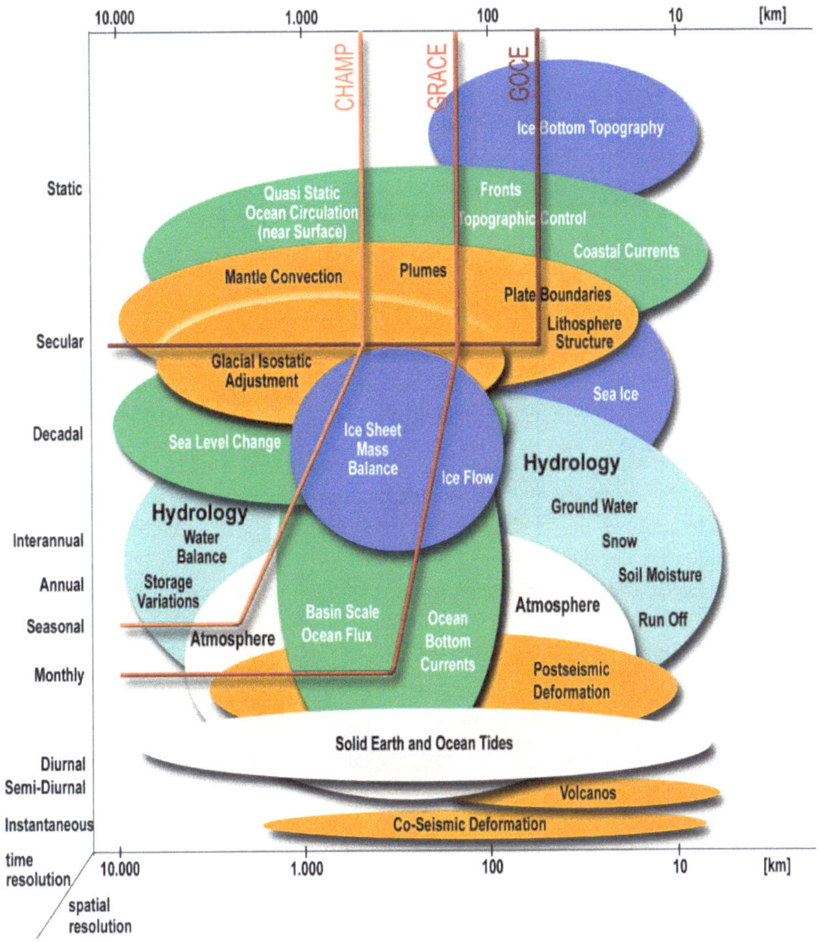

Fig. 3.5. Resolvability of mass transport by satellite missions. Mass transport phenomena and mass distribution characteristics and its resolvability by the satellite missions CHAMP, GRACE and GOCE. From Ilk et al. (2005).

be derived by combination with traditional hydrographical measurements. New insights in global and basin related heat and mass transport can be expected. Temporal variations of the dynamic ocean surface are caused mainly by temperature related volume changes where the mass column remains unchanged. Mass changes by fluid dynamics causes more problems and can be detected only by changes of the gravity field as expected by the GRACE mission. This will lead to a clear progress in the understanding of ocean circulation. Future topics of research include the determination of large-scale heat and mass transport, the investigation of circulation systems such as the Antarctic circum polar current, Weddell- and Ross eddies, the separation of steric and mass-related changes of the global sea surface and of the dynamics of currents, and the interaction of temporal and quasi-static circulations (eddies, fronts).

Hydrological cycle: The time dependent gravity field as expected from the GRACE mission enables for the first time to detect continental mass changes with a resolution of 1 cm water column in monthly snapshots. This allows to close the hydrological cycle at different scales in time and space. Topics of research include the global water balance and water transfer between atmosphere, continents, oceans and ice shield, the large-scale variations of terrestrial water storage under characteristic conditions, the large scale temporal variations of evapotranspiration, the evaluation and development of large-scale hydrological models, the water balances in difficult accessible regions, the long term trends of continental water storage as a consequence of environmental changes, and identification of hydrological problem zones with respect to water management and the availability of water resources.

Ice mass balance and sea surface: The polar ice caps play a key role in Earth system because imbalanced masses and resulting changes of the sea surface are global. Of central relevance is the precise determination of the mass balance of the complete ice shields by the actual gravity field missions CHAMP, GRACE and GOCE. Altimetry enables the precise measurement of the topography and of ice height changes, with the missions CryoSat and ICESat also in the climate sensitive ice shield regions. Interferometric SAR (ENVISAT) enables the area-wise determination of ice motions which can be compared to balance velocity models. The precise measurement of the thickness of the sea ice (CryoSat) provides new insight in the actual climate development. Topics of research are: registration of mass changes of the polar ice caps and the consequences for sea level rise; investigation of the changes in the border areas of ice masses; validation and improvement of glaciological models as important component in coupled climate models; determination of ice mass induced recent crust deformations (glacial isostasy); additional data sets for validation, densification and interpretation of satellite data; modeling of sea ice dynamics based on new remote sensing data.

Crust and mantle: The new gravity field missions open new dimensions in the research for geodynamic mass transport within the Earth: GOCE will improve the resolution of the static gravity potential and its gradients in the medium and short wavelength range by more than one order of magnitude; GRACE will provide for the first time the temporal variation of the potential down to a resolution of 400 km; it can be expected that mass distribution and mass transport will become directly observable. Topics of research are: glacial-isostatic adjustment processes and lateral variations of mantle viscosity; global mass transport in the mantle and dynamic topography based on new seismic tomography data and 3D-distributions of mantle viscosity; sub-lithosphere mantle convection and deviations of seismic discontinuities in 410 and 600 km depth; models of active and passive continental margins based on high-resolution gravity data, decoupling processes at active subduction zones; episodic mass redistributions at plate margins; improvement of global and regional crust and lithosphere models.

3.10 Earth rotation: understanding Earth system dynamics

3.10.1 Earth rotation measurements

Changes in the Earth's rate of rotation become apparent when comparing time kept by the rotating Earth, known as Universal Time, to uniform time scales based either upon atomic clocks or upon the motion of the Sun and other celestial bodies. Prior to the development of atomic clocks, the most accurate measurements of changes in the Earth's rate of rotation were obtained by timing the occultations of stars by the Moon. With the advent of atomic clocks in 1955, a uniform atomic time scale became available that could be used as a reference when measuring the transit times of stars as they pass through the local meridian. Changes in the Earth's rate of rotation could then be determined more accurately from optical astrometric measurements of star transits than they could from measurements of lunar occultations. And prior to the development of space-geodetic techniques, optical astrometric measurements of changes in the apparent latitudes of observing stations yielded the most accurate estimates of polar motion. The space-geodetic techniques of VLBI, GNSS like GPS, and SLR and LLR are now the most accurate techniques available for measuring changes in both the Earth's rate of rotation and in polar motion.

An integral part of geodesy has always been the definition and realization of a terrestrial, body-fixed reference frame, a celestial, space-fixed reference frame, and the determination of the indexEarth orientation parametersEOPs (precession, nutation, spin, and polar motion) that link these two reference frames together. But with the advent of space geodesy — with the placement of laser retro-reflectors on the Moon by Apollo astronauts and Soviet landers, the launch of the LAGEOSs, the development of very long baseline interferometry, and the development of global navigation satellite systems like the global positioning system — a quantum leap has been taken in our ability to realize the terrestrial and celestial reference frames and to determine the Earth orientation parameters.

The only space-geodetic measurement technique capable of independently determining all of the EOPs is multibaseline VLBI. All of the other techniques need to either apply external constraints to the determined Earth orientation parameters or can determine only subsets of the EOPs, only linear combinations of the EOPs, or only their time rates-of-change.

Ring laser gyroscopes (see Section 2.5.2) are a promising emerging technology for determining the Earth's rotation. In a ring laser gyroscope, two laser beams propagate in opposite directions around a ring. Since the ring laser gyroscope is rotating with the Earth, the effective path length of the beam that is co-rotating with the Earth is slightly longer than the path that is counter-rotating with it. Because the effective path lengths of the two beams differ, their frequencies differ, so they interfere with each other to produce a beat pattern, the frequency of which is proportional to that component of the instantaneous angular velocity $\Omega(t)$ of the Earth that is parallel to the normal of the plane of the ring. Ring laser gyroscopes measure the absolute rotation of the Earth in the sense that, in principle, just a single measurement is required to determine the Earth's instantaneous rotation. All of the other techniques discussed above are relative sensors because they infer the Earth's rotation from the change in the orientation of the Earth that takes place between at least two measurements that are separated in time.

Earth orientation parameters can be determined from measurements taken by each of the techniques discussed above. But each technique has its own unique strengths and weaknesses in this regard. Not only is each technique sensitive to a different subset and/or linear combination of the Earth orientation parameters, but the averaging time for their determination is different, as is the interval between observations, the precision with which they can be determined, and the duration of the resulting EOP series. By combining the individual series determined by each technique, a series of the Earth's orientation can be obtained that is based upon independent measurements and that spans the greatest possible time interval.

3.10.2 UT1 and Length-of-Day Variations

Length-of-day observations show that it consists mainly of: (1) a linear trend of rate -1.8 ms/cy, (2) decadal variations having an amplitude of a few milliseconds, (3) tidal variations having an amplitude of about 1 ms, (4) seasonal variations having an amplitude of about 0.5 ms, and (5) smaller amplitude variations occurring on all measurable time scales.

Linear trend: Tidal dissipation causes the Earth's angular velocity and hence rotational angular momentum to decrease. Since the angular momentum of the Earth-Moon system is conserved, the orbital angular momentum of the Moon must increase to balance the decrease in the Earth's rotational angular momentum. The increase in the orbital angular momentum of the Moon is accomplished by an increase in the radius of the Moon's orbit and a decrease in the Moon's orbital angular velocity. The tidal acceleration of the Moon can be determined from observations of the timings of transits of Mercury as well as from satellite and lunar laser ranging measurements. Tidal forces distort the figure of the Earth and hence its gravitational field which in turn perturbs the orbits of the Moon and artificial satellites. Lunar and satellite laser ranging measurements can detect these tidal perturbations in the satellites' orbits and can therefore be used to construct tide models and hence determine the tidal acceleration of the Moon. Using LLR measurements, a value of 25.73 ± 0.5 "/cy for the tidal acceleration of the Moon is found, which by Kepler's law corresponds to an increase of 3.79 ± 0.07 cm/yr in the semimajor axis of the Moon's orbit, and which should be accompanied by a 2.3 ms/cy increase in the length of the day.

By *a priori* adopting a value for the tidal acceleration of the Moon, lunar and solar eclipse observations can be used to determine the secular increase in the length of the day over the past few thousand years. Using eclipse observations spanning 700 BC to 1600 AD, lunar occultation observations spanning 1600 to 1955.5, and optical astrometric and space-geodetic measurements spanning 1955.5 to 1990, it was found that the length-of-day has increased at a rate of 1.80 ± 0.1 ms/cy on average during the past 2700 years.

Since the observed increase in the length of the day is 1.8 ms/cy, whereas the observed tidal acceleration of the Moon implies a 2.3 ms/cy increase in the length of the day, some other mechanism or combination of mechanisms must be acting to change the length of the day by 0.5 ms/cy. One of the most important mechanisms acting to cause a secular trend in the length-of-day on time scales of a few thousand years is glacial isostatic adjustment (GIA). Other possible mechanisms include present-day changes in glacier and ice sheet mass and the accompanying change in nonsteric sea level, tectonic processes taking place under non-isostatic conditions, plate subduction, earthquakes, and deformation of the mantle caused by pressure variations acting at the core-mantle boundary that are associated with motion of the fluid core.

Decadal variations: The most important mechanism acting to cause decadal variations in the length of the day is core-mantle coupling. It has been recognized for quite some time that the core is the only viable source of the large decadal LOD variations that are observed, and current models of Core Angular Momentum (CAM) predict decadal length-of-day variations that agree reasonably well with those observed.

While the exchange of CAM with the solid Earth can clearly cause decadal LOD variations of approximately the right amplitude and phase, the mechanism or mechanisms by which the angular momentum is exchanged between the core and solid Earth is less certain. Possible core-mantle coupling mechanisms are viscous torques, topographic torques, electromagnetic torques, and gravitational torques.

Viscous coupling is caused by the drag of the core flow on the core-mantle boundary, with the strength of the coupling depending on the viscosity of the core fluid. Given current estimates of core viscosity, it is generally agreed that viscous torques are too weak to be effective in coupling the core to the mantle.

If the core-mantle boundary is not smooth but exhibits undulations or "bumps", then the flow of the core fluid can exert a torque on the mantle due to the fluid pressure acting on the boundary topography. The strength of this topographic coupling depends on the amplitude of the topography at the core-mantle boundary. Because of uncertainties in the size of this topography and a controversy about how the topographic torque should be computed there is as yet no consensus on the importance of topographic coupling as a mechanism for exchanging angular momentum between the core and mantle.

Electromagnetic torques arise from the interaction between the magnetic field within the core and the flow of electric currents in the weakly conducting mantle that are induced by both time variations of the magnetic field and by diffusion of electric currents from the core into the mantle. The strength of this electromagnetic torque depends on both the conductivity of the mantle and on the strength of the magnetic field crossing the core-mantle boundary. If the conductivity of the mantle, or of a narrow layer at the base of the mantle, is sufficiently large, then electromagnetic torques can produce decadal length-of-day variations as large as those observed. But because of uncertainties in the conductivity at the base of the mantle, the importance of electromagnetic coupling, like that of topographic coupling, as a mechanism for exchanging angular momentum between the core and mantle remains unclear.

Gravitational attraction between density heterogeneities in the fluid core and mantle can exert a torque on the mantle, leading to changes in the length of the day. The strength of this gravitational torque depends upon the size of the mass anomalies in the core and mantle, which are poorly known. As a result, there have been few quantitative estimates of the magnitude of the gravitational torque. However, Buffett (1996a,b) has suggested that the inner core may be gravitationally locked to the mantle. If so, then any rotational disturbance of the inner core, possibly caused by electromagnetic torques acting on the inner core, will be transmitted to the mantle, causing length-of-day changes. Mound & Buffett (2005) consider this last mechanism to be the most viable mechanism for exchanging angular momentum between the core and mantle.

Tidal variations: Tidal forces due to the gravitational attraction of the Sun, Moon, and planets deform the solid and fluid parts of the Earth, causing the Earth's inertia tensor to change and hence the Earth's rotation to change. Dissipation associated with mantle anelasticity causes the deformational and hence rotational response of the Earth to lag behind the forcing tidal potential. As a result, not only does mantle anelasticity modify the in-phase rotational response of the Earth to the tidal potential, but out-of-phase terms are introduced as well. Anelastic effects are found to modify the elastic rotational response of the Earth by a few percent.

Dynamic effects of long-period ocean tides on the Earth's rotation can be computed using ocean tide models based upon Laplace's tidal equations. But the accuracy of ocean tide models and hence of the predicted effect of ocean tides on the Earth's rotation greatly improved when satellite altimetry, in particular ERS-1 (for high latitudes) and TOPEX/Poseidon sea surface height measurements became available. Dynamic ocean tide effects are larger at the fortnightly tidal frequency than they are at the monthly frequency.

Ocean tides in the diurnal and semidiurnal tidal bands also affect the Earth's rate of rotation. Comparisons of observations with models show the dominant role that ocean tides play in causing subdaily Universal Time 1 (UT1) and LOD variations, with as much as 90% of the observed UT1 variance being explained by diurnal and semidiurnal ocean tides. Apart from errors in observations and models, the small difference that remains may be due to nontidal atmospheric and oceanic effects.

The diurnally varying solar heating of the atmosphere excites diurnal and semidiurnal tidal waves in the atmosphere that travel westward with the Sun. These radiational tides are much larger than the gravitational tides in the atmosphere, with the amplitude of the surface pressure variations due to the radiational tides being about 20 times larger than the amplitude due to the gravitational tides. While gravitational tides in the atmosphere have no discernible effect on the Earth's rotation, the radiational tides do have an effect. Since the oceans respond dynamically to tidal variations in the atmospheric wind and pressure fields, the oceans also contribute to the excitation of UT1 and LOD by the radiational tides. In fact, the effect of radiational tides on UT1 and LOD is typically included in tables of the effects of diurnal and semidiurnal ocean tides on the Earth's rate of rotation.

Seasonal variations: Seasonal variations in the length-of-day are primarily caused by annual and semiannual changes in the angular momentum of the zonal winds. Although only 1% of the atmospheric mass is located in the region of the atmosphere above 10 hPa, the strength of the zonal winds there is great enough that they have a noticeable effect on seasonal length-of-day variations. Apart from errors in observations and models, the residual that remains after modeled atmospheric and oceanic effects have been removed from the observations may be caused by hydrologic processes.

Interannual variations: Like seasonal variations in the length of the day, variations on interannual time scales are also predominantly caused by changes in the angular momentum of the zonal winds. The most prominent feature of the climate system on these time scales is the El Niño Southern Oscillation (ENSO) phenomenon. Numerous studies have shown that observed LOD variations on interannual time scales, as well as interannual variations in the angular momentum of the zonal winds, are (negatively) correlated with the SOI, reflecting the impact on the length-of-day of changes in the zonal winds associated with ENSO.

Studies of the effects of oceanic processes show that they are only marginally effective in causing interannual length-of-day variations. The interannual LOD signal that remains after atmospheric and oceanic effects are removed may be caused by hydrologic processes. Like seasonal variations, better atmospheric, oceanic, and hydrologic models are needed to close the LOD budget on interannual time scales.

Intraseasonal variations: Like the seasonal and interannual variations in the length of the day, variations on intraseasonal time scales are also predominantly caused by changes in the angular momentum of the zonal winds. The Madden-Julian oscillation with a period of 30 to 60 days is the most prominent feature in the atmosphere on these time scales and a number of studies have shown that fluctuations in the zonal winds associated with this oscillation cause the length-of-day to change. Studies of the effects of oceanic processes show that they are only marginally effective in causing intraseasonal length-of-day variations. Hydrologic effects on intraseasonal length-of-day variations are thought to be relatively insignificant, although the monthly sampling interval of current hydrologic models makes it difficult to study such rapid variations.

3.10.3 Polar Motion

Observations of polar motion show that it consists mainly of: (1) a forced annual wobble having a nearly constant amplitude of about 100 mas, (2) the free Chandler wobble having a period of about 433 days and a variable amplitude ranging from about 100 to 200 mas, (3) quasi-periodic variations on decadal time scales having amplitudes of about 30 mas known as the Markowitz wobble, (4) a linear trend having a rate of about 3.5 mas/yr and a direction towards 79°W longitude, and (5) smaller amplitude variations occurring on all measurable time scales.

Linear trend: One of the most important mechanisms acting to cause a linear trend in the path of the pole on time scales of a few thousand years is glacial isostatic adjustment (GIA). The isostatic adjustment of the solid Earth as it responds to the decreasing load on it following the last deglaciation causes the figure of the Earth to change, and hence the pole to drift. Models of GIA show that its effect on the pole path is sensitive to the assumed value of lower mantle viscosity, to the assumed thickness and rheology of the lithosphere, to the treatment of the density discontinuity at 670 km depth, and to the assumed compressibility of the Earth model.

However, GIA is not the only mechanism that will cause a trend in the pole path. The present-day change in glacier and ice sheet mass and the accompanying change in nonsteric sea level will also cause a linear trend in polar motion. But the effect of this mechanism is very sensitive to the still unknown present-day mass change of glaciers and ice sheets, particularly of the Antarctic ice sheet. Other mechanisms that may cause a linear trend in the path of the pole include tectonic processes taking place under non-isostatic conditions, plate subduction, mantle convection, upwelling mantle plumes, and earthquakes.

Decadal variations: Since optical astrometric measurements are known to be corrupted by systematic errors, there has always been some doubt about the reality of the decadal variations evident in these measurements. But since the highly accurate space-geodetic measurements are less susceptible to systematic error than are optical astrometric measurements, decadal variations seen in the space-geodetic measurements can be considered to be reliable.

The cause of the decadal-scale polar motion variations evident in space-geodetic measurements is currently unknown. Since core-mantle processes are known to cause decadal variations in the length of the day, they may also excite decadal variations in polar motion. But electromagnetic coupling between the core and mantle appears to be two to three orders of magnitude too weak and topographic coupling appears to be too weak by a factor of three to ten. Like the decadal variations in the length-of-day, invoking the inner core when modeling core-mantle processes may ultimately provide the long-sought explanation for the cause of the decadal variations in polar motion.

Tidal variations: Tidally induced deformations of the solid Earth caused by the second-degree zonal tide raising potential cause long-period changes in the Earth's rate of rotation. But since this potential is symmetric about the polar axis, tidal deformations of the axisymmetric solid Earth cannot excite polar motion. However, due to the nonaxisymmetric shape of the coastlines, the second-degree zonal tide raising potential acting on the oceans can generate polar motion via the exchange of nonaxial oceanic tidal angular momentum with the solid Earth.

Better observations of the effect of long-period ocean tides on polar motion are needed, as are better models for these effects. Observations of these effects are internally incosistent, and predictions from the available theoretical ocean tide models do not agree with each other or with the observations.

Comparisons of observations with models show the major role that ocean tides play in causing subdaily polar motion variations, with as much as 60% of the observed polar motion variance being explained by diurnal and semidiurnal ocean tides. Apart from errors in observations and models, the difference that remains may be due to nontidal atmospheric and oceanic effects.

Chandler wobble: Any irregularly shaped solid body rotating about some axis that is not aligned with its figure axis will freely wobble as it rotates. The Eulerian free wobble of the Earth is known as the Chandler wobble in honor of Seth Carlo Chandler, Jr. who first observed it. Unlike the forced wobbles of the Earth, such as the annual wobble, whose periods are the same as the periods of the forcing mechanisms, the period of the free Chandler wobble is a function of the internal structure and rheology of the Earth and its decay time constant, or quality factor Q, is a function of the dissipation mechanisms acting to dampen it. The observed values for the period and Q of the Chandler wobble can therefore be used to better understand the internal structure of the Earth and the dissipation mechanisms, such as mantle anelasticity, that dampen the Chandler wobble causing its amplitude to decay in the absence of excitation.

While there is growing agreement that the Chandler wobble is excited by a combination of atmospheric, oceanic, and hydrologic processes, the relative contribution of each process to its excitation is still being debated.

Seasonal polar motion: The annual wobble is a forced wobble of the Earth that is caused largely by the annual appearance of a high atmospheric pressure system over Siberia every winter.

A rather large residual remains after the effects of the atmosphere and oceans are removed from the observed seasonal polar motion excitation. This residual is probably at least partly due to errors in the atmospheric and oceanic models, but could also be due to the neglect of other excitation processes such as hydrologic processes.

Interannual polar motion: Like the seasonal wobbles, the wobbling motion of the Earth on interannual time scales is a forced response of the Earth to its excitation mechanisms. For periods between 1 year and 6 years excluding the annual cycle, oceanic processes are much more important than atmospheric in exciting interannual polar motions.

Intraseasonal polar motion: Like the seasonal and interannual wobbles, the wobbling motion of the Earth on intraseasonal time scales is a forced response of the Earth to its excitation mechanisms. For periods between 5 days and 1 year excluding the seasonal cycles, atmospheric processes are more effective than oceanic in exciting polar motion.

3.11 Earth rotation: understanding processes in the solid Earth

The key questions that are addressed in this note are the following: What can we learn from Earth rotation about the Earth's interior (core, mantle)? What can we learn about the exterior fluids (global mass balance, hydrological cycle, global change)? Similar techniques can be applied to the other terrestrial planets and help us to better understand their interior, their evolution, and their external and internal fluids. The Section 6.1 will address that question.

3.11.1 Earth's interior from Earth rotation

The Earth rotation and orientation are measured by VLBI, satellite observations using DORIS, GNSS, as well as SLR, and LLR. The precise use of reference frame, terrestrial and celestial reference frames, allows a very good determination of the Earth position and orientation in space as a function of time. This allows a very good precision on the EOPs as a function of time that comprise precession, nutation, length-of-day variations, and polar motion. Precession and nutation allow getting information on phenomena related to the deep Earth interior such as core-mantle coupling, mantle anelasticity, or inner core coupling with the liquid core and the mantle. Figure 3.6 presents the geophysical parameters that are determined from nutation.

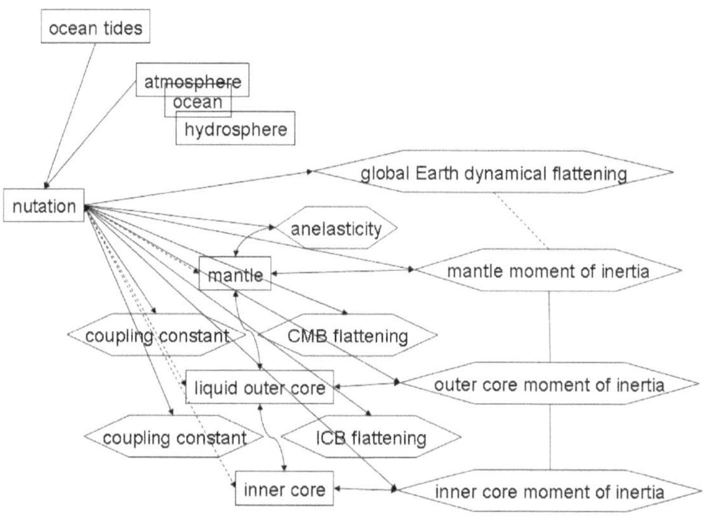

Fig. 3.6. Geophysical parameters obtained from nutation.

At this step it must be noted that the very good constraints that are obtained are constraints on the coupling constants but not really on the coupling mechanisms involved. In other words, the coupling constant can be determined from the EOPs but for explaining these coupling constants, one needs to consider the physical mechanisms such as inertial coupling, electromagnetic coupling, topographic coupling, or viscous coupling. The choice of the mechanism is related to other considerations. The evaluation of the relative influence is related to further computation and improvements are still necessary in that field. The situation for nutation is a good example. There is a trade-off between the flattening of the core (in particular the non-hydrostatic contribution) and the real part of the coupling constant at the Core-Mantle Boundary (CMB). The presently adopted nutation theory has considered electromagnetic coupling as coupling mechanism at the CMB. Consequently, with the help of a complicate theory, it was possible to relate the imaginary part of the coupling constant and its real part. The flattening of the core could be determined to correspond to an increase of about 350 m of the equatorial radius with respect to the polar radius, in addition to the hydrostatic contribution computed from the core rotation inside the Earth. But if the coupling mechanism is different or is a combination between electromagnetic coupling and topographic coupling, this may not hold. Consequently, it is true that the observation of EOPs may help to understand the interior of the Earth a but further step in the theory is still necessary. Additionally, the external geophysical fluids influence the nutation. Although the most important influence is at the nutation corresponding to one solar day as expected, there are other non-well determined influences on other nutation periods. The interpretation of the data in terms of physics of the interior may be influenced by that. For example, a semi-annual modulation of the diurnal atmospheric effect on Earth may have a considerable contribution on the retrograde annual nutation, which is the most important nutation for determining the resonance parameters and the core coupling. GGOS will aim at better determining the geophysical fluid effects on the Earth and in particular on Earth rotation. This will further help to better determine the internal geophysical contribution to nutation and to better constrain the physics of the Earth interior. This determination has some limitation from the modeling point of view but should certainly be one of the most important advances in our understanding of the Earth.

3.11.2 Geophysical fluids from Earth rotation

The geophysical fluids have large effects on Earth rotation. They are related to the global behavior of the atmosphere, of the ocean, or hydrosphere. For the nutation, for instance, the influence mentioned above can be determined as resented in Figure 3.7.

Additionally, the dynamical flattening can be determined from precession and nutation. On the other hand, geophysical fluids contributions to it can also be deter-

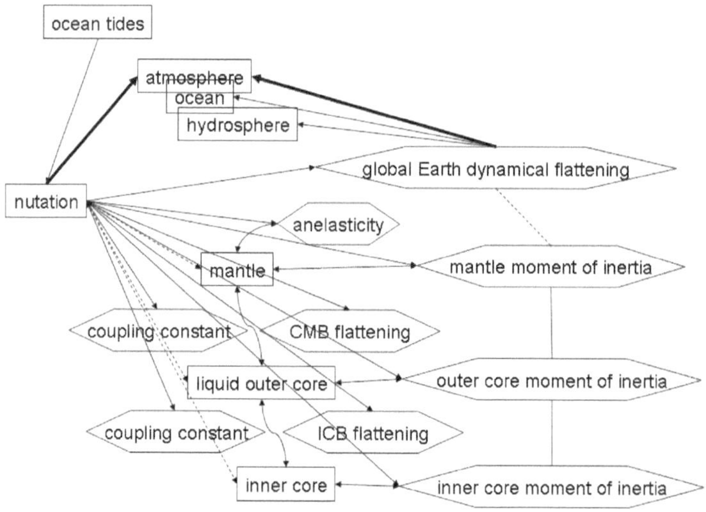

Fig. 3.7. Determination of atmospheric global properties from nutation and from the parameters determined from nutation.

mined and we will benefit from a comparison of both kinds of observations. This is presented on Figure 3.8.

Concerning the other Earth orientation parameters, the geophysical fluids such as the ocean, the atmosphere and the hydrosphere are the most important contributions. Observations of polar motion and length-of-day variations provide access to the global contributions from these geophysical fluids. The only remaining problem will be to separate the relative contributions. The interaction between these geophysical fluids is important to consider for that matter. We have represented our view in Figure 3.9.

3.11.3 General remarks

The high precisions of observation in different fields (orientation, gravity, geophysical fluids) are very necessary to better understand our planet as well as to better understand the other planets (see Section 6.1). GGOS aims at refining the consistency and determination of all these and will consequently greatly enhance our understanding of the interior of the Earth. Similar projects of synergism are desirable and its concept is already applied to other planets. The consequences of the GGOS effort will be huge and have a great potential for a better understanding of the interior of the Earth.

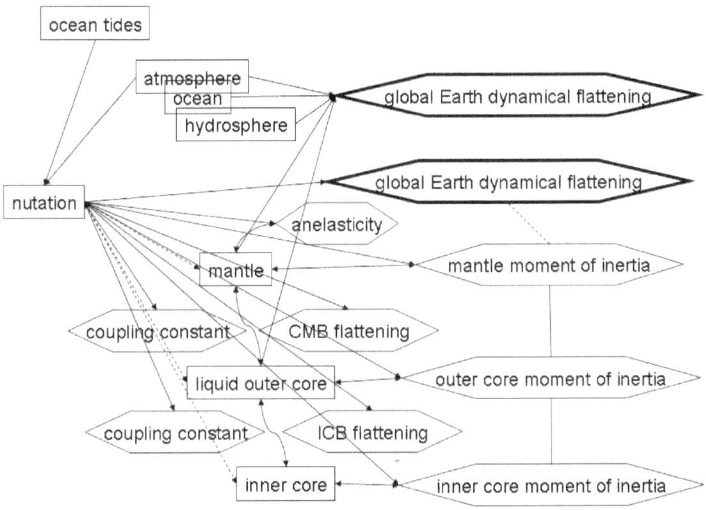

Fig. 3.8. Comparison of the dynamical flattening obtained from precession and nutation to the contributions determined from the geophysical fluids.

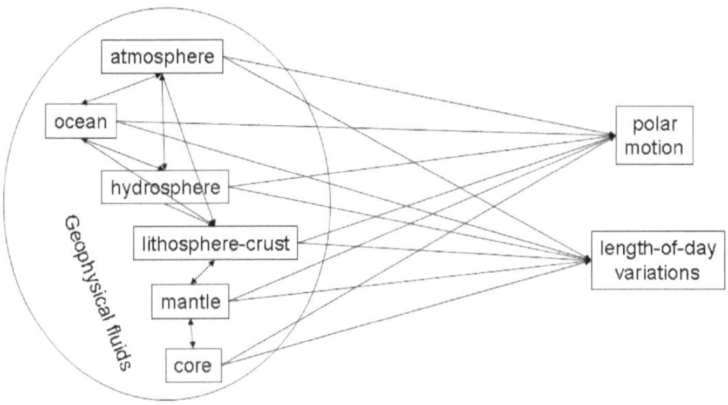

Fig. 3.9. Geophysical fluid effects on polar motion and on Length-Of-Day variations.

Chapter 4
Maintaining a modern society

C. Rizos, D. Brzezinska, R. Forsberg, G. Johnston, S. Kenyon, D. Smith

In innumerable ways, geodesy contributes to the functioning of a modern society. While that contribution is critical, it is not necessarily well known or understood by most outside the geodetic community. Geodesy defines the coordinate infrastructure underlying many of the functions of modern society. Like the wooden frame of a house, that infrastructure is the unseen framework upon which different "layers" of spatial information (e.g., the geometric data and thematic description of spatial elements such as points, polygons, lines, 3D objects, and their topologies – i.e., how they relate to each other –, and imagery from space and airborne platforms) are constructed so that they align with each other perfectly. This infrastructure is known as SDI, and geodesy defines the foundation of the SDI. In this chapter the terms "spatial" and "geospatial" will be assumed to be inter-changeable. In many countries, the importance of SDI has risen to the level of government mandate. In parallel with this SDI development, Positioning, Navigation and Timing (PNT) is increasingly needed in many aspects of life, in many business and engineering applications, and to aid decision-making at all levels of government and private enterprise. Due to the globalization and interoperability requirements, spatial data and positioning are increasingly required with respect to a global reference frame. Both, spatial data and (geo)positioning rely on geodesy and increasingly on global geodesy. "Georeferencing" is defined here as the process of assigning coordinates (or "positioning") to an entity (point, line, polygon, etc.).

4.1 Spatial data infrastructure

In the U.S., for example, the government's Office of Management and Budget (OMB) issued circular A-16 ("Coordination of Geographic Information and Related Spatial Data Activities"). This document establishes that the Federal Geographic Data Committee (FGDC) be in charge of establishing the National Spatial Data Infrastructure (NSDI). Furthermore, it identifies the National Oceanic and Atmo-

H.-P. Plag, M. Pearlman (eds.), *Global Geodetic Observing System*,
DOI 10.1007/978-3-642-02687-4_4, © Springer-Verlag Berlin Heidelberg 2009

spheric Administration (NOAA) as the "lead agency" in providing "geodetic control" for the federal government in support of the NSDI. The geodetic control is specifically referred to as the National Spatial Reference System (NSRS). Because no other agency inside of NOAA is responsible for geodetic control (nor specifically the NSRS), the naming of NOAA as the lead agency effectively translates into the National Geodetic Survey (NGS), an office within NOAA. This means that in the United States, NGS must define, maintain and provide access to the NSRS - that is, make sure that everyone who needs to reference anything spatially (e.g. to "position" anything), can do so consistently with others.

One important aspect for georeferencing for military and civilian activities all over the world is the use of a globally consistent reference frame such as the ITRF. As the most accurate realization of ITRS, the ITRF provides a single, accessible 3-D reference frame for geospatial data from a variety of sources. Another global reference system currently still used for many applications is the WGS 84. As pointed out in Section 2.2, the realizations of WGS 84 (through GPS) today are closely aligned to ITRF and supported by ITRF.

Local maps and geodetic control are still commonly used worldwide and the conversion of this information into a common system provides users with the ability to unambiguously georeference with respect to locations on or above the Earth's surface. It also minimizes possible errors when using modern satellite-based geopositioning technologies such as GNSS. Hence increasingly national, and even local, datums are "compatible" with ITRF, i.e. these datums are Earth-centered, Earth-fixed, and their relationship to the very accurate, high integrity ITRF is determined to a high level of confidence. In many countries and regions, the relations between the national or regional frames are monitored on a continuous basis by the national agency responsible for operational geodesy in that country.

In Australia, SDIs are being developed largely within individual government jurisdictions: federal, state and territory and, increasingly, local government. The Australia and New Zealand Land Information Council (ANZLIC), the peak Spatial Information Council comprising senior government officials from the Australian Government, eight State and Territory governments, and New Zealand, coordinates the development of the Australian Spatial Data Infrastructure (ASDI). ANZLIC has a number of standing committees to advise it on technical issues, amongst which the Intergovernmental Committee on Surveying and Mapping (ICSM) develops and promotes data standards across the jurisdictions. The most tangible component of the ASDI is the Australian Spatial Data Directory (ASDD), comprising 25 geographically distributed and independently maintained nodes, collectively containing over 40,000 metadata records. Groups of agencies are coming together to develop coordinated SDIs in response to major national priorities. Australian Government agencies, for example, have formed the Australian Ocean Data Center Joint Facility to coordinate marine data. The Australian SDI comprises a diverse set of organizations and locations and relies heavily upon the Australian National Geospatial Reference System (NGRS) to link them together and provide seamless coordinate sets for the entire continent. For this reason the responsibility for maintenance of the national geodetic infrastructure, and provision of the linkage to the international geodetic infrastructure, falls to the agency Geoscience Australia (GA). The national datums are maintained cooperatively by GA and the respective state government agencies.

A similar federated model exists in Canada. The Canadian Geospatial Data Infrastructure (CGDI) contains all of the technology, standards, access systems and protocols necessary to harmonize all of Canada's geospatial databases, and make them available on the Internet (Geoconnection, 2007). Geospatial databases include: topographic maps, air photos, satellite images, nautical and aeronautical charts, census and electoral areas, forestry, soil, marine and biodiversity inventories. Geospatial information plays an important role in the everyday lives of Canada's citizens, echoing the drivers of the NSDI and ASDI referred to already. In essence, geospatial information provides details or characteristics (i.e. buildings, roads, demographics, water, soil, weather, topography, wildlife, farming, etc.) regarding a geographic location, on land or water, and at a street, local, regional, provincial, national, or global level. As in the case of the US, Australia and Europe (see below), the Canadians have recognized that developments in information technology over the past decade have made both the access to and the need for geospatial information expand exponentially. In 1999, the Government of Canada invested Canadian $60 million in a national partnership initiative to improve access to geospatial information over the Internet. This initiative, known as "GeoConnections" (http://www.geoconnections.org), was led by Natural Resources Canada and was mandated to accelerate the development of a CGDI.

In Europe the SDI program is known as the Infrastructure for Spatial Information in Europe (INSPIRE) (see http://inspire.jrc.it/home.html). The general situation regarding spatial information in Europe is one of fragmentation of data sets and sources, gaps in availability, lack of harmonization between data sets at different geographical scales, and duplication of information collection. It was concluded that "these problems make it difficult to identify, access and use data that is available." Awareness has been growing at both the national and EU level regarding the need for quality geo-referenced information to support, in the case of one quoted example, "understanding of the complexity and interactions between human activities and environmental pressures and impacts." On the 29 January 2007, the European Council adopted a directive aimed at establishing INSPIRE, in accordance with the joint text agreed by the Council and the European Parliament on 21 November 2006 (see http://inspire.jrc.it/home.html for details).

While SDI initiatives are much more than about reference frames and coordinates, there is a trend towards ever higher accuracies in the SDI. This means that a corresponding improvement in the accuracy of the geodetic infrastructure generally one order of magnitude higher is required. The SDI can be visualized as many layers of spatial information resting on a strong geodetic "foundation" (see Figure 4.1, an example taken from the ASDI). Hence this foundation must be defined and maintained to a high level of integrity. Any crustal motion impacts the realization of the national reference frame, and must therefore be monitored so that a valid 4D reference system can always be reconstructed. Furthermore, all geospatial data sets must be referenced to the correct 4D reference frame or datum. The transformations between different data, some historical (e.g. contained within analog or digital cartographic data), some contemporary, must be defined to the appropriate level of accuracy. Finally, the quality and integrity of the GNSS technology (the geopositioning "workhorse") and the associated high accuracy techniques, must be consistent and quantifiable if the crucial connection between geopositioning and SDI is to be maintained for the benefit of so many applications.

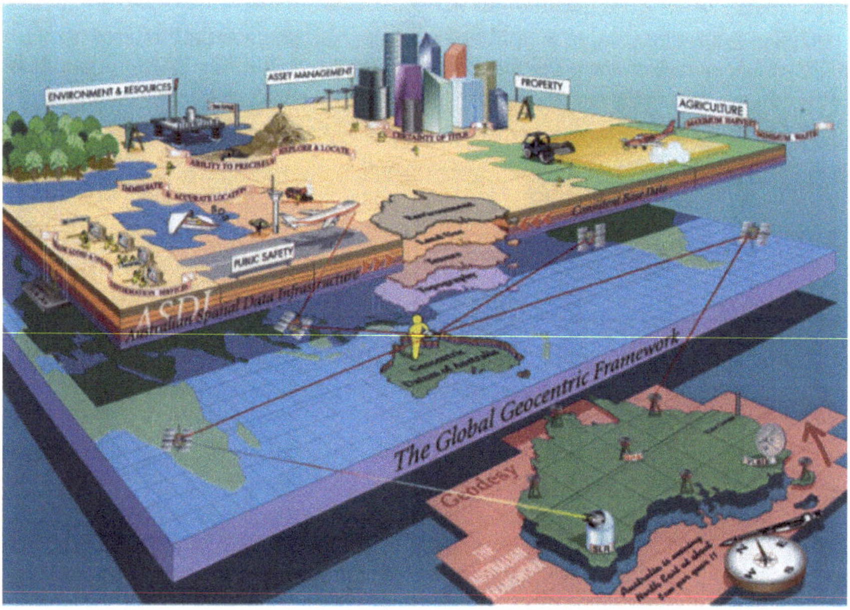

Fig. 4.1. A Model of the Australian Spatial Data Infrastructure.

The potential economic, social or environmental impacts of inconsistent geopositioning or misaligned spatial data sets are illustrated by the following examples:

1) Two engineering crews building a bridge from opposite sides of a bay; failing to align properly when they meet in the middle.
2) Floodplain maps, levee construction, and ocean storm surge models all using different height systems; thus failing to adequately identify which city areas are vulnerable to flooding during hurricanes or storm surges.
3) Road construction and car navigation systems using different coordinate systems; failing to safely determine which lane a car is driving in.
4) Automated airline navigation systems which facilitate takeoff and landing at airports using different, inconsistent geopositioning for their runways; thus potentially leading to safety failures.
5) Geospatial products to support disaster relief efforts, such as in the case of Hurricanes Katrina and Rita in 2005 in the U.S. Gulf Coast Region.

These examples are but a handful that demonstrate the critical nature of a well-defined and widely-adopted SDI. The use of geodesy to define the spatial data infrastructure of a city, region, country or even of the entire planet (the ultimate goal) impacts the functioning of modern society in so many ways. Without an appreciation of SDI, the applications discussed here would appear disjointed and unconnected, yet the reality is very different. This chapter will highlight the economic and social impact of geodesy in different application areas by focusing on the common theme – that geodesy is the foundation for all high fidelity spatial data, and the very different applications of geopositioning/georeferencing and spatial information are merely industry-specific examples.

4.2 Navigation

Navigation is the act of guiding a person or moving platform (ship, aircraft, spacecraft) in a safe and expeditious manner from one point to a destination. A crucial characteristic of navigation is that geopositioning is required in real time, while underway. For many centuries navigational science has been a driver for advances in geodesy, applied mathematics, cartography, mechanical (and in the 20th century electronic) position determination technology, to name but a few. Initially navigation was intimately related to maritime commerce and warfare, but in the 20th century navigation has been applied first to air flight, later to spacecraft operations, and increasingly in the last decades to land and personal navigation. The next frontier is navigation as an indispensable capability of robotic or autonomous vehicles (Section 4.3). Modern navigation is reliant on geopositioning technology (these days principally GPS) and spatial data (i.e., digital maps and geospatial databases).

The GPS of the U.S. has truly revolutionized geodesy, surveying and navigation in the last two decades. Remarkably, the same space hardware and control facilities allow geodesists to determine the ITRF to sub-cm accuracy, and navigators at sea to position their ships to ten meter accuracy. Hence GPS (and soon other GNSSs) is one of the "threads" that links geodesy to vital applications such as navigation. Although maritime navigation is subject to guidelines issued by the International Maritime Organisation (IMO), and aircraft navigation to standards and procedures approved by the International Civil Aviation Organisation (ICAO), both have identified GPS/GNSS as the critical geopositioning technology. Furthermore, national and international charts (and other forms of spatial information) are increasingly based on a reference frame defined through geodesy. Therefore, another benefit of the SDI is to ensure the layer of spatial information vital for safe navigation on the surface of the Earth is "aligned" with all other data, such as the natural topography, transport infrastructure, the built environment, etc. The WGS 84 datum, although not of the same fidelity as ITRF, is important as it is the reference system of broadcast GPS data and thus is the "default" reference frame of non-augmented real-time GPS point positioning. Most international maritime and aeronautical charts are based on the WGS 84 datum.

4.2.1 Marine navigation

The ever increasing reliance of navigation on GPS/GNSS, and spatial information georeferenced by the geodetic framework or datum, means that society benefits enormously from the increased efficiency and safety of maritime, air, space and land navigation. This can be illustrated with many examples, but consider one from maritime navigation. In the U.S., the NGS participated in a navigation demonstration in the San Francisco Bay region. Bathymetric charts were converted to an ellipsoid height system (to be totally consistent with GPS heights, which are also ellipsoidal) and GPS receivers were placed around the Bay, on buoys and at three locations on a ship (to determine roll, pitch and yaw). These sources of data, combined with detailed information about the size and shape of the ship, were used with kinematic positioning software to determine, in real time, the location of the ship's underkeel location to a few centimeters of accuracy in real time. Because the bathymetric charts had been converted from a hydrographic to a geodetic (ellipsoid) frame, they were consistent with the GPS positions of the ship. This allowed for a accurate location of the bottom of the ship relative to the dredged channel in the Bay to be determined to a few centimeters accuracy (requiring the bathymetric charts to be of equal accuracy). Normally, a ship's knowledge of its underkeel clearance is on the order of a meter or so and this often means that ships must maintain greater than a meter of clearance as a safety buffer, simply because of vertical uncertainties. And what that, in turn, means is that ships intentionally carry less cargo so as to float higher to maintain this larger safety buffer. What the demonstration proved was that this buffer could be significantly decreased while maintaining safe passage. Specifically for the shipping industry this means an increase in cargo capacity, increased shipping efficiency, lower freight charges (and increased profit), without the need for expensive port and channel upgrades. This application has been identified as crucial for improved marine navigation in many other countries.

4.2.2 Air navigation

Similar examples may be quoted from civil aviation, which has made GNSS the cornerstone of the Future Air Navigation System (FANS). According to ICAO, at its simplest level, FANS-equipped aircraft use GPS to determine their location and altitude. GPS in this context replaces older and less accurate navigation systems. The aircraft transmits its location using Very High Frequencies (VHF) or satellite communications so that Air Traffic Control (ATC) can learn the locations of aircraft without the need to use radar. ATC radar is often absent over oceans and certain continental areas, so the first benefit is an increase of safety - decreasing the risk of midair collisions. Prior to the advent of FANS, pilots had to speak their location over voice links, typically HF radio. Given the inaccuracy of Inertial Navigation Systems (INS) and the noise present on High Frequencies (HF) links, ATC would insist on quite large separations between aircraft. By increasing accuracy and thus decreasing separations, FANS allows aircraft to fly closer to their preferred routes or ideal routes. An ideal route is typically: a) that great circle route (the shortest distance between two locations), and b) the route best suited to the prevailing winds. The optimal route is to position the aircraft with a maximum tail wind, certainly doing the utmost to avoid strong head winds. However, for air navigation more importance is often placed on the reliability and integrity of the geopositioning technology (and the charts/maps) than on navigational accuracy. For this reason ICAO has encouraged the development of independent GNSSs (e.g. Russia's GLONASS, and the planned EU GALILEO), as well as new transmitted GNSS signals, in order to provide for sufficient redundancy. Geodetic monitoring of the quality of GNSS, as well as the datum upon which the terrain features are mapped, are critical contributions that geodesy makes to air navigation.

The potential use of GNSS positioning to assist aircraft landings, particularly in low visibility conditions, is also currently being considered. Clearly this is one example where consistency of SDI is imperative, ensuring the airfield information is compatible with the datum used by the aircraft navigation system.

4.2.3 Land navigation

Land navigation is a rapidly growing segment of what is now generally known as the "telematics" market, and it also relies on the "twin" geodesy contributions of a precise geopositioning technology such as GNSS on the one hand, and digital road/map data on the other. New Location-Based Service (LBS), for consumer and enterprise users, are valued at tens of billions of dollars per year, and represent a new industry segment with massive potential for growth (expected to grow at over 10% per year over the next decade or so). Increasingly the SDIs must expand to accommodate ever more spatial data sets of interests to LBS users, such as points-of-interest, satellite and ground-level imagery, and even information with short temporal relevance.

4.3 Engineering, surveying and mapping

This field has traditionally been the preserve of experts in measurement and geopositioning. However, as in the case of navigation, with the advent of GPS, and more generally GNSS, the ability of more and more people to quickly generate their own accurate geopositioning information has expanded significantly. There is no doubt that field mapping/surveying (and construction/mining engineering that are dependent upon the products of mapping/surveying) has seen significant improvements in productivity, reliability and accuracy – leading to associated savings in construction and mapping costs. Increased automation of machinery, advances in Information and Communication Technology (ICT), increased integration of different geospatial measurement and imaging technologies, and the trend to more real time operations (including sophisticated field-to-finish surveying systems that exchange geospatial information between office and field units, across data networks, and even between field units), all will place ever greater demands on geodesy and geopositioning technologies. That includes offshore engineering and hydrographic mapping operations.

There is unfortunately a downside. In mere seconds, users of GPS can obtain 3-D coordinates. However, detailed knowledge of what reference frame or datum the positions are referred to, the type of height obtained, and the accuracy of the coordinates may not be obvious to those without professional education. This situation is exacerbated when new technologies such as Real-Time Kinematic (RTK)-GNSS are used in engineering and other precise applications. The RTK systems allow sub-centimeter differential positioning between a "base station" and a "rover". The most alarming problem with this situation is that a base station whose true location is not properly known will yield rover coordinates that are equally in error. Consider, for example, the situation where two local governments (say a municipality and a state) operate RTK base stations and offer an RTK subscription service. If those two overlapping services do not have their base stations consistent with one another (e.g. through ties in some form to the SDI), then all engineering projects built with one service will not align properly with those of the second service. This coordinate inconsistency could lead to massive economic losses and is nowadays a critical challenge to governments at all levels. The issue of overlapping RTK networks already exists in many places in the world. In the U.S., this issue has spurred NGS to begin investigating the creation of an RTK accreditation system to ensure that overlapping RTK networks will be "NSRS compatible."

4.3.1 Machine guidance

For example, the guidance of construction machinery is primarily based on centimeter level RTK-GNSS techniques, possibly augmented with inertial, laser scanners, close-range radar, optical guidance, and other positioning sensors. The construction vehicle driver is "guided" so that trenching, grading and other operations are carried out according to a design model. This results in greater certainty in construction, higher productivity and less waste (e.g. over-excavating), and less labor, with commensurate improvements in safety. In addition, when the track of where the machinery has been is determined and transmitted back to an office a permanent digital record of the new terrain surface becomes available. The trend is toward fully-automated machinery, "controlled" by precise GNSS navigation and its augmentations (inertial, etc.). Similar trends are expected with the automation of farming (see Section 5.8.3), container port loading/unloading operations and open-cut mining, to name but a few potential machine automation applications. Initially such automation will take place for machinery operations on the surface of the Earth, where GNSS signals can be tracked. Indoor positioning at such high levels of accuracy is far more challenging, and advances in technology and its implementation will be slower than in the case of "open sky" applications. However, it must be emphasized that for many of these applications what is required is accuracy and reliability of geopositioning, both in a coordinate/reference frame sense (i.e., the coordinates must be expressed in the correct datum, to which other spatial data are referred) and relative to the "real world" (i.e., objects and terrain that the machinery must navigate around and over).

4.3.2 Land titling and development

Land development and infrastructure engineering requires integrity of coordinates (both those derived from GNSS, as well as implied from previously mapped features). Imagine the uncertainty of land boundaries combined with ambiguous geopositioning in the field: designs for a new road or other infrastructure would be wrong, and this mistake would be compounded in the field when the geopositioning technology guides construction machinery to the wrong location. Land ownership remains to this day one of the cornerstones of national and personal wealth. Current trends are towards a cadastre where parcel boundaries are described by coordinates rather than length. This trend is in response to the need to overlay a series of other land management data sets onto the cadastre, and to implement efficiencies when surveyors relocate existing boundary corners. Land owners and governmental agencies can now correctly interpret land use patterns on a parcel-by-parcel basis, including the location of underground services like sewage, communications, water and electricity. A unique challenge occurs in mining operations: how to relate underground operations (where GNSS cannot be used) to the surface (where increasingly GNSS is the critical geopositioning technology). Furthermore, with increasing interest in the "marine cadastre", and the definition of offshore boundaries in general (Exclusive Economic Zone, state jurisdiction, continental shelf, fisheries, mining/drilling, conservation zones, etc.), coordinates on the ocean surface and those at the sea bottom must be seamlessly connected to a nation's land datum. Again, the SDI layer model is pertinent.

Another concept is the inclusion of cadastral survey observations into regional geodetic adjustments constrained by space geodetic techniques. This is commonly referred to as a "geodetic cadastre", and has the impact of propagating geodetic coordinates at a far greater density than most geodetic networks could hope to achieve. In countries like New Zealand this is used to monitor the temporal deformation in network geometry caused by localized tectonics.

4.3.3 Engineering geodesy and structural monitoring

Engineering geodesy is the term that describes the use of very precise measurement and analysis models to monitor the deformation of anthropogenic structures such as tall buildings, stadiums, bridges, breakwaters and other port engineering, offshore platforms, dams, etc., as well as ground subsidence due to a underground mining, soil compaction and fluid extraction (e.g., oil, water, gas). Increasingly, GNSS receivers, accelerometers and a variety of geotechnical sensors (inclinometers, strainmeters, fiber-optic cables, etc.) are being installed at the time of construction. Such a whole-of-life "structural health monitoring" approach to infrastructure management is in contrast to the current practice of retrofitting structures to monitor loads (and subsequent deformations) after they have been built, for example to ensure safe continued operation of a structure as it nears the end of its life, or when/after it is subjected to loads well beyond what it was designed to withstand. A precise, consistent and time-stable geodetic reference frame (which may be a local datum, and not necessarily a global one) is essential, as well as high integrity geopositioning technologies such as GNSS and differential InSAR (in the case of ground subsidence).

4.3.4 Geographic information systems

Mapping technologies such as scanning and imaging sensors need direct georeferencing (e.g. using GNSS/INS). These days geospatial data are stored, manipulated, analyzed and visualized within special database and computer graphics systems known as Geographic Information Systems (GIS). GIS is designed to allow the display and query of spatial data in the form of "layers". To "stack" the spatial data layers so that they are aligned correctly requires that they be carefully referenced to the underlying geodetic layer or datum (Figure 4.1). Geographic elements within each layer include points, lines and polygons, each with "attribute" information (e.g., describing what it is, and other pertinent textual/contextual information). A special example of point data are terrain models such as ground-level DEM, or tree/building-top Digital Surface Model (DSM). It is this combination of spatial and text information that gives GIS its unique abilities to provide both graphical and quantitative answers to "what if?" scenarios and queries framed in terms of Boolean operations, as well as to create "custom" thematic maps or graphical outputs. GIS is increasingly being used as a decision-making tool, assisting policy advisers, government agencies and businesses to determine the consequences of a certain action or impact on land, communities, infrastructure, biota, etc. GIS technology, geopositioning and spatial information are all necessary for natural and built environmental monitoring, natural resource management, land development, disaster and emergency management (see Section 4.5), transport planning, epidemiological studies, and much more.

Increasingly GIS capabilities are being offered via the World Wide Web (WWW) as "web-GIS", so that users can create custom "views" of spatial data by selecting from many spatial data sets those that are of interest to them. Queries can then be made as in the case of traditional desktop GIS (see above). Again, it is crucial for correct decision-making that the data sets are integrated within the SDI, so that there are no ambiguities with respect to the datum.

Mapping technologies such as LIDAR, and airborne and satellite digital cameras/sensors, are generating enormous volumes of data. The growth of spatial databases, especially those containing remote sensed high resolution (i.e. sub-meter ground object resolution) satellite images, is very rapid. Direct georeferencing ensures that this data can be correctly integrated with historical spatial data and data generated from ground-level surveys. The management of these spatial databases is an increasingly challenging problem.

4.3.5 Height systems

The vertical datum is a critical issue for many engineering, surveying, and mapping operations. Extremely accurate airborne and helicopter LIDAR systems are being used to support many engineering projects and to develop new DEMs and DSMs, but in many cases they depend on the quality of the vertical reference surface to meet the requirements of the project. The reason is LIDAR systems, as in the case of digital and traditional photogrammetric systems, derive their georeferencing information from GNSS/INS. Hence the height information is in relation to a reference ellipsoid, and not to the geoid (or other physical/mean sea level datum). EGMs can define the vertical surfaces to which heights are referenced, but being global in nature, are incapable of capturing local geoid structures. This can lead to errors of decimeters to meters over areas tens to hundreds of km across. In the case of the NASA/NGA SRTM, all heights are referenced to a geoid model such as EGM96, while many heights used for engineering and surveying projects in the continental U.S. use the detailed geoid models developed by NGS (using a combination of EGMs, DEMs, local surface gravity measurements as well as leveling and GPS data), the latest referred to as GEOID03. In Australia the geoid model, developed by GA, is known as AUSGEOID98, and involves a combination of geopotential model, surface and oceanic gravity anomalies, leveling data and satellite altimetry data. In Europe, the European Gravimetric Geoid 1997 (EGG97) is being updated because significant new and improved data sets have become available since the last computation in 1997. These improvements include better global geopotential models from the CHAMP and GRACE missions, better DEMs in some regions (e.g., new national DEMs, SRTM3, GTOPO30), updated gravity data sets for selected regions, updated ship and altimetric gravity data including improved merging procedures, the use of GPS/leveling data, as well as improved modeling and computation techniques. In Canada the vertical heights derived from GNSS can be converted to the system known as CGVD28 using a geoid-based transformation model (http://www.geod.nrcan.gc.ca/software/gpsht_e.php). Every nation maintains their own vertical reference system, frequently (though not exclusively) tied to mean sea level (at least locally) and therefore "connected" to a regional model of the geoid by some means. Increasingly it is being recognized that geodesy must unify these vertical datums in order to ensure interoperability, in much the same way that most nations have migrated to an Earth-centered, Earth-fixed (ECEF)-datum such as an ITRS realization.

4.4 Timing applications

GPS is used for time transfer between precise "clocks" (see Section 2.9.4), as well as an inexpensive and readily available time-base for synchronization of computer networks, telecommunications switches, energy grids, etc. The integrity of GPS and in the future GNSS, which can be considered a byproduct of modern geodesy, is therefore vital.

4.5 Early warning and emergency management

Search and rescue operations rely on accurate geopositioning and up-to-date spatial information. For example NGA developed a host of GIS products for emergency management of Hurricanes Katrina and Rita in 2005, and the tsunami that struck Indonesia in 2004. Efficient rescue operations depend on maps and GIS information that are of high quality, and increasingly developed from the latest spatial data sources, e.g. commercial high-resolution satellite imagery, or in future from "rapid mapping systems" based on Unmanned Aerial Vehicle (UAV)-mounted and terrestrial imaging/scanning systems. The ability of a host of agencies to then utilize high quality GIS information provides superior situation-awareness and allows first-responders the means to make informed decisions. Emergency crews then "navigate" the disaster area, and relief supplies and equipment are delivered, using GPS/GNSS technology.

Risk management requires modeling the effects of severe natural events such as cyclones and tsunamis, which require a uniform and accurate understanding of the topography, including the transition from land-based DEMs to bathymetry. This allows engineers to accurately estimate inundation zones and wind speed variations in the case of cyclones, and inundation and run-up zones for tsunami events. The only system capable of providing accurate heighting for both is based on the ellipsoid, since the relationship between Lowest Astronomical Tide (generally used for bathymetry) and Mean Sea Level (generally, though erroneously used for DEMs) is not well understood at all locations. Geodetic infrastructure provides access to accurate height information.

Another emerging application of geopositioning in the emergency management area is vehicle and personal location systems in emergencies, particularly bush or forest fires where visibility is much reduced. A knowledge of the location of assets directly impacts on fire management, and evacuation processes. This is analogous to minimizing occurrences of "friendly fire" in the military environment.

Many high-profile events held around the world, including the Olympics, Super Bowl, World Series, Soccer World Cup, political conventions, G8 and APEC leaders meetings, etc., depend on high quality geospatial information in preparation for possible emergencies that may arise from an act of terrorism or other threat. These GIS products are built from imagery and other geospatial data, and need to be referenced to a common datum, and be accessible to emergency management officials if and as they are needed. The rapid deployment and response of emergency personnel can only happen if they know precisely the geospatial "picture" of the situation.

4.6 Infomobility

Modern society is mobile, requiring timely answers to such queries as: "Where am I?", "Where are you?", "Where is that?", "How do I get from here to there?" while on the move. The availability, on the one hand, of wireless communications and sophisticated mobile information devices (the cell or mobile phone, personal digital assistant, etc.), together with GPS/GNSS and GIS on the other hand, has spawned a new ICT field of "infomobility". This new cross-disciplinary field (also known as location-aware or context-aware computing, pervasive or ubiquitous computing, telematics, telegeoinformatics, etc.) links information technology and wireless communications (together representing the ICT discipline), and geopositioning systems to digital maps and geographically-referenced data delivers information-rich services appropriate to the location of a person or object, via a mobile device. Infomobility is part of the broader field of geotechnology (also known as geoIT, geospatial information science, etc.) identified by the US Department of Labor in 2004 as one of the three most important emerging and evolving fields, along with nanotechnology and biotechnology (Gewin, 2004). The characteristic which most distinguishes infomobility from other geotechnology applications is the small interactive mobile device having both wireless connectivity (to web servers via the mobile Internet) and location determination capability (via GPS or some other means), running special client software.

Infomobility applications are also often referred to as LBS, although it could be argued that LBS is a subset of infomobility as it implies some sort of "transaction" between the mobile user and a service provider. Infomobility may be partitioned into three major categories:

- Vehicle-based applications, sometimes also known as "telematics" services, comprising essentially of car navigation systems, driver and passenger assistance services, and vehicle management systems (Rizos & Drane, 2004). These rely primarily on GNSS, augmented with wheel sensors and map-matching to provide continuous location information, even in urban areas and tunnels where GPS/GNSS signals do not penetrate.
- Personal services, many of them similar to vehicle telematics services, but delivered to a portable device to aid people on the move. The requirements to deliver geopositioning is far more stringent than for vehicles, as people spend a lot of

their time inside buildings, where GNSS is not well suited (though Assisted-GNSS techniques do increase the sensitivity of receivers to weak satellite signals attenuated by buildings, walls, trees, etc. Therefore pedestrian navigation requires a "mix" of geopositioning technologies, including cellphone signals, WiFi, inertial, pedometer, and so on.
- The tracking of inanimate objects, in logistics and warehousing applications. These may be for individual high value assets, or container/pallet tracking. As with pedestrian applications, the mix of geopositioning technologies may have GPS/GNSS as a core, but augmented with short range communications sensors, such as Radio-Frequency (RF) tags known as Radio-Frequency Identification (RFID) and WiFi to give indoor coverage.

What sets infomobility or LBS-type applications apart from "information on the move" (e.g., m-commerce, e-commerce) is the central role played by geopositioning and geospatial data in general. GNSS is a key technology, but there are other geopositioning sensors that may be used in combination with GNSS (or even on their own), such as inertial sensors, pedometers, altimeters, WiFi, cellphone, UWB (UWB), RFID, and vision systems. It must be emphasized that these applications have commercial (or business) counterexamples to personal (or consumer) ones, such as:

- The tracking of commercial fleets of vehicles (taxis, couriers, public transport, etc.) is an important telematics application.
- Positioning/tracking of emergency service workers (e.g., firefighters, police, rescue workers) and soldiers, are examples of personal infomobility applications. The environments in which they are expected to operate challenge standard GNSS technology, which has led to considerable innovation in geopositioning.
- The massive uptake of RFID "tagging" systems will displace barcodes as the primary means of keeping track of stock in warehouses, in transit, in retail shops, etc. Within a decade many household items will be "tagged", and hence "trackable".

It is predicted that there will be a massive growth in such applications, and listed here are just some of these in order to give an impression of the future infomobility applications:

- cyber-geography, geo-blogging, GeoSearch, georeferencing web pages, the Geospatial Web;
- spatialized (media) art, location-based games;
- context-sensitive (location-based) advertising, spatial searches (e.g., Google Earth, NASA's World Wind, Microsoft's Virtual Earth, Yahoo), tourism guides;
- augmented and mixed reality for work and pleasure;
- supranet, smart dust, sensor nets, etc.;
- information portals for mobile workers (answering questions such as "Where are you?", "Are you all right?");
- tracking and location privacy (rights and freedoms in a modern "spatially-enabled" society);

- RFID-tracking, enterprise databases and navigation queries (answering questions "Where is it?" about inanimate objects); and
- telematics and Intelligent Transport Systems.

Geospatial information is the Information Technology (IT) "content" for most, if not all, of these applications. As with many other geospatial applications, geoIT requires that different spatial data (the coordinates of a mobile device, the location of a place-of-interest, the road network linking different parts of the city, and so on) must all be consistent, i.e. on the same geodetic datum.

4.7 Management of and access to natural resources

4.7.1 Water management and hydrology

The SDI of geodesy provides a framework for basic water management - river monitoring relies on a fundamental reference system, and the assessment of water rights is fundamentally tied to the properties of area and geography. The assessment of aquifers is aided by new technology: satellite and absolute gravity measurements can infer changing volumes of water stored in subterrain aquifers, and repeated GPS/GNSS observations and InSAR can detect surface movements caused by depletion of such aquifers. Geodesy therefore can play a role in monitoring the rate of fluid extraction (including illegal groundwater mining), providing information to the relevant authorities.

The new satellite missions measuring gravity change (currently, GRACE) promise major improvements of the quantification and understanding of the global hydrological cycle. Whereas traditional hydrological models have focused on details, using models of snow and rain falls, groundwater changes, river runoff, etc., on local to regional scales, the measurement of gravitational changes gives overall integrated bounds on the hydrological cycle on continental scales. The proven monitoring of the yearly hydrological cycle over major tropical regions such as the Amazon Basin and South-East Asia are spectacular examples of the success of the new class of gravity field change-measuring satellites, and with likely new technological breakthroughs (e.g., laser interferometry missions) within the next decade, the gravitational monitoring of major drainage basins on a global scale may be feasible.

Major engineering structures used for water management (dams, channels, etc.) are obviously heavily relying on the geodetic SDI. The level of water is dependent on the gravity field, and accurate 3D-networks with associated gravitational information (geoid models) are a necessity for major construction projects. The monitoring of such projects after construction is another domain where geodesy plays a key role, with the long-term stability of reference frames and monuments being essential for the measuring any movements or subsidence of the structures and artificial lakes.

4.7.2 *Energy resources*

Energy, in the form of electricity, is derived from a variety of renewable and non-renewable sources. Examples of the former include photovoltaic technology, wind, hydrodynamic and tidal power. The latter include coal, nuclear and gas-fired power stations. All forms of energy generation require massive investments in infrastructure: power stations, pipelines, transmission lines, power grids, etc. From a strictly engineering point of view, such infrastructure must be designed and built in the right location, the structures may need to be monitored (episodically or continuously), and the components must "fit" together (not just in a physical sense, but operations need to be synchronized to microsecond levels of accuracy - see Section 4.4).

Traditionally the geodesy and surveying disciplines have assisted exploration geophysicists in locating coal, ore bodies, gas and oil fields, through the provision of land and marine gravity data, including marine gravity derived from satellite radar altimetry. There is still considerable activity in seeking out non-renewable sources of energy from offshore locations, in deeper water, further away from shore. Offshore engineering is a discipline that is becoming increasingly important. However, apart from the challenge of working in a watery environment, the SDI off shore is very much wanting. Geospatial data quality is very patchy, and it is generally conceded that the quality of maps of the Moon's surface is better than for many parts of the Earth's deep oceans. The challenge in ocean bottom geodesy has long been recognized (Spiess, 1990, e.g.,), and despite the fact that the IYPE has the goal to extend the success of geodesy on land onto the ocean bottom(Chen et al., 2005), it is not likely that considerable progress will be made soon.

Increasingly the geodesy and surveying skills (and the underlying geodetic reference frame) must support new power generation technologies, such as designing new wind farms, tidal power systems, etc. In these cases use is made of GIS to assess the likely impact of new infrastructure on the land, communities and biota. A new field of mapping is in support of Carbon Credits, e.g. in the planting and monitoring of forests.

4.8 Monitoring the environment and improving predictability

Vulnerability of society to extreme weather events, disturbances in critical infrastructure by environmental disasters, and potential impacts of climate change necessitate improved predictions of weather, climate, and other environmental events. Geodetic techniques, in particular, GNSS, sense the atmosphere with electromagnetic waves, which can be used to extract information on ionospheric electron content, tropospheric water vapor content, and, potentially, carbon dioxide. Below, GNSS meteorology and space weather are identified as examples. Other applications such as monitoring of sea level, ice sheets, lake levels, and carbon dixoide are addressed in other chapters.

4.8.1 GNSS meteorology

Water vapor plays a crucial role in the dynamics and thermodynamics of many atmospheric processes that act over a wide range of temporal and spatial scales, covering both the global hydrological and energy cycles, which effectively define the local and global climate change, contributing largely to the greenhouse effect, and playing a critical role in the vertical stability of the atmosphere and in the structure of the evolution of atmospheric storm systems. The scarcity of traditional meteorological observations, especially over the Southern Ocean and Polar Regions, as well as the shortcomings of the traditional methods over the land, have contributed greatly to uncertainties in global and regional weather analysis. GNSS offers a more economical, and in principle near-real time, method of measuring atmospheric water vapor.

A number of studies conducted in the 1990s have shown that the amount of precipitable water contained in the neutral atmosphere can, in fact, be retrieved using ground-based GNSS receivers (see Section 2.9.1, Figure 2.38 on page 74). In addition, it has been demonstrated that a GPS receiver aboard a microsatellite in a low Earth orbit, supported by a ground-based network of receivers, can be used to determine the atmospheric refractive index as a function of altitude during the event of satellite occultation by the Earth (see Figure 2.39). Thus, the availability of remote sensing observations from GNSS radio occultation sensors provides a unique opportunity to improve the quality of regional meteorological analysis, particularly over the traditionally under-sampled regions, as well as promise of higher spatial and temporal resolutions, if a sufficient number of sensors are launched and supported by an adequate ground-based tracking network.

4.8.2 Space weather

Space weather refers to conditions on the sun and in the solar wind, magnetosphere, ionosphere and thermosphere that can influence the performance and reliability of space-borne and ground-based navigation and communication systems, and can even endanger human life or health. Thus, ionospheric irregularities are one of the key components of the space weather that require special attention. Advances in space-weather forecasting require improved understanding of the ionosphere/thermosphere system. Any forecast model must begin with a detailed specification of the current state of the system, which is provided either by empirical models or by assimilative models. The ionosphere is one of the most detrimental error sources in satellite navigation and communication systems. While spatial and temporal distribution of ionospheric disturbances are of primary interest in their own scientific context, they are also of special interest to communication, surveillance and safety-critical systems supporting air navigation, as they affect the skywave signal channel characteristics.

A large number of globally distributed permanently tracking GNSS stations can deliver large volumes of data suitable for continuous, near-real time ionosphere monitoring during the disturbed and quiet geomagnetic conditions, and offers an attractive alternative to the traditional methods. GNSS provides high-resolution TEC measurements. Traditional ionosphere monitoring methods based on ionosondes and Incoherent Scatter Radars (IRS) have many disadvantages as compared to GNSS. Hence GNSS geodesy offers the best option for providing detailed information on ionospheric conditions, an essential component of any space weather monitoring system.

Chapter 5
Earth observation: Serving the needs of an increasingly global society

D. Sahagian, D. Alsdorf, C. Kreemer, J. Melack, M. Pearlman, H.-P. Plag, P. Poli, S. Reid, M. Rodell, R. Thomas, P. L. Woodworth

In this chapter, we examine the potential of geodesy from the point of view of what society needs in terms of Earth observations to ensure the security of people and resources, and to achieve a sustainable utilization of ecosystem goods and services. Many of these applications depend on Earth observations, both *in situ* and remotely from space or airborne sensors. The framework for these observations is dominated by a few international programs, such are implemented and supported by GEO, IGOS-P, Committee for Earth Observation Satellites (CEOS), the European Commission (EC), United Nations (UN) agencies, and national contributions.

GEO's visions for GEOSS, in brief, is to enable a future where decisions can be based on information (see GEO, 2005a, for the full text of the vision; also on page 280). GEO is focused around nine Societal Benefit Areas (SBAs) of Earth observation as identified by EOS-II (see Table 5.1). Thus, we start here from the information needs of these SBAs, which then can be translated into observational needs. Finally, the potential contribution of geodesy can be extracted from these needs. In the next section, we will briefly review the Earth observation framework and introduce the nine SBAs. Subsequently, we devote separate sections to each of the SBAs that can benefit from geodetic observations and/or products.

5.1 The current and future framework of global Earth observations

Major early milestones towards integration of the global Earth observing systems were the definition of the Integrated Global Observing Strategy (IGOS), and the establishment of the Global Three Observing Systems (G3OS) in the context of the United Nations Framework Convention on Climate Change (UNFCCC) in the mid-1990's. Initially, IGOS was of particular importance within Earth monitoring based on remote sensing (see, e.g., Williams & Townshend, 1998), and it was developed in the framework of the G3OS (see, e.g., Dahl, 1998). The drivers for IGOS are

H.-P. Plag, M. Pearlman (eds.), *Global Geodetic Observing System*,
DOI 10.1007/978-3-642-02687-4_5, © Springer-Verlag Berlin Heidelberg 2009

the scale of the issues (global climate change, sustainable development) to be addressed, the cost of space components for remote sensing of the Earth environment, the logistics especially for *in situ* data, and the need for data integration from multiple sources for products of use to decision makers, science, and society at large. For key variables of the Earth system, IGOS attempts to provide long-term continuity, adequate data archives and accessibility, consistency of data records, and the ancillary data required for data quality assessment. IGOS provides the framework for a coherent response of the monitoring system to the integrated user requirements. IGOS intends to build upon existing strategies for international observation programs, focusing on the identification of areas where the existing systems can be improved, where duplication of observations can be reduced, and gaps in observations and data sets can be identified. If effectively implemented, IGOS appears to be the strategy for providing the observational basis for a future Earth information system. A key issue identified in IGOS is the need to transform many observational activities from their research states into operational monitoring.

In 1998 the further development and implementation of IGOS was put into the frame of the IGOS-P (see, e.g., Smith, 1998, for the early development of IGOS-P). IGOS-P is a partnership of organizations that are concerned with global environmental change issues. IGOS-P links research, long-term monitoring and operational programs. IGOS-P seeks to provide a comprehensive framework to harmonize the common interests of the major space-based and *in situ* systems for global observations of the Earth. Its aim is to provide an over-arching strategy for conducting observations relating to climate and atmosphere, oceans and coasts, the land surface and the Earth's interior. The Partners, through IGOS, build upon the strategies of existing international global observing programs, and upon current achievements, in seeking to improve observing capacity and deliver observations in a cost-effective and timely fashion. Main efforts of IGOS-P are directed to those areas where satisfactory international arrangements and structures do not currently exist. Most of the IGOS-P efforts are concentrated in a small number of so-called Themes with strong linkages to critical social issues.

The last few years have seen a rapid programmatic development in Earth observations on global scale, stimulated in part by activities in Europe. There, the Global Monitoring of Environment and Security (GMES) initiative was launched in May 1998 and adopted by European Space Agency (ESA) and the European Union (EU) Councils in June and November 2001, respectively. The overall aim of GMES is to support Europe's goals regarding sustainable development and global governance by providing timely and quality data, information and knowledge (European Commission and ESA, 2003).

Following up the recommendations of the World Summit on Sustainable Development (WSSD) held in Johannesburg in 2002, the first EOS-I was held in Washington, DC, in July 2003. This summit initiated an unprecedented global effort towards coordination of global Earth observation. Through its declaration (see Annex 1 in GEO, 2005b), EOS-I established the *ad hoc* GEO with the task to draft a 10-Year Implementation Plan for GEOSS. Subsequently, this *ad hoc* GEO met six times, and supported by several Subgroups, drafted the requested plan (GEO, 2005a) together

Table 5.1. The nine Societal Benefit Areas of Earth observations as identified by EOS-II. From GEO (2005b).

Area	Objective of GEO
Disaster	Reducing loss of life and property from natural and human-made disasters
Health	Understanding environmental factors affecting human health and well being
Energy resources	Improving management of energy resources
Climate	Understanding, assessing, predicting, mitigating, and adapting to climate variability and change
Water	Improving water resource management through better understanding of the water cycle
Weather	Improving weather information, forecasting, and warning
Ecosystems	Improving the management and protection of terrestrial, coastal, and marine ecosystems
Agriculture	Supporting sustainable agriculture and combating desertification
Biodiversity	Understanding, monitoring and conserving biodiversity

with a reference document containing many details of the vision for GEOSS (GEO, 2005b). It is noted here that in less than two years of its existence, the membership of the *ad hoc* GEO had grown from initially about 30 countries to more than 60 countries.

The work of GEO was guided by the Framework Document adopted by the EOS-II, which was held in Tokyo in April 2004 (see Annex 2 in GEO, 2005b, for the full text). This Framework document identified nine major SBAs of Earth observations (see Table 5.1) and emphasized strongly the importance of coordinated global Earth observations.

The GEOSS Implementation Plan was adopted by EOS-III, which took place in February 2005 in Brussels. The same meeting transitioned the *ad hoc* GEO into a permanent group. The presence is dominated (and so will be the next few years) by the first steps towards an implementation of GEOSS. IAG is involved in this process in order to ensure that the geodetic observing system is developed consistently with the needs and progress of GEOSS for a maximum benefit.

GEO (2005b) provides for each of the benefit areas an overview of the requirements in terms of quantity and status of the observational capacity. Extracting the quantities potentially coming or benefiting from GGOS results in the list compiled in Table 5.2.

Geodetic observations contribute to Earth observation in two very distinctive ways, namely (1) geodesy provides the reference frame in which all Earth observations can be associated with coordinates, and (2) geodesy provides observations of quantities related to relevant processes (see Section 2.3).

All measurements depend upon a suitable reference frame in which positions can be determined and against which changes in position can be measured. Based on the available tools (see Chapter 2) geodesy provides this reference frame for the Earth on a global scale in the form of the ITRF as well as for space in form of the ICRF (see Section 2.2). With these reference frames, geodesy serves a common need for all SBAs and society at large (see Chapter 4). In the following, we will not address these requirements to any detail. Rather, we will focus on the requirements for geodetic observations.

Geodetic observations can provide insights into a number of critical areas that impact human society. These range from understanding earthquake processes, assessments of hazards, detecting and tracking tsunamis to monitoring the effects of climate change, and even prediction of volcanic eruptions (see Chapter 3). The following sections consider the observational needs for seven out of the SBAs listed in Table 5.1.

Table 5.2. Requirements for geodetic observables for the nine Societal Benefit Areas as reported in GEO (2005b). The fields and their status are extracted from the discussion of the User Requirements (URs) for the nine benefit areas in GEO (2005b). There, the status is indicated with the follow classes: 0: ok; 1: marginally acceptable accuracy and resolution; 2: could be ok within two years; 3: could be available in six years; 4: still in research.

Observable quantity	Status
Deformation monitoring, 3-D, over broad areas	3
Subsidence maps	3
Strain and creep monitoring, specific features or structures	2
Gravity, magnetic, electric fields - all scales	3
Gravity and magnetic field anomaly data	2/3
Groundwater level and pore pressure	4-1
Tides, coastal water levels	1
Sea level	2-1
Glacier and ice caps	2
Snow cover	2
Moisture content of atmosphere/water vapor	2
Extreme weather and climate event forecasts	3
Precipitation and soil moisture	3-1

5.2 Disasters: Reducing loss of life and property from natural and human-made disasters

One of the most important services that science can provide to society is understanding, predicting, and reducing of vulnerability to natural hazards. These can be divided into those stemming from the dynamics of the fluid envelope of the Earth such as storms, storm surges and floods, those stemming for the dynamics of the solid Earth, such as earthquakes, volcanoes, sinkholes, subsidence, precarious rocks, rockslides, and landslides, and those resulting from interaction of the solid Earth with its fluid envelope, in particular tsunamis triggered by earthquakes, rockslides, volcanic eruptions, and submarine landslides.

In disaster prevention and mitigation, Earth observations are pivotal in at least three aspects: (1) understanding the processes causing these hazards and assessing their risks for planning and mitigation, (2) monitoring the development of hazardous situations and providing a basis for a decision on early warnings, and (3) determining the extent of a disaster as support for rescue and damage assessment. The first two aspects are central for early warning systems. A comprehensive and effective early warning system requires four elements, namely

- risk knowledge: *a priori* knowledge of the likely risk scenarios a community might be faced with;
- monitoring and warning service: the capacity to monitor risks and rapid and reliable decision mechanisms for early warning;
- communication: the ability to disseminate understandable warnings to those at risk;
- response capability: knowledge and preparedness capacity by all partners of the information chain to act appropriately.

Geohazards are intimately connected to displacements and deformations of the Earth's surface. Consequently, geodetic observations play a crucial role in all three aspects of disaster prevention and mitigation, including risk knowledge and the monitoring of hazardous situations required for the implementation of early warning systems. The importance of geodetic observations for these hazards has been emphasized by many (e.g. UNAVCO, 1998; Solomon & the Solid Earth Science Working Group, 2002; Raymond et al., 2003). Marsh & the Geohazards Theme Team (2004) state that "*Geohazards driven directly by geological processes all involve ground deformations. Their common observational requirements are for global, baseline topography and geoscience mapping, against which surface deformations ... can be monitored.*" Thus, the observations provided by existing global and regional geodetic networks have already transformed our understanding of geohazards, and it is likely that these networks will play an even more important role in the future as their coverage and precision improve. In many regions, observing systems dedicated to geohazards would also have to be flexible in spatial and temporal resolution, as well as readiness on demand. Therefore, in many parts of the world, dedicated ground-based geodetic networks are needed. In addition to the classical, point-oriented geodetic techniques, 2-dimensional imaging techniques such as InSAR are also needed. These techniques allow the monitoring of relevant areas with high spatial resolution, although currently not with the low latency and temporal resolution required for some geohazards applications.

5.2.1 Landslides, rock falls and subsidence

Landslides are a major hazard causing many fatalities and significant damage in many locations. In the past century (1903-2006), landslides killed more than 60,000 people globally, affected more than 10 million people (many of them homeless after the event), and caused damage on the order of 5 billion dollars (Salichon et al., 2007). Many landslides take place in widespread areas of slope instabilities caused by severe storms, earthquakes, volcanic activity, coastal wave erosion, and wildfires.

Landslide danger may be high even as emergency personnel are providing rescue and recovery services. Often, earthquakes are accompanied by landslides, rockfalls, and other surface disruptions that can cause as much or more damage to anthropogenic structures and systems than the earthquakes themselves. These events are difficult to predict, but depend on recent weather conditions (i.e., precipitation and soil moisture), as well as land cover, topography, and earthquake recurrence interval. Steep topography near lakes and fjords has the potential of large waves caused by rockslides into the water below and pose a potential threat in some areas. Moreover, in many mountainous areas, the steep hill sides are a potential threat for the people living at the base of these slopes or for the infrastructure at the bottom of such hills. In many areas, slop slides or slow landslides pose a problem, too.

In order to provide accurate landslide hazard maps, forecasts of landslide occurrence, and information on how to avoid or mitigate landslide impacts, several questions must be considered: Where and when will landslides occur? How big will they be? How fast and how far will they move? What areas will they affect or damage? How frequently do they occur in a given area?

In known unstable areas, networks of campaign-type or permanent GNSS stations can be used to detect a change in the motion and thus indicate a potentially perilous situation. However, the recurrence period of land- and rockslides can be long and in many areas the risk is not obvious. InSAR is an emerging technology that allows the determination of surface deformation with high spatial resolution and accuracy in many regions (see Section 2.4.5). InSAR is expected to play a leading role in the detection of geohazards and the monitoring of hazardous areas. InSAR has been successfully applied to the mapping of coseismic displacements (e.g., Massonet et al., 1993), deformation at volcanoes, silent landslides (Ferretti et al., 2004), and anthropogenic subsidence. In particular, the combination of permanent GNSS stations with InSAR is expected to improve the time series of deformation measurements considerably.

Potentially hazardous landslides and slow landslides associated with human activities, as well as anthropogenic soil subsidence caused by groundwater, oil, and gas extraction and mining activities, can increasingly be detected by using InSAR. Ferretti et al. (2004) analyzed an InSAR-based time series of surface displacements and detected several unstable areas in the San Francisco Bay area. In order to reveal such areas at an early stage of the development of landslides or larger deformations, an accuracy of 1 mm/yr and high spatial resolution (< 100 m) are required.

The monitoring of anthropogenic subsidence also requires high spatial resolution and the determination of changes in the secular velocity of vertical land motion on the level of 1 mm/yr. In areas with active mining and groundwater extraction, changes in secular land motion have to be available with low latency in order to detect potential hazards in a timely manner.

5.2.2 Volcanic eruptions

Volcanic eruptions are comparable to landslides in number of fatalities and extent of damage (Salichon et al., 2007). Major volcanic eruptions have local to global impacts, and are typically presaged by directly observable events, including seismicity and infra sound (Johnson et al., 2008), gas release, surface deformation, and small precursor eruptions. Modern volcano monitoring systems integrate localized monitoring components and remote sensing.

GNSS and gravity measurements are integral parts of any monitoring system of potentially hazardous volcanoes. The combination of these measurements provides a basis for understanding the dynamics of subsurface magma movements and the development of hazardous situations. Surface displacements can indicate magma movements not necessarily associated with increased seismicity.

Increasingly, InSAR is applied to the monitoring of volcanoes (see Section 2.4.5). However, for early warning purposes, the combination with local GNSS networks is crucial. Unfortunately, many hazardous volcanoes are not sufficiently monitored. The development of relatively cheap disposable GNSS stations would be an advantage at hazardous volcanoes.

5.2.3 Earthquakes

Earthquakes are a major causes of disasters which, over the last hundred years (1903-2006), killed nearly 2 million people, affected nearly 100 million people, and caused damage of more than 300 billion U.S. dollars (Salichon et al., 2007). Increasingly, megacities are developing in areas prone to experience major earthquakes, thus making disasters more likely. As in the case of volcano monitoring, local *in situ* observation systems are increasingly supplemented by continuous and broad scale networks such as the Plate Boundary Observatory (PBO) in the U.S. The GNSS networks provide fundamental observations of the deformation process during the complete earthquake cycle from preseismic to co- and postseismic deformations. Hence, strain rates determined from geodetic observations are increasingly used in hazard assessments. Moreover, image techniques such as InSAR are increasingly supplementing the ground-based techniques.

Much of the geodetic infrastructure is currently focused on research related to the processes causing earthquakes. Increasingly, the geodetic networks also support the rapid detection of earthquakes for early warning and damage reduction response (e.g., by rapid shutdown of gas pipelines, stalling or diversion of traffic on roads and railroads, shutting down of nuclear power plants, etc. This application requires real-time detection of ground motion with reaction times in the range of a few seconds), as well as rapid damage assessment in support of rescue.

Seismic hazards can also result from mining, filling of reservoirs, and extraction of oil and gas. In order to detect seismic hazards induced by mining, monitoring of the strain rates in the mining area is the appropriate tool. The seismic hazard associated with the filling of large reservoirs is thought to be caused by changes in the subsurface pore pressure and not the loading-induced stress (Roeloffs, 1988; Talwani & Acree, 1985).

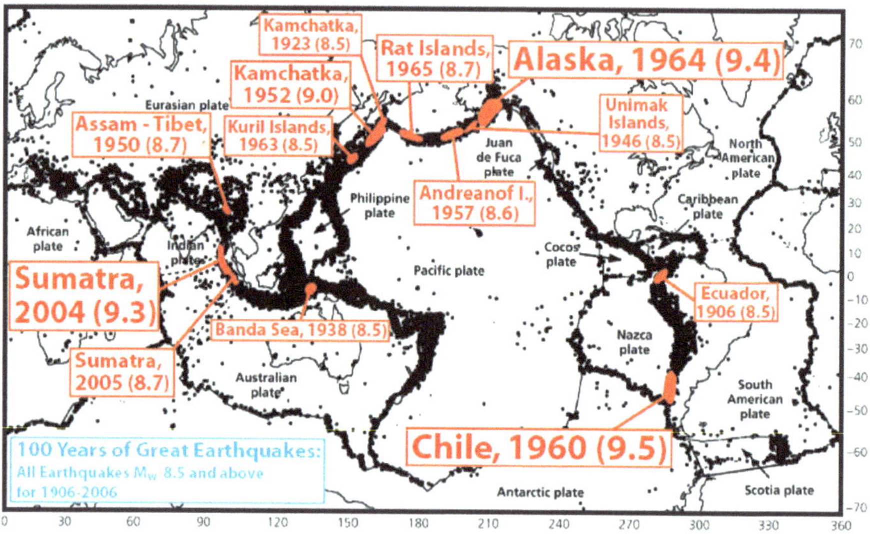

Fig. 5.1. Location of the largest earthquakes since 1900. Indicated are the locations of earthquakes with $M_w > 8.5$.

5.2.4 *Tsunamis*

Tsunamis are generated by submarine earthquakes, landslides and volcanic eruptions. Although tsunamis are frequent, most have small amplitudes (a few centimeters) and do not pose any danger for coastal areas. Only large earthquakes (moment magnitude greater than 7.5) with an epicenter at shallow depth can excite tsunamis which can result in dangerous coastal wave heights larger than a few meters. Generation of tsunamis by earthquakes is therefore restricted to submarine seismogenic regions with shallow and potentially large earthquakes. However, knowledge of the location of these faults is not sufficient to identify all potentially hazardous areas. Therefore, an important task is the detection of potentially hazardous regions in the ocean.

Over the last hundred years, most of the large earthquakes with magnitudes of 8.5 and larger, which are potentially responsible for devastating ocean-wide tsunamis, were located around the Pacific Ocean (Figure 5.1). However, large tsunamis can also originate in other regions. Smaller tsunamigenic earthquakes occur in many other regions (e.g., the Mediterranean and the Caribbean), and pose a danger for nearby coastal areas.

Submarine landslides can happen in large areas of the continental shelves, where sufficient sediments have accumulated to allow turbidity currents to form. Moreover, in some areas of steep topography and appropriate geology, rockslides can occur, as in the Mona Rift area north of Puerto Rico (Grindlay & Hearne, 2005). Coastal landslides can result in large waves if sufficient material is involved. Examples are the landslides on Stromboli on 30 December 2003, where the ash deposits slid and created a tsunami traveling around the coast of the island and causing considerable damage, with waves reaching maximum heights of 5 to 10 m (Bonaccorso et al., 2003). Much larger slides have happened at this island over the last 13,000 years, which are likely to have caused large tsunamis affecting the Aeolian Islands and the coasts of South Italy (see La Rocca et al., 2004, and the references therein).

The major difference between tsunami generation by a landslide and an earthquake is in the movement of the source region. For an earthquake, a tsunami is generated mainly by vertical movement in the source region (usually 100 kilometers or more wide), causing a sea surface anomaly. However, in a landslide, a much smaller source region moves mainly horizontally. As a consequence, the long-wave approximation is valid for tsunamis generated by earthquakes, but not for landslides. Moreover, directivity of the tsunami is generally more expressed for those created by landslides.

Submarine volcanic eruptions are mainly associated with mid-ocean ridges, hotspots, and back-arc basins. However, in most cases, the mid-ocean ridge and hotspot volcanoes are not likely to be explosive, and these eruptions are not likely to create tsunamis. Similarly, earthquakes at mid-ocean ridges are normally not large enough to trigger significant tsunamis.

Knowledge of the tsunamigenic source locations is only a first step in establishing the tsunami hazards for a given coast. Most tsunamigenic sources have strong anisotrophy in the propagation of tsunami energy away from the source (e.g. Titov et al., 2005b). Consequently, the tsunami hazard at any point on a coast depends not only on the distance to potential sources, but also the direction with respect to the propagation pattern for a particular source. Moreover, the shape of the coast, its topography and the bathymetry of the ocean basin between the coast and the source are important factors determining the tsunami hazard.

Tsunami hazard maps, and more generally, sea level hazards maps, are necessary for planning of a reliable and economically feasible sea level hazard observing system. This has also been acknowledged in the definition of a task for the GEO Work Plan focusing on a Global Tsunami Hazard Map (GTHM). The methodology will have to take into account problems that the incomplete record of events causes for the direct application of a probabilistic analysis, comparable to probabilistic seismic hazard analysis (see e.g. Wang & Ormsbee, 2005). The experience gained in establishing the Global Seismic Hazard Map (GSHM) (Shedlock et al., 2000), the Global Stress Map (Heidbach et al., 2004) and the Global Strain Map (Kreemer et al., 2003) can help in developing the methodology for the GTHM, and the information contained in these maps is of direct relevance. Thus, geodetic observations of the kinematics of the Earth's surface that allow the determination of the strain field near subduction zones, are an important input to this hazard assessment.

Geodesy also plays a role in the monitoring required for any early warning system. A rapid and precise quantification of earthquake sources is central to tsunami warning systems, because tsunami models are initialized by assuming a displacement field of the ocean floor. The early prediction of tsunamis on the basis of detected earthquakes is currently limited due to two shortcomings:

- First estimates of the magnitude of large earthquakes often prove to be too low (Kerr, 2005; Menke & Levin, 2005) due to saturation of the near-real time methods, leading to an underestimation of the tsunamigenic potential.
- Tsunami propagation models are sensitive to the initial conditions (i.e., the model of the seismic rupture process, Titov et al., 2005a). However, establishing the rupture processes of particularly large earthquakes for a specific earthquake in near-real time is difficult. Currently, it takes considerable studies before the relevant details are investigated (as well illustrated by the sequence of papers addressing the magnitude and rupture process of the 2004 Sumatra-Andaman earthquake, see, e.g., Plag et al., 2006b, for references).

After the 2004 Sumatra tsunami, at least seven large undersea earthquakes occurred. Large-scale tsunami warnings were issued for five of them (Nias, March 2005, M 8.7; West California, June 2005, M 7.1; Tonga, May 2006, M 7.8; Kuril Islands, November 2006, M 8.3; and Kuril Islands, January 2007, M 8.1). However, most of these events did not generate significant tsunamis in the areas for which warnings were issued. However, the July 2006 West Java event with a magnitude of 7.7 and the April 2007 Solomon earthquake with magnitude 8.0 each generated unexpectedly large local tsunamis that killed more than 600 and 30 people, respectively. These incidents illustrate that the currently used method for early warnings based on earthquake-magnitudes from seismometers alone is not reliable to accurately predict the size and impact area of tsunamis. In the case of an earthquake, there are two steps in the prediction of tsunami impact in a specific region: (1) determination of the tsunami potential of the event based on the magnitude and rupture process, (2) prediction and/or detection of the tsunami propagation towards the specific region.

Static coseismic displacements determined from GPS stations in the near-field of earthquakes agree well with the displacements determined through integration of strong motion records (Larson et al., 2003; Miyazaki et al., 2004). Blewitt et al. (2006b) demonstrated that a relatively sparse GPS station network with a radius of about 2000 km around the epicenter of the 2004 Sumatra earthquake was sufficient to determine the magnitude of this event accurately using the GPS data up to 15 minutes after the earthquake origin time. Their results indicate that if GNSS data from a sufficiently dense network around the source of a large earthquake were available in real time, GNSS ground-based stations could yield realistic low-latency (order 15 minutes) estimates of the seismic moment and displacement field of the event. This information would be a valuable addition to tsunami warning systems for devastating oceanwide tsunamis. However, even with such networks in place and an operational processing set up, it would still take about 30 minutes for reliable tsunami predictions. Although such predictions would be of very high value for oceanwide tsunamis, this time lag is unacceptable for coastal areas close to the epicenter.

In summary, the current state-of-the-art in the low-latency detection of tsunamigenic sources and the subsequent low-latency prediction of the tsunami propagation and impact do not warrant a tsunami monitoring system based solely on source detection and numerical propagation predictions (Defra, 2005). Moreover, some tsunamigenic events are not associated with strong or clear seismic signals (e.g. landslides, impacts) and cannot be detected easily. Therefore, a key element in a monitoring system for early warning purposes will directly focus on the propagation phase of the tsunami and aim to detect a tsunami at an early stage of this propagation phase.

In most cases, sea level information gathered by a suitable monitoring network will help to reduce the uncertainties in the source detection and classification, and thus provide a sound basis for the decision to issue timely warnings. In many cases, only an observing system aiming at the tsunami itself will provide the necessary observations to identify a tsunami. This monitoring system in principle will also be applicable to other sea level hazards, in particular large storm surges. However, sea level observations alone may not be sufficient since most existing sea level sensors currently are located at coastal locations, while the sea level events need to be detected before they reach the coastal parts under risk. Tsunami monitoring systems in the Pacific therefore apply a number of ocean bottom sensors both for the detection of tsunamigenic events and the monitoring of the tsunami itself (see Section 2.9.3). But these systems are expensive and demanding in maintenance, and it is unrealistic to assume a dense-enough network in many parts of the global oceans.

In the aftermath of the Sumatra-Andaman earthquake, a number of groups have developed algorithms and software for detecting tsunamis in tide gauge data for confirmation of previous alerts and passing warnings along (Woodworth, 2008, private communication). Moreover, a number of studies have suggested other approaches that may help in tsunami warning. For example, Garcés et al. (2005) indicated the detection potential of deep infra-sound associated with tsunamis. Plag et al. (2006b) pointed out that GPS networks and gravimeters, in principle, could sense the loading deformations induced by the tsunami and the associated gravity perturbations, respectively, and Nawa et al. (2007) confirmed this by showing that the tsunami signal was present in observations of superconducting gravimeters. Tilt meters and horizontal seismometers can observe loading induced tilt. GNSS buoys can measure the sea level variations caused by a passing wave, and in some cases, i.e. in shallower areas, also currents associated with the orbital motion of the particles in these waves. Bao et al. (2005) showed that the Sumatra tsunami affected GRACE observations. Coupled ocean-ionospheric signals have also been studied with respect to their early warning potential (Occhipinti et al., 2006). Gower (2005) studied the signal of the 2004 Sumatra tsunami in satellite altimetry and found these observations valuable for the post-event validation of model predictions. After that, tsunami source models inferred from satellite altimetry data have reported by others (e.g., Hirata et al., 2006; Song et al., 2005; Fujii & Satake, 2007). Ablain et al. (2006) applied more advanced technique to extract tsunami signals from sea level anomaly data by satellite altimetry to raw data and produced high quality tsunami signals. Hayashi (2008) studied the detectability of tsunami in satellite altimetry observations and constructed tsunami-height profiles with only 4- to 5-cm root mean square errors. However, real-time applications are limited by satellite locations at the time of an event (inappropriate observation timing), insufficient observation frequency, and a lack of real-time data processing capabilities.

Storm surges and tsunamis are barotropic waves associated with transport of large masses that load and deform the solid Earth's surface, similar to ocean tides, and thus produce surface displacements, tilts, and gravity signals. For example, large storm surges in the German Bight induce surface displacements of up to 50 mm. The tsunami excited by the 2004 Sumatra earthquake resulted in maximum vertical displacements of the Earth's surface of the order of 20 mm (Plag et al., 2006b) with a deformational signal of the order of 10 mm arriving in Ceylon and India about 20 minutes before the tsunami. These geodetic signals in principle can be measured with GNSS, sensitive gravimeters, and tiltmeters, respectively. While the direct measurement of the vertical displacement at single GNSS sites may be difficult to achieve against the typical noise level of high-resolution GNSS time series, it is likely that GNSS networks can sense the deformational signal, particularly if these networks extend sufficiently far in-land. Free oscillations of the Earth, which have periods in the same range as the tsunami waves, are not expected to create significant problems for the GNSS detection of the loading signals, as the geometric amplitudes of the free oscillations are extremely small and expected to be at the maximum at the 1 mm level. Single gravimeter stations equipped with superconducting gravimeters in principle are able to measure the gravity signal resulting from the sum of the mass relocation and the induced deformation. This application requires procedures which allow the separation of the non-tidal part of the signal in near-real time. For large earthquakes, which generate free oscillations of the solid Earth, the separation of these free oscillations from the loading induced signal in gravity constitutes a particular challenge. The advantage of all signals induced by the loading is that they propagate well in advance of the load. However, most of the available sensors have a relatively high noise level at the time scale important for the detection of an event from minutes to several hours. It is expected that the noise level in displacements determined from GNSS data with high sampling rates (1 to 30 s) can be reduced considerably in the near future through improved processing algorithms. Nevertheless, all these techniques are currently being researched.

Most of the approaches mentioned above require feasibility studies quantifying the tsunami and storm surge signals as well as ambient noise levels. Moreover, development of observational techniques capable of detecting the signals in those quantities that turn out to be promising is required.

The occurrence of devastating tsunamis and extreme storm surges is relative rare. As pointed out in a recent U.K. report (Defra, 2005), only a system used more or less continuously can be expected to be operational in the case of a rare event. Thus, a dedicated tsunami detection system is likely not to be operational in the case of a rare but devastating tsunami, while a multi-hazard and multi-application system is far more likely to ensure continuous operation. GNSS networks with their many applications are therefore well suited for integration in early warning systems for rare events, and their potential for low-latency detection of displacement fields, loading signals, and ionospheric signals should be exploited.

5.2.5 Storm surges

Many of the World's coasts have a long history of disasters caused by storm surges, with large loss of lives and property. Though many of these areas are today protected by advanced systems of dikes, barriers and levies, these protective devices can fail under extreme storm surges. Preparedness for extreme events is low and the resulting disaster can be expected to be of regional scale, affecting the economic development of whole countries or regions. The potential and long-lasting effect of events leading to failure of coastal protections has been sadly demonstrated by the 2005 New Orleans disaster.

There is considerable knowledge and understanding of storm surge hazards for many coastal areas based on those experienced over the last few hundred years. However, recent findings indicate that storm surge statistics based on the observational records might severely underestimate the risk of extreme storm surges exceeding those observed in the last few hundred years. For some coastal areas combined meteorological and hydrodynamic models predict storm surges with considerable accuracy. However, because extreme storm are likely to be underestimated by the current models, it appears reasonable to include detection of extreme storm surges into a monitoring system for tsunami detection.

In coastal regions, there is growing concern about the impacts of hurricanes and other major storms. Although these events are frequent, their courses and landfall positions are challenging to predict before the storm develops, at which point present system do increasingly well. However, precise gravimetric measurements of ocean thermal structure both horizontally and vertically, make it possible to forecast the development and intensity of major storms along various path scenarios dependent on synoptic atmospheric circulation systems and location of air masses in the vicinity of the storm. More precise elevation mapping can better characterize coastal vulnerability to such events. Similar to the case of tsunamis (see Section 5.2.4 above), geodetic observations can also play a role in detecting a moving surge as part of a warning system.

5.2.6 Flooding

River floods in continental interiors lead to devastation of infrastructure, loss of crops, and often loss of life. While it is relatively straightforward to predict the frequency of floods on a statistical basis (100-yr floods, etc.), typical land uses in floodplains (urban, agricultural, etc.) are too valuable for complete and permanent abandonment in anticipation of rare events. Consequently, there is great value in the ability to predict floods using runoff and soil moisture observations upstream of a locality of interest. For major drainage basins, gravimetric techniques (e.g., GRACE, see Chapter 2) can be applied to the monitoring of available and mobile surface and near-surface water masses and their variability in time and space, so that drainage network models can be reliably applied to prediction of flooding in key locations, such as St. Louis (1993) or New Orleans (2005). In the interest of preparing for the impacts of flooding, more precise elevation maps of the soil surface (cm resolution), and ultimately of the water table (m resolution) would be beneficial to disaster preparation and identification of vulnerable areas.

Floods due to failure of natural or anthropogenic reservoir dams can be disastrous. Therefore, geodetic monitoring of major reservoir dams should be considered in order to detect any instabilities at an early stage.

5.2.7 The slowly developing disasters: sea level rise

A slowly changing Local Sea Level (LSL) by itself need not constitute a severe sea level hazard. Many coastal areas cope with secular LSL changes of up to ± 10 mm/yr, and some locations with large extraction of groundwater, oil or gas, with considerably larger rates. In many cases, LSL changes of the order of a few mm/yr are easily accommodated by slow adaptations through coastal engineering. However, slow LSL changes affect the statistics of extreme sea levels and can lead to significant changes in hazards and risks. A recent example is New Orleans, where rapid subsidence combined with a LSL rise increased the vulnerability of the area and contributed to the disaster caused by Hurricane Katrina (Dixon et al., 2006).

Moreover, changes in atmospheric conditions also affect the statistics of the extremes and in particular the maximum sea levels that can be expected in a specific location. Consequently, in assessing the sea level hazards at a given location, scenarios of future LSL on all relevant time scales (for storm surges, tsunamis, and slow LSL changes) will have to be considered.

In the recent past, LSL changes caused by increased ice sheet melting has gained considerable public attention. A rapid melting of the Greenland ice sheet, as considered by Zwally et al. (2002), would have severe global consequence including wide-spread societal impacts due to migration of coastal population. GEO considers secular LSL changes as a potential, slowly developing disaster. The film 'An inconvenient Truth' also focuses on the implications of future sea level rise. The severe consequence of a significant rise in sea level for coastal zones in the world were illustrated by Rowley et al. (2007). As an example, the effect of a regional sea level rise of four meters around Florida is illustrated in Figure 5.2. Day et al. (2007) discuss the potential contribution of a relatively stable sea level after the last deglaciation for the emergence of complex societies. The changes in coastline expected as a consequence of a rapid sea level rise could have the reverse effect and significantly impact the stability of the global society.

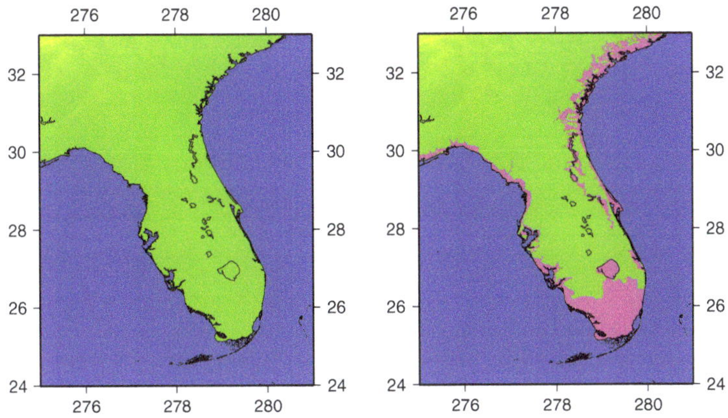

Fig. 5.2. Effect of a regional LSL rise of 4 m on the coastline in Florida. Although a regional rise in LSL of 4 m is not predicted by any of the IPCC scenarios (Bindoff et al., 2007), a catastrophic disintegration of parts of the Greenland or Antarctic ice sheet, as discussed for example by Zwally et al. (2002) for Greenland, could lead to changes of this order of magnitude. Left: present day coast line. Right: Coast line after a regional sea level rise of 4 m.

At any location, the position of the sea surface is determined by a number of processes in the atmosphere, ocean, and solid Earth. Its position with respect to the underlying land surface, i.e., LSL, is the output of numerous Earth system processes acting on a wide range of spatial and temporal scales. For low frequencies, this leads to a complex equation of LSL as a function of the heat and salinity distribution in the ocean, ocean currents and atmospheric circulation, mass changes in the ocean, large ice sheets, continental glaciers, and the terrestrial hydrosphere, postglacial rebound, geodynamic and anthropogenic vertical land motion and geoid changes, as well as changes in shape and extent of the ocean basins (Plag, 2006a, , and Figure 5.3). As a consequence, local and regional LSL changes show large deviations from a global average. Over the last century, a global average rise in sea level of 1 to 2 mm/yr has been determined (see Section 5.5.4). However, in many regions and locations, secular trends in LSL of the order ± 10 mm/y and more are seen.

Understanding sea level variations requires observations from a very carefully designed observing system providing all quantities in a long-term stable reference frame well tied to the gravity field (i.e., the CM). In fact, understanding and predicting LSL changes may be one of the most demanding applications for geodetic observations. Because the sea surface adjusts closely to an equipotential surface of the Earth's gravitational field, the RFO needs to be tied to the CM. A potential secular translation of the RFO with respect to the CM of the order of 2 mm/yr is expected to bias estimates of global sea level trend on the order 0.2 to 0.3 mm/yr (see Section 2.2). However, locally this translation can result in biases of vertical land motion on the order of ± 2 mm/yr and more. In order to reduce the uncertainties, the tie between RFO and CM should be constrained to ± 0.5 mm/yr or better.

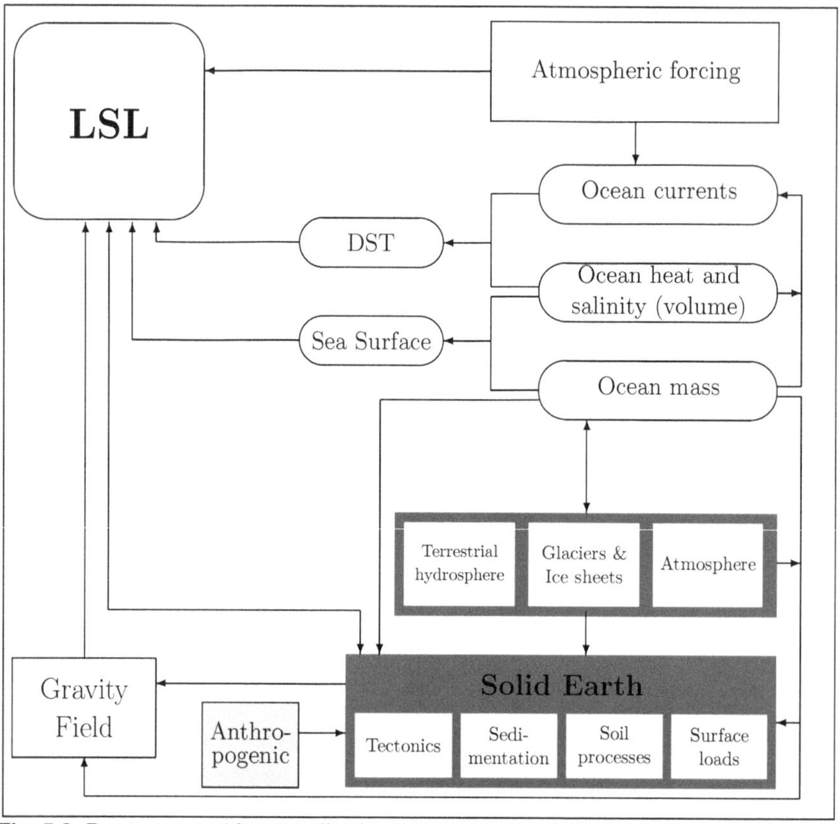

Fig. 5.3. Processes and factors affecting long-period local sea level. Mass movements in the terrestrial hydrosphere (groundwater, rivers, lakes, and reservoirs) and land-based cryosphere (glaciers and ice sheets) and mass exchange with the ocean load and deform the solid Earth and affect the gravity field. The deformations and the associated gravitational changes result in LSL changes, depending on where mass has been relocated. Ocean mass changes as well as ocean volume changes caused by heat and salinity changes affect the sea surface position. Heat and salinity changes also affect the ocean currents and thus change the Dynamic Sea Surface Topography (DST). Atmospheric circulation forces regional wind-driven currents affecting the DST. DST and sea surface changes caused by regional and global processes change LSL in any location. The atmosphere also acts locally on the sea surface and thus changes sea level. Past changes in the ice sheets and glaciers lead to postglacial rebound, which affects sea level through vertical land motion and geoid changes. Tectonic processes in the solid Earth both result in vertical land motion, changes in the size of the ocean basins, and changes in the geoid. In areas where sedimentation takes place, the compaction of the sediments and their load on the solid Earth introduce vertical land motion. Moreover, changes in LSL feed back on the solid Earth and can cause the destruction of peat through oxidation and thus lead to subsidence. Anthropogenic vertical land motion associated with exploitation of groundwater, oil and gas as well as changes in sedimentation can change the Earth surface position. Variations in sedimentation due to river regulation (reduction) or land use (increase) also affect LSL, particularly near river deltas. Figure modified from Plag (2006a).

Scenarios of plausible future trajectories of LSL require realistic global, regional and local assumptions. These include assumptions concerning changes in the global ice sheets, ocean and atmospheric circulation, water storage on land, and local vertical land motion. For present-day, past and future changes in the water mass stored on the continents, the cryosphere and ocean, the fingerprints in sea level (Plag & Jüttner, 2001; Mitrovica et al., 2001, 2009) can be computed using the so-called "sea level equation" (Farrell & Clark, 1976; Milne et al., 1999; Mitrovica & Milne, 2003). These fingerprints of LSL changes induced by mass transports are spatially variable with the local changes exceeding by far −100% of the global average close to the changing load and reaching up to +140% in the farfield. For other contributions such as changes in ocean and atmospheric circulation, a global Earth system model with sufficient spatial resolution is required. Finally, vertical land motion resulting from natural and anthropogenic causes need to be based on observational evidence.

In some coastal areas, anthropogenic subsidence can combine with LSL changes to constitute a severe threat to the coastal population and infrastructure. For example, in the northern part of the Gulf of Mexico, a combination of sediment loading and oil extraction has caused LSL in Galveston to rise nearly 1 cm/yr over the last 50 to 100 years. In Porto Corsini in the Adriatic, excessive groundwater extraction has caused large subsidence of the soil and a local sea level increase reaching peak values of several cm/yr. Another example is the city of Venice and the Lagoon, where pumping of groundwater during the first half of the 20th century led to significant anthropogenic subsidence, which was superimposed on a natural subsidence of the Lagoon due to tectonic and sediment processes. In these cases, monitoring of the Earth's surface with a combination of GNSS station networks and InSAR appears to be capable of providing the accuracy and high spatial resolution required to assess and predict LSL changes.

5.3 Energy Resources: Improving management of energy resources

Traditionally, geodesy has contributed to the exploration of energy sources such as oil and gas resources in particular by measuring gravity anomalies (see Section 4.7.2). Airborne gravimetry (see Section 2.6.4) has contributed substantially to our knowledge of the geographic location of potential resources.

Exploitation of energy resources such as oil, gas and coal is inevitably associated with impacts on the Earth surface and the infrastructure implemented for the exploitation. Space-geodetic techniques allow the monitoring of surface displacements in the vicinity of mines or in areas of oil and gas extraction. These displacements are indicative of the reservoir dynamics and the observations help to enable a controlled exploitation.

Stability of infrastructure such as offshore oil and gas platforms is intimately related to the exploitation of the underlying resources. GNSS measurements on platforms allow the determination of instantaneous subsidence rates, which can be used to regulate the extraction rates. However,the current stability of the global geodetic reference frame is not sufficient to provide reliable velocities on monthly to yearly time scales, particularly for sites far off-shore, where no nearby stable reference sites can be found(Plag, 2005). In cases, where platform settlement is observed, the subsidence of the platform measured by GNSS provides information on the vertical displacement of the ocean floor, which in turn is directly related to reservoir properties (Plag, 2005).

In open-pit mining, the steering of heavy equipment increasingly depends on geodetic techniques (see Section 4.3). Currently, the steering is mainly based on local augmentation systems, requiring a high level of local technical skills. It can be expected that improved access to a global reference frame would reduce the level of local skills required.

The use of renewable energy sources also benefits from geodetic techniques, observations and tools. Efficient management of forests is greatly eased by having precise positioning available, for example, to registered infested trees and to monitor spreading of tree disease. Wind fields are increasingly derived from SAR observations. Mapping tidal and wave energy also benefits from satellite altimetry and SAR, respectively.

The improved capability to measure surface displacement with GNSS and from these observations to deduce strain fields has led to new applications of geodetic techniques related to energy resources. One example is in the area of geothermal energy. Geothermal activity in places such as Iceland and New Zealand is generally associated with magmatic processes and has an evident impact on the surrounding landscape.Non-magmatic geothermal activity, on the other hand, is often much less evident at the Earth's surface, yet its potential as an energy source can be significant.For example, most of the geothermal resources in the Great Basin in the western United States are non-magmatic.Of particular interest is northwestern Nevada, which finds itself tectonically between crustal extension in the East and shear deformation in the West.Deformation that results from tectonic activity can generally be expressed in terms of a strain rate tensor field and can be quantified through the use of geodetic velocities.

A GPS network has been installed in northwestern Nevada to make semi-continuous measurements in order to obtain a station velocity solution that is then used to map the 2D strain-rate tensor field. Crustal strain is correlated with the locations of current geothermal systems and details of the tensor style, and its spatial variation are explored (Blewitt et al., 2003, 2005; Kreemer et al., 2006a). Current findings suggest a strong correlation between locations of existing systems and the level of transtensional strain.Transtensional strain is a combination of shear and dilatation and can be accommodated through strike-slip and normal faulting, respectively. Conceptually, shear strain would continuously fracture fault planes through the entire crust, whereas dilatation tends to pull fault planes apart, allowing pathways for the movement of fluids. The combination of shear and dilatation can thus create fluid conduits to great depths, and sustain them through continued stress.

If geothermal output is controlled at a crustal scale, the geodetic stations need to be separated no further than the thickness of the seismogenic crust (~15 to 20 km in Nevada) to best map crustal strain. Smaller spatial scaling adds redundancy, because the measured strain distribution is the result of slip at depth on a fault that is locked at the surface, and hence is spatially smoothed by an amount related to the seismogenic thickness. Thus geodetic networks can be used to characterize regional strain partitioning, but cannot determine whether partitioning occurs along a single fault.

The network deployed in Nevada consists of stainless steel pins as markers, which allows for the antenna to be re-mounted with sub-millimeter precision at each measurement campaign. Sites are occupied about 30 to 50% of the time. The combination of monument set-up and site occupation history has provided velocities that after 2.5 years of data collection overlap within 95% confidence with those determined at available co-located deep-braced continuous monuments over more than 6 years. The average strain rate in northwest Nevada is about 20 nanostrain/yr. Thus for a network of about 20 km spacing, the differential station velocity is ~0.4 mm/yr. Formal uncertainties in velocity after 2.5 year data collection are an order of magnitude smaller than this suggesting that we can adequately resolve velocity variations on the scale of individual crustal blocks.

Because the aim is to relate geothermal resources to the (often slow) interseismic strain accumulation, InSAR techniques may not be useful as a direct application tool. However, the exploration of geothermal resources often requires substantial water pumping which can affect the local deformation field near existing facilities. InSAR is a very effective tool for quantifying the extent and depth of such deformation. As such it can benefit the interpretation of potential local strain anomalies derived from the GPS measurements.

5.4 Climate change: Understanding, assessing, predicting, mitigating, and adopting to climate variability and change

One of the major consequences of climate change is the propensity of all systems to generate feedback which may operate in concert or in opposition. A clear example of a positive feedback mechanism arose from the anthropogenic depletion of the ozone layer. A reduction in ozone led to weaker absorption of ultraviolet radiation in the middle stratosphere, which in turn led to stratospheric cooling which helped to sustain the heterogeneous processes responsible for destroying ozone, thereby cooling the stratosphere still further.

The ozone loss problem has been largely resolved by banning (or at least reducing) the emissions of chlorofluorocarbons (CFCs), substituting species which tend to break down in the troposphere (hydrochlorofluorocarbons, or HCFCs) which are, themselves, in the process of being phased out. In mitigating the stratospheric ozone loss problem, we have unfortunately contributed to tropospheric warming by replacing ozone destroying catalysts with strong greenhouse gasses.

The long-term effects of altering atmospheric composition, such as elevating the concentrations of greenhouse gases, are only partially understood. Viewed against the backdrop of the large natural variability in the Earth system, it may seem to some as if the relatively small anthropogenic alterations in planetary radiative balance are insignificant and do not warrant mitigation. However, the impacts of sea level rise, melting permafrost, and an increase in extreme atmospheric events such as hurricanes (like Katrina and Rita hitting the USA in 2005, and Gustav and Ike hitting the USA in 2008) or cyclones (like Nargis hitting Myanmar in May 2008) offer sobering reminders of the devastating influence of extreme atmospheric events, and that it would be wise to tread with caution.

Recent research (e.g., Emanuel, 2003; Pielke Jr. & Landsea, 1998) suggests that there might be some causality between hurricane intensity and rising ocean temperatures, this sensitive, they claim, to global warming. Whether warming is due to natural or to anthropogenic forcing is immaterial, but it does emphasize the need for a better understanding of the complexities of the Earth system before further modifying the composition of the atmosphere.

Evidence for climate change has been gathered by scientists in several areas, such as ice caps melting, sea level rise, modification of migrating species habitats, and others. These changes have a strong impact on some particularly exposed communities (e.g. low-level islands). The global society needs to prepare for these proven effects which will eventually concern all of us either via the food chain or via population migration. Identification of future risks is required for proper preparation. Such identification requires that climate change be better understood and quantified, and then to a certain extent, forecast. Forecast cannot be dissociated from the ability to first observe the phenomena associated with climate change.

The attribution of climate trends in the current atmospheric observations is complicated by natural atmospheric variability and large-scale oscillations such as the ENSO or the eleven-year solar cycle. However, reanalyses enable us to infer climate signals from a combination of model and ensemble of measurement systems. It has been suggested that reanalyses would be more robust and reliable than single instrument records. This superiority has emerged only recently as new methods such as variational bias correction are being used to detect and correct instrumental drifts as well as instrument problems, provided some reference observations are available. These climate reanalysis models can be used to simulate the past and can also be used to project future climate to help us prepare for changes.

Critical to that process is the availability of long-term records from single instruments, free of breaks and instrumental biases, and to which the entire data assimilation system runs can be anchored. The instruments that are part of GGOS can provide such observations of atmospheric-induced delays in regions away from the lower boundary (ground) where other effects (urban heat islands, land use changes, etc.) may interfere with the atmospheric trends observed. Measurements of delays in GNSS propagation signals between transmitters and receivers placed in low-Earth orbit can provide such so-called radio occultation measurements (see Section 2.9.1) and thus offer a way to monitor the stratospheric mass field in climate reanalyses.

Geodetic observations are also valuable for the validation of reanalyses. Atmospheric and oceanic mass transport induce signals in the geodetic observations. Mass transport changes the gravity field and, through interaction with the solid Earth impacts Earth rotation. The surface loading associated with the mass transport in atmosphere and ocean loads and deforms the solid Earth. The improved ability to predict these signals will allow validation of climate models based on geodetic observations.

The metric of choice most often used for assessing climate change is the rate of change of atmospheric temperature near the surface, because it is easily measurable and because it controls other environmental parameters. In order to monitor that temperature, particular emphasis has been given to designing instruments and methods to collect measurements with precision of 0.1 K per decade within the suggested climate trends. While it is important to plan and obtain such measurements with the necessary precision, it is equally important to measure the consequence of such trends on the static and the dynamic structures of the atmosphere. As such, magnified effects of climate change are also to be considered.

For a static illustration, referring to the hydrostatic equilibrium of the atmosphere, there is a magnifying effect of temperature change on air density and hence atmospheric layer thicknesses. Assuming for example a 0.1 K homogeneous warming throughout any given atmospheric layer whose boundaries are defined by fixed pressure levels, that atmospheric slab would expand by about 0.04% of its original thickness. In real terms and with a tropospheric average temperature of 250 K, this would amount to raising the mid-latitude near-tropopause level of 200 hPa by 5 m, all other parameters held constant. With GGOS ensuring a reference frame with centimeter accuracy over a decade, positioning upper-air atmospheric pressure *in situ* sensors, accounting also for possible changes in height at the Earth surface in the same time frame, this trend could be identified with high reliability.

The dynamic impact of climate change is reflected in alterations in the patterns of atmospheric circulation. As horizontal temperature gradients change, the cells that make up the general circulation system are affected in their strength and shape (including extent and position). Vecchi et al. (2006) have found evidence of a weakening in the tropical Pacific Walker circulation over 130 years (between 1861 and 1992) based on sea level pressure observations in that region. Using climate models to elaborate on the origin of that decline, they found that anthropogenic changes in the atmosphere could explain the observed decrease in sea level pressure gradient. Similarly, on the basis of climate simulations made at Meteo France for the 4-th assessment report of the IPCC, a weakening of the Hadley cell circulation system is also predicted under IPCC climate scenarios. The total mass of air flowing through the upper branch of that cell at latitude 15°N and between 200–100 hPa pressure levels is currently about 50 Megatons per second in January. A shift in the location of that upper branch of the Hadley cell in the climate runs mentioned here would mean that by January 2030 this atmospheric flow would decrease by up to 5 Megatons per second. With an ability to measure mass displacements, GGOS could help in monitoring such changes. This would complement the atmospheric efforts to measure detailed structural changes such as air density and wind measurements.

Moreover, changes in ocean and atmospheric circulation will affect the angular momentum transfer between ocean and atmosphere on the one hand and the solid Earth on the other. This will affect Earth rotation. Therefore, observations of Earth rotation variations are a data set providing constraints for general circulation models.

Geodesy thus has the potential to bring to climatologists data sets that will help anchor and validate climate models from which forecasts of atmospheric trends can be made for the purpose of preparing for the impact of climate change on the global society.

In terms of monitoring climate change, geodetic observations are pivotal in several aspects. Changes in the dynamic sea surface topography are derived from satellite altimetry observations and can be compared to those changes derived from temperature and salinity data.

Because of the sensitivity of the ice sheets and glaciers to global warming, monitoring of the mass and volume of the Earth's ice bodies are a key activity, also fundamentally relying on the geodetic reference frame and being facilitated by new measurement techniques. Satellite and airborne measurements of ice surface heights by laser and radar provide direct measurements of changes. However, glaciers are very dynamic bodies, and local elevation changes are often a consequence of changing ice dynamics, which is not always representative of larger regions. Therefore repeated large-scale monitoring is required, notably by a combination of laser, radar and gravity satellite missions. Such missions are complementary and will eventually give the full picture of change both on local and continental scales.

To asses the results of such ice monitoring missions, knowledge of crustal uplift associated with melting ice sheets is needed. Such assessment requires data from permanent GNSS stations and repeated absolute gravity measurements from networks spanning wide zones around the ice sheets (Wahr et al., 1995; Plag et al., 2007c). Currently such uplift models are the limiting factor for gravitational change monitoring of Antarctica, whereas the melting of the margins of the Greenland ice sheet is clearly demonstrated with GRACE.

Geodesy is fundamental in monitoring sea level changes, one of the most serious impacts of climate change. Global sea level changes are derived from satellite altimetry observations, which pose the most stringent requirements on the stability of the geodetic reference frame (e.g., Blewitt et al., 2006a). Projected scenarios of local and regional sea level rise provide a basis for planning of adaptation strategies, but require detailed understanding of trends in oceanic and solid Earth contributions. The latter poses high demands on the tie between RFO and the CM.

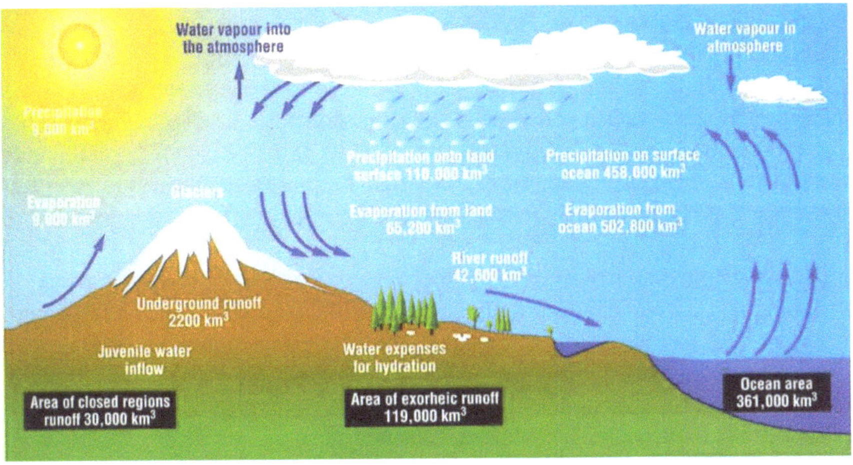

Fig. 5.4. The large-scale features of the global water cycle. Numbers are fluxes in km^3.

5.5 Water: Improving water resource management through better understanding of the water cycle

5.5.1 The global hydrological cycle

Earth is a unique, living planet due to the abundance and vigorous cycling and replenishing of water throughout the global environment. The water cycle operates on a continuum of time and space scales and exchanges large amounts of energy as water undergoes phase changes and is moved from one part of the Earth system to another. Water is essential to life and is central to society's welfare, progress, and sustainable economic growth. However, global water cycle variability which regulates flood, drought, and disease hazards is being continuously transformed by climate change, erosion, pollution, salinization, and agriculture and civil engineering practices. The water cycle delivers the consequences of climate change while responding directly to the drivers of that change. The most visible manifestation that could be expected from climate warming would be changes in the distribution of precipitation and evaporation, and the exacerbation of extreme hydrological events, floods and droughts. From both scientific and practical perspectives, the key question is whether projected climate change will entail significant changes in the Earth's global water cycle.

The water cycle plays the following key roles in the Earth system:

- Water exists in all three phases in the climate system and the phase transitions are a significant factor in the regulation of the global and regional energy balances.
- Water vapor in the atmosphere is the principal greenhouse gas and clouds at various levels and composition in the atmosphere represent both positive and negative feedback in climate system response to anthropogenic perturbations —

hence the water cycle and its dynamics represent a major source of predictive uncertainty about global change.

- Process such as ocean, ice-sheet, soil moisture, and groundwater dynamics represent the slow water cycle components that form the basis for understanding and predicting global and regional climate, while processes such as precipitation, cloud dynamics, water vapor, and evaporation represents the fast components of the water cycle and forms the basis for prediction of hydrological extremes.
- Water is an excellent solvent and global biogeochemical and element cycles are mediated by the dynamics of the water cycle.
- The variability and changes in the global cycling of water is linked to variability and changes in cycling of carbon, methane, nitrogen, and other nutrients at regional and global scales.
- In total, water is the element of the Earth system that most directly impacts and constraints human society and its well-being.

Despite the fundamental role of the coupled water and energy cycle for the Earth system, the knowledge of key quantities is still associated with large uncertainties (e.g., Dooge, 2004; U.S. Climate Change Research Program, 2007). For example, the fluxes between the main reservoirs in the global water cycle published in literature over the last three decades (for an example, see Figure 5.4) have changed considerable indicating potentially large uncertainties in these numbers. In particular, fluxes between terrestrial surface waters, cryosphere, and groundwater are largely unknown. The determination of the continental water storage in space and time is not possible nowadays with sufficient accuracy. However, as discussed in Section 2.6.5, the gravity missions particularly if combined with changes in Earth's geometry and rotation have already provided new insight in monthly and submonthly changes in continental water storage, and a continuation of these missions is likely to provide a monitoring of these changes on spatial scales down to a few hundred kilometers and temporal scales down to a few days.

5.5.2 Water for life: the challenge of water management

The importance of natural resources to modern society has never been greater, nor have resources ever been more threatened by global change, human population increase, and anthropogenic activity in general. The importance of the management of natural resources is probably best illustrated by the example of water. In many areas of the world, current demands exceed the supply (as indicated by the water scarcity index, Figure 5.5), and water has to be transported over great distances. This situation is expected to become more severe over the next decades (e.g., EEA, 1999; Lawford & the Water Theme Team, 2004; Bernasconi & others, 2005; Oki & Kanae, 2006; United Nations, 2006).

Clean, fresh water is arguably the most important resource to human society, as it controls our ability to produce sufficient food to support the burgeoning human population. Usable water resources reside in lakes, streams, artificial (dammed) reser-

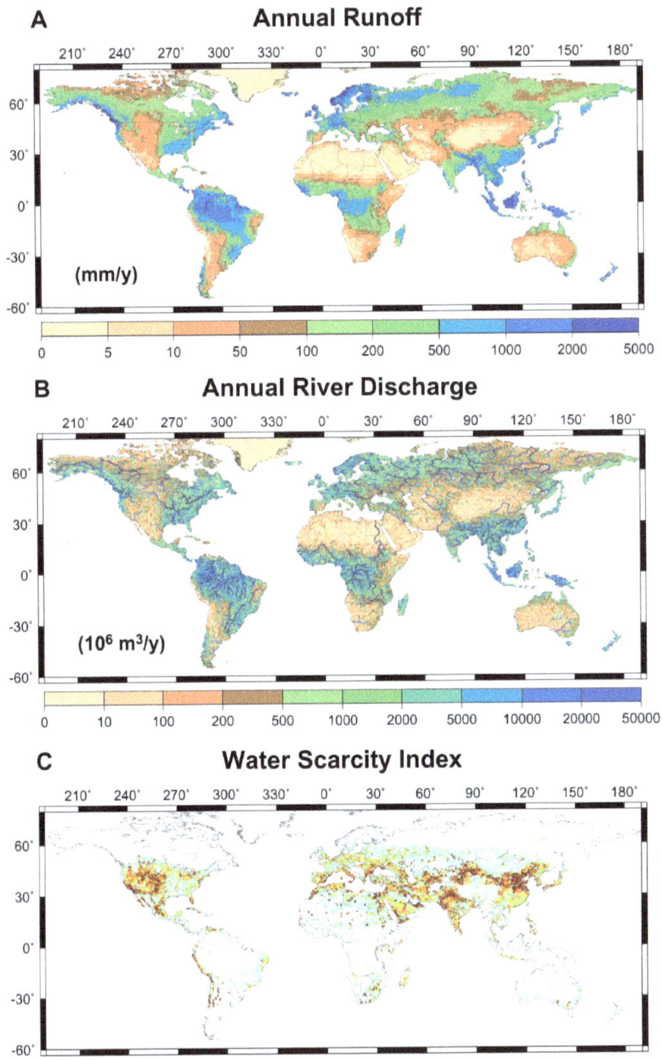

Fig. 5.5. Earth's water resources: relation of supplies to demands. From Oki & Kanae (2006). Reprinted with permission from AAAS.

voirs, and groundwater. Of these, groundwater represent the greatest volume, and is also the most vulnerable to long-term contamination. The level of highly variable internally draining lakes must be consistently monitored in order to track changes in available irrigation water as a result of surface water diversion (commonly for irrigation) in areas such as the Aral Sea and Lake Chad. Further salinization of surface waters, rendering them useless or irrigation and other uses should be monitored so that water use policy upstream can be more effectively developed.

The discharge of rivers into the global ocean controls coastal zone water chemistry and ecosystem function, yet is highly modulated by water use throughout the drainage basin. Geodetic observations that track river stages globally can complement river stage gauges and discharge stations to follow changes in water utilization as well as provision of fresh water to the coastal zone. This information would be useful to those concerned with water resource depletion by one state or nation before it reaches the region or coast of another. In many cases, flow control through dams in river systems is the subject of controversy between those upstream and downstream. Reservoir levels can readily be monitored using geodetic techniques to inform and support short and medium-term resource planning.

The vast majority of liquid fresh water on the planet resides underground, and is easily accessible through wells. However, in semi-arid to arid regions, where the stress on water resources is most acute, aquifers do not recharge at a significant rate relative to rate of withdrawal. Utilization of such water resources is considered "mining" as this water is a non-renewable resource. Consequently, water tables drop, and the aquifers are assigned limited lifetimes before depletion. The changing mass distribution due to water withdrawal, whether in a confined aquifer (leading to land surface subsidence), or unconfined aquifer (leading merely to lowering of the water table and deepening of the unsaturated zone), can be detected geodetically, and can provide global assessments of groundwater alterations, previously unavailable due to either lack of data, unwillingness to share such information, or the impracticality of concatenating thousands or millions of local to regional aquifer reports.

The water crisis is largely a crisis of governance (United Nations, 2006), brought about by water management obstacles such as sector fragmentation, poverty, corruption, stagnated budgets, declining levels of development assistance and investment in the water sector, inadequate institutions and limited stakeholder participation, but the lack of detailed knowledge of the global water cycle from local to global scales is contributing and enforcing this crisis. Therefore, Earth observations can improve the knowledge base and thus help to mitigate this crisis. As discussed in Section 2.6.5, on regional to global scales, the mass transports observed by GGOS are already improving the database concerning the motion of water through the hydrological cycle, and future combined analysis of the variations in Earth's gravity field, shape and rotation will help to reduce the uncertainties.

5.5.3 *Observations of the Global Water Cycle*

The path forward for observing the global water cycle must be based on integrated observations, as opposed to isolated observations that focus on a single flux or state. The components of the water cycle need to be measured simultaneously in order to allow the estimation of fluxes between the components of the climate system. Precipitation and evapotranspiration over land and ocean surfaces require that the state of the system at the surface and in the atmosphere be monitored simultaneously. By integrated observations, we mean the simultaneous retrieval of related water cycle variables. From a technical perspective, this suggests a satellite platform with sensors for multiple frequencies, combining passive and active sensors, and perhaps LIDAR. From an Earth science perspective, the water cycle variables and the required spatial and temporal observation requirements to satisfy the science and applications should drive the sensor package and not (as is often done traditionally) the inverse.

It would be most useful to develop the water cycle observational perspective considering that the water cycle can be divided into slow and fast branches. The 'slow branch' would consist of measurements relevant to the retrieval of soil moisture, groundwater, snow and ice, freeze-thaw states, ocean dynamics, ocean salinity and perhaps water body extent and river discharge. These components do not have a regular diurnal cycle. The 'fast branch' would consist of precipitation (liquid and solid), evapotranspiration, clouds and water vapor. The dynamics of the components can vary significantly within a day. Table 5.3 provides a summary of the measurement requirements for a complete monitoring of the water cycle and the capability to retrieve fluxes at interfaces of the land, atmosphere, and ocean components of the water cycle.

Table 5.3. Key variables required for monitoring the Earth system water cycle and fluxes.

Variable	Role in the Water Cycle	Orbit	Measurement Horiz. Spatial Resolution	Measurement Temporal Revisit
Precipitation Rate/Type	Diabatic heating, surface forcing	GSO	1-5 km	0.5 Hrly
Soil Moisture	Link water, energy, biogeochemistry	LEO	1-10 km	Daily
Surface Freeze/Thaw and Sea-Ice	Climate, Carbon cycle, Ocean Dynamics	LEO	0.1-1 km	Daily
Open Ocean and Coastal Salinity	Density flows in Oceans	LEO	10 km	Weekly
Snow Cover Extent	Surface energy balance	LEO	0.1-10 km	Daily
Snow Water Equivalent	Water storage dynamics	LEO	0.1-10 km	Daily
River and Lake Elevation	Water transport, biogeochemistry	LEO	0.1 km	Daily
Water Vapor	Water and energy transport	GSO	5-10 km Horiz.; 0.5 km Vert.	0.5 Hrly
Cloud Properties	Water and energy transport, radiation balance, precipitation genesis	GSO	1-5 km	0.5 Hrly
Land and Sea Temperature	Energy balance	GSO	1-10 km	0.5 Hrly
Ocean Height	Ocean currents and vertical mixing	LEO	10-100 km	Daily
Evaporation (Land and Ocean)	Water, energy, and carbon cycle	LEO	1-10 km	0.5 Hourly
Ocean Salinity	Ocean currents and vertical mixing	LEO	10-100 km	Daily
Water Quality	Environmental and human health	LEO	0.1-10 km	Daily

From an observational and technological perspective, the 'slow branch' would be observed from LEO, with the technological focus primarily on improved spatial resolution and 'fast branch' having the technological challenge of obtaining the needed resolution at a GeoStationary Orbit (GSO). Below we lay out some challenges that should be met over the next decade or so.

5.5.4 Slow branch challenges

The challenge for observing the slow branch of the water cycle is to progress from single-variable isolated water cycle instruments to multi-variable integrated water cycle instruments. It is most likely that the desired integration will progress along the lines of extending and integrating the capabilities of sensor technologies in given electromagnetic band families. For example, we have several current sensors that make observations in different microwave wavelengths - our suggestion is that we progress towards integrating these capabilities into fewer more capable instruments that can simultaneously observe multiple components of the water cycle. Because the slow branch of the water cycle generally changes on timescales longer than 1 day, LEO orbits are appropriate. Below we summarize a few of the current and planned measurements of the slow branch of the water cycle that should be progressively integrated over the next few decades.

Soil moisture and freeze-thaw state: Given its critical role in the terrestrial hydrosphere as the 'regulator' between the water and energy cycles, it is clear that improvements in weather and seasonal climate forecasting will depend on improved soil moisture observations. Soil moisture (including its freeze/thaw state) is a key variable that links the water, energy, and biogeochemical (carbon, nutrients, and elements) cycles. It has long been recognized that this state of the terrestrial hydrosphere needs to be monitored at high resolution and with good fidelity in order to make significant advances in Earth system science as a whole. It is expected that the availability of soil moisture data will link the science communities in water, climate-energy, and biogeochemistry. In addition, soil moisture serves as the memory of the terrestrial hydrosphere and it has significant impacts on operational weather and seasonal predictability of the climate system. Further, there are long-standing needs of these primary observations for decision makers, especially in the area of drought and flood management. Currently soil moisture is being estimated from the NASA Aqua AMSR-E sensor at 10.7 GHz, with a nominal resolution of about 50-km and posted at a 25-km spacing based on over-sampling. However, the high microwave frequency is severely limited by low penetration depth and scattering by vegetation. A soil moisture mission providing a 40-km product using radio-brightness measurements from 1.4 GHz (L-band), a 3-km product based on an active L-band sensor and a combined passive-active 10-km product is possible. Airborne campaigns have definitively shown that combination of sensors and frequencies are optimal for soil moisture monitoring.

Seasonal snow: Snow plays two important roles within the terrestrial water cycle. Its cover and seasonal duration provides significant albedo contrast that has been shown to affect hydrological and climate variability at global scales, and its amount is a fundamental source of moisture that transfers wintertime precipitation into spring and summer soil moisture and river discharge. Its measurement is synergistic to soil moisture. Since radio-brightness measurements at higher frequencies are used in the retrieval of snow, the same antenna used for soil moisture will provide higher resolution for the higher frequencies, perhaps leading to improved retrievals in areas with large terrain. Therefore there is a natural synergy between soil moisture and snow. The low frequency (L-band) active radar is the observational sensor of choice regarding freeze-thaw state, and is also synergistic to the measurement needs for soil moisture.

Surface water extent and hydraulics: The ability to measure, monitor, and forecast the supply of fresh water, is of high-priority and concern to GEO, the World Climate Research Programme (WCRP) strategic framework 2005-2015 Coordinated Observation and Prediction of the Earth System (COPES), and the relevant UN agencies. The seasonal extent of wetlands and the extent of flooding of tropical rivers like the Amazon is a critical variable for understanding the biogeochemical cycle within the earth system. Currently estimating the discharge of highly braided Arctic rivers is problematic, yet critical to the understanding of the changes being observed in the Arctic and their predictive consequences. At present, the satellite-based measurement of surface water extent and stage has occurred through 'observations of convenience' from altimeters designed for ocean applications. Nonetheless, these measurements have demonstrated that monitoring water bodies from space is feasible and the information provides critical new insights into the terrestrial hydrological cycle in environments not well-monitored *in situ*. Additionally, it is the only feasible measurement platform to provide consistent, global measurements. The technological challenge is to provide an antenna-sensor package to image water bodies with an intrinsic resolution on the order of ten meters, so that both its elevation and slope can be estimated, allowing for both water extent and (for rivers) discharge estimation. A design based on interferometric Ka-band radar with a baseline of $\sim$10 m would allow for these retrievals. Such a satellite system is needed to determine the spatial and temporal variability in freshwater stored in the world's terrestrial water bodies – a most critical water cycle issue.

Groundwater mass: A promising measurement concept is the gravimetric determination of changes in groundwater storage, based on extremely precise observation of time-dependent variations in the Earth gravity field, using space-based gravity gradiometer systems. Groundwater constitutes ninety-six percent of Earth's unfrozen fresh water (Shiklomanov, 1993 Shiklomanov, 1993). It is a vital resource which provides for irrigation, industry, and domestic usage. In many parts of the world it is being depleted due to unsustainable rates of pumping, which may lead to future conflicts and human hardship. Groundwater varies slowly relative to soil moisture, surface water, and non-permanent snow cover, but it is dynamic on seasonal to interannual timescales (Alley et al., 2002). Indeed, changes in terrestrial water storage, particularly groundwater storage, have contributed to observed changes in sea level during the past two decades (Milly et al., 2003; Sahagian, 2000). Quantifying groundwater storage variations is critical for improving large scale water balance assessments (see also above, Section 5.5.2). Groundwater maintains streams between storms by supplying baseflow, and, with soil moisture, it determines the infiltration to runoff ratio and thus the timing, duration, and intensity of floods. Groundwater also feeds back to atmospheric processes and the carbon cycle by enabling phreatophytes to continue to transpire during droughts.

Despite its importance, groundwater's natural variability and vulnerability to overproduction and climate change have not been adequately characterized by the scientific and water resources communities. It is often assumed that over the course of a year, a zero net change in groundwater storage will occur. In fact, the interannual variability of aquifer storage can be substantial, of the same magnitude as root zone soil water storage variability (e.g., Eltahir & Yeh, 1999; Rodell & Famiglietti, 2001; Seneviratne et al., 2004). Groundwater may be an important indicator of natural and human induced climate variations, if the effects of pumping and injection can be removed.

Inadequate monitoring, political boundaries, and the absence of centralized, digital archives of measured groundwater levels have restricted the number and quality of aquifer storage and flux assessments, even in developed nations. Indeed, two major conclusions of National Research Council (2004) were, 1) "*Our ability to quantify spatial and temporal variability in recharge and discharge is inadequate and must be improved given the importance of groundwater in the hydrological cycle, the contribution of groundwater to base flow in streams and inflow to lakes, and society's reliance upon groundwater for water supply*", and 2) "*The roles of groundwater storage, and recharge and discharge fluxes in the climate system are under-appreciated and poorly understood*".

Geodetic measurement systems can be valuable to groundwater resources assessments and scientific investigations because they enable data to be obtained through non-destructive means (i.e., without digging). In particular, satellite based monitoring of Earth's time variable gravity field has the potential to revolutionize the study of hydrology providing global observations of water mass redistribution. Whereas the current generation of radar and radiometer based remote sensors only provide data on water stored in the upper few centimeters of the soil column, satellite gravimetry has an unlimited penetration depth.

The GRACE mission (see Section 2.6.5), which is the first twin satellite gravimetry mission, is now being used to generate time series of total terrestrial water variations, among other applications (e.g. Tapley et al., 2004a). Rodell & Famiglietti (2002); Rodell et al. (2006) have shown how groundwater storage variations can be isolated from the GRACE derived water storage fields. However, the resolution and accuracy of the gravimetry technique must be improved before it is fully embraced by the hydrological community. The sensitivity of this first demonstration of "photon-less" remote sensing is expected to allow detection of changes in mass distribution equivalent to ± 1 cm variation in water storage over a 500×500 km^2 area. Current GRACE measurement uncertainties are on the order of 10 kg H_2O/m^2 (1 cm equivalent height of water) over a 750,000 km^2 region, and they degrade rapidly as the spatial resolution increases (Wahr et al., 2006). While this is sufficient for many large scale hydrological and oceanic investigations, most water resources, meteorological, agricultural, and natural hazards applications require higher resolution data. Furthermore, GRACE launched in 2002 with an expected lifetime of nine years, while climate variability assessments require a longer, nearly continuous record. This emphasizes the importance of developing a follow-on gravimetry mission with advanced technology to increase spatial resolution while decreasing uncertainty. The monthly temporal resolution of GRACE is an issue for many applications, but it should be sufficient for groundwater assessments. Moreover, recently introduced new analysis methods of GRACE data have yielded submonthly temporal resolution (see Section 2.6.5 and Luthcke et al., 2006).

Cryosphere: Ice in the boreal and polar latitudes shows significant interannual variability in the instrumental record. The ice cover has a significant effect on surface albedo and hence it is a source of diabatic heating anomalies on large scales. The ice is also a significant insulator, and subsurface thermal regime and heat fluxes into the atmosphere are affected by variability in seasonal ice cover. Recent studies in atmosphere and ocean dynamics have demonstrated that sea ice could be a significant source of interannual memory in the climatic system. Its extent is also a major determinant of polar amplification of global change. Major melting of polar cryosphere could also be the trigger for catastrophic climate change if the ocean thermohaline circulation is disrupted by major flow of freshwater into polar saline oceans. Paleoclimate indicators show that the ocean thermohaline circulation can change regimes if freshwater inputs are dramatically changed — an example is the freshwater input into the North Atlantic due to the break-up of the Laurentide ice sheet. All these considerations indicate that the monitoring of the cryosphere is important for understanding and predicting the role of the water cycle on the Earth system. The age of ice (first-year versus multi-year ice) and ice extent at high-resolution need to be estimated. New multi-frequency active and passive microwave systems will be required in order to overcome the confounding effects of snow cover and melt pools that limit current capabilities.

The contribution of geodesy to the investigation of ice sheets, glaciers, and sea ice is discussed in Chapter 2 and in Section 3.4, and here we consider the relevance of such investigations to society. Since the mid 1990s, new observation techniques have shown that nearly all ice on Earth is undergoing rapid change: Arctic sea ice is shrinking, both in extent and thickness; low-latitude glaciers and ice caps are losing mass at rapidly accelerating rates; and even parts of the vast ice sheets in Greenland and Antarctica are shrinking (although there is considerable uncertainty in mass balance, especially for Antarctica). Although some of these changes may simply represent natural variability that we are only seeing now because of the new observation techniques, some undoubtedly represent recent changes resulting from substantial local warming. These changes have clear societal impacts: shrinking low-latitude glaciers affect tourism and local water availability; shrinking Arctic sea ice affects regional weather, climate, and living conditions for Arctic flora and fauna, including humans, and potentially opens new ship routes; and increasing losses from glaciers and ice sheets certainly affect the rate of sea-level rise, and potentially affect ocean circulation patterns, and hence global climate. Thus, apart from the scientific motivation to improve understanding of our planet, there are important practical reasons for monitoring the behavior of these ice bodies.

As discussed in Chapter 2 and Section 3.4, the recent improvement in our knowledge of what is happening to ice on Earth results largely from remote-sensing measurements from aircraft and, increasingly, from satellites. This includes the measurement of parameters that are of obvious relevance to glaciology, such as ice-surface elevation and velocity and ice thickness, but also measurements of less obvious relevance. These include measurements of temporal changes in Earth's gravity field, which give insights into the rates of change of the mass of entire ice sheets, estimates of the rate of sea-level change from tide gauges and satellite altimeters, and estimation of the rate of crustal motion beneath the ice. All of these measurements share a heavy dependence on geodesy to provide an accurate framework within which to set the measurements, and on very accurate aircraft and satellite trajectories to ensure that the accuracy of the measurements and their locations within the framework are known, with a good understanding of associated errors.

There are three ways to measure the mass balance of the big ice sheets in Greenland and Antarctica: comparison of total snowfall with total losses; measurement of volume changes, using altimetry of the ice surface; and measurement of temporal changes in gravity, indicative of mass changes. Of these, all but the first require correction for changes in the elevation of rock beneath the ice. This is particularly so for interpretation of gravity changes, because rock is so much denser than ice. By necessity, the required estimates of crustal motion beneath ice sheets come from models, which become progressively more reliable as more information becomes available on actual vertical motion to constrain the models. This in turn depends heavily on highly accurate geodetic measurements.

In addition to the approaches described above for measuring ice-sheet mass balance, changes in length of day and in the direction of the Earth's rotation axis also reveal mass redistribution. Precise geodetic observations of these changes are valuable constraints for mass redistributions on regional to global scales, particularly if combined with observations of gravity changes and surface displacements (see Section 2.6.5).

Earth's climate is changing, with temperatures increasing almost everywhere. In most regions, the increase is slow and accompanied by periods of cooling, but in others, such as the Antarctic Peninsula and parts of Greenland, it is remarkably fast. Ice is responding to these changes, but it also has the potential to affect them. As the spatial extent of Arctic summer sea ice shrinks, it is replaced by dark ocean, capable of absorbing far more solar radiation than the sea ice, and thus amplifying the warming. More subtly, drainage of meltwater from shrinking glaciers and ice sheets, affects the density "layering" of the sea, and this affects the deep ocean currents responsible for transferring enormous amounts of heat from low to high latitudes. Possible consequences are the subject of ongoing research, and there is little agreement on whether they may be serious, but in view of the very large amounts of heat transferred by the ocean, this is an issue that cannot be ignored. Required research will include a detailed monitoring of Earth's ice.

The reservoirs of ice on Earth are vast enough to raise sea level by 55 m if all ice melted. Although this would take many thousand of years, it is clear that even the melting of just a small percentage of this ice could be catastrophic for tens of millions of people and impact hundreds of millions, with enormous costs both to global finances and to global security. Continuing to monitor ice on Earth, as well as ocean density, structure, and continental water impoundment distribution will play a key role in future studies of sea-level change, and this in turn requires continuation and enhancement of geodetic measurements.

Climate change and global ecology: Climate affects all life on Earth, with small changes having serious impact on some life forms, some of which may play important, and perhaps yet unrecognized, roles in processes affecting human welfare. Thus, even those humans with little interest in global ecology would be well advised to preserve ecosystems and the environment that sustains them to the extent possible. Already, climate change is affecting many aspects of global ecology, including: growing season for crops and forests; fish and sea mammal migrations; patterns of insect migration, with associated effects on the distribution of diseases; and the viability of polar bears in a world of diminishing summer sea ice. On a more frivolous, yet of local economic concern level, climate change is affecting our recreation habits, with ski slopes closing from lack of snow, and fewer glaciers to visit as they retreat to higher and higher elevations. Far more significant than impacts on tourism, such changes have the very serious effect of substantially altering the timing and amount of meltwater available to sustain nearby agriculture.

Sea-level change: Results from the measuring techniques largely made possible by accurate geodesy show that changes in ice mass since the mid 1990s are responsible for ~30% of a total sea-level increase of > 3 mm/yr, and that this contribution is increasing with time. Clearly, as a progressively increasing percentage of humanity shifts to homes near the coast, this is a source of increasing concern. Only a decade or two ago, it was generally accepted that sea level was rising by about 1 mm/yr over the 20th century, and, although this was probably an under-estimate (1 - 2 mm/yr may be a better average), we are now experiencing triple this rate. Some of the observed increase has been caused by ocean warming, with the remainder likely to be caused by melting ice. There is a third component in the sea level equation that may have been confounding attempts to project future sea level rise from the observed 20th century sea level records from tide gauges. This component stems from direct human activities that transfer water between reservoirs (e.g. between continents and ocean). While some activities such as groundwater mining, deforestation, and surface water diversion serve to transfer water from continents to oceans, thus increasing the rate of sea level rise, these contributions were counteracted and possibly overwhelmed in the 20th century by the construction of new dams and water impoundment on the continents (Sahagian et al., 1994; Chao, 1994). With dams "holding back" water at a rate of at least 0.5 mm/yr sea level equivalent (and perhaps up to 2 mm/yr), tide gauge measurements over the 20th century would not have reflected the entire contribution of ocean warming and glacial melting. As such, if the construction of new dams in the 21st century is not continued at as great a rate as it was in the 20th (as it is not expected to for various reasons), we should expect to see an increase in observed sea level rise, as indeed, we are already seeing.

Ocean height: The data from the early ERS and TOPEX/Poseidon ocean height missions provided oceanographers with unprecedented means to constrain the models of ocean circulation. They transformed the discipline and allowed oceanographers to model and predict case situations in ocean climate. The success of the early missions motivated follow-ons, including the current Jason missions. These ocean height missions provide data at fairly coarse resolution. Development of future capability to perform high-resolution and high-repetition mapping of ocean height will enable oceanographers to address the ocean weather challenge in support of coastal hazards and biogeochemical cycle applications.

Ocean salinity: Surface ocean salinity affects the density of surface waters and the extent of vertical mixing. The vertical mixing is a significant determinant of ocean heating gradients and circulation. The vertical mixing rate across the oceans is also a significant factor in the biogeochemical cycles. The Aquarius Earth System Science Pathfinder (ESSP) is scheduled to provide open ocean salinity measurements. Follow-on capabilities should include sensors that can map coastal waters at high resolution. Mixing in the coastal zone has significant implications for water quality monitoring and biogeochemical cycles science.

5.5.5 Fast branch challenges

In addition to the integration of sensor technologies to enable simultaneous multi-variable observations of the fast branch of the water cycle, we are also challenged with providing observations at a sufficiently high temporal resolution that processes such as storms can be tracked. The ability of the Geostationary Satellite Server (GOES) to track water vapor from a GSO should motivate the extension of these capabilities to precipitation, cloud properties and highly accurate measurements of air temperature, humidity, clouds, and surface temperature: A GSO version of Global Precipitation Measurement (GPM), CloudSat, Calipso and Atmospheric Infra-Red Sounder (AIRS) . Without such sensor systems, it is unclear whether the critical advances in cloud resolving parameterizations, and subsequent advances in weather forecasting, can be achieved. While studies that evaluate the trade-offs between temporal and spatial resolution need to be carried out, we need to become creative in developing technology to achieve these goals. Below we summarize a few of the current and planned measurements of the fast branch of the water cycle that should be progressively integrated and moved to geostationary platforms over the next few decades.

Precipitation: The experimental Tropical Rainfall Measuring Mission (TRMM) demonstrated the capability to combine the advantages of the active and passive microwave techniques for precipitation observation. Precipitation radar-data can be used to better constrain the cloud models incorporated in retrieval algorithms, thereby considerably improving the accuracy of retrievals based on passive microwave observations only. The seminal work of the TRMM team forms the scientific and technological basis for a global precipitation measuring-system, combining observations from at least one active precipitation radar in inclined orbit, a constellation of several (6-8) passive microwave imaging radiometer-spacecraft in staged polar orbits, and surface-based rain gauges. This GPM mission constellation concept , together with more detailed characterization and improved modeling of cloud structure and properties, constitutes the best currently feasible approach for quantifying the rate of the global water cycle. GPM is one of the next generation of systematic measurement missions that will be launched around 2010 by a consortium of international space agency partners.

Water vapor and clouds: Water vapor in the lower troposphere directly impacts precipitation forecasts, and water vapor (principally) in the upper-troposphere is the largest contributor to the atmospheric greenhouse effect. NASA has made a major scientific and technological investment in the development of the experimental AIRS instrument. Imaging multispectral radiometers, such as the Moderate-Resolution Imaging Spectroradiometer (MODIS) on Earth Observation Satellite (EOS) Terra and Aqua provide measurements of a variety of basic water and energy cycle variables, from sea- and land-surface temperature to cloud amount and optical properties and radiation fluxes. In addition, the planned experimental missions Cloudsat and CALIPSO will provide measurements of global cloud properties and their vertical structure. CloudSat is designed to measure the vertical structure of clouds and precipitation from space. A measurement and algorithm approach is used that combines radar information with radiance data obtained from other sensors of the EOS constellation. Information derived from this combination includes detailed vertical profile information about the water and ice contents of clouds, the occurrence of precipitation and quantitative information about precipitation (solid and liquid precipitation are readily detected by 94 GHz radar). CloudSat will provide new knowledge about clouds and precipitation and the connection of clouds to the large-scale motions of the atmosphere, offering tests of global climate and weather forecast models as well as cloud resolving models and related parameterizations. Finally, the AIRS/Advanced Microwave Sounding Unit (AMSU)/Humidity Sounder for Brasil (HSB) instrument suite observes surface temperature, cloud fraction, cloud top pressure and temperature, profiles of atmospheric temperature and water vapor, plus a rain flag. All are directly or indirectly relevant to the hydrological cycle. Challenges remain, however, before the AIRS observations reach their full potential for forecasting precipitation events.

Evaporation: Evaporation from continents and ocean surfaces serves as the crucial link between the surface water and energy budgets. To date, efforts have aimed to provide estimates of ocean evaporation from remotely-sensed data of the Special Sensor Microwave/Imager (SSM/I) measuring near-surface humidity and winds. However, the veracity and utility of these estimates are limited by the quality of the retrieved data, the absence of other crucial data (such as near-surface temperature), and assumptions regarding the algorithmic formulas. Thus improvements can be made in two ways, through improved satellite data (such as QuickScat which could provide improve near-surface winds and Aqua with higher quality humidity retrievals) as well as the advancement of existing algorithms that use not only current space-based instrument measurements but also adapt to forthcoming satellite retrievals. The feasibility of measuring continental evaporation via remotely sensed data (from the SSM/I and the Advanced Very High Resolution Radiometer (AVHRR)) has been demonstrated. However, most of these pilot studies are limited in space and time and therefore currently possess no capability to globally estimate continental evaporation. Our only current capabilities for providing global estimates of continental evaporation rest upon global land modeling efforts - such as the Global Soil Wetness Project (GSWP) and the GLDAS. A two-pronged effort is needed to improve our capabilities to remotely sense evaporation. First, the surface and near-surface atmospheric quantities which are required as input for algorithms must be advanced, either by revisiting previous data or by exploiting data from future missions. This not only includes data used for algorithmic expressions, but also in-situ data that can verify the veracity of the remotely sensed measurements. Equally important is to revisit previous efforts and advance them through further development of the theoretical framework upon which the retrieval algorithms are based.

Innovative technology solutions: There are potentially many solutions to the above challenges, but long-term technological development is needed. Taking the required resolutions from Table 5.3, we need to convert these into antenna (and related hardware) requirements, and develop a technology strategy with end dates related to space demonstration. Below three areas are briefly presented:

- **Carpet Sensors:** One potential area for innovative technology is the development of 'carpet satellites'. Here a 'mat' of 9 to 16, small-scale sensors would fly in formation-like a sensor mat-resulting in an effective large-scale antenna, not unlike the 2-D array of the ESA Soil Moisture and Ocean Salinity (SMOS) stick antenna. Location among the sensors would be done through laser ranging, and the result would be an effective antenna perhaps a few kilometers in size. Different sensors may be on different micro-satellites, with a 'central' satellite having a more complete set of sensors than on some of the other components. This idea is perhaps the most innovative (risky) and a multi-decadal effort of Observing System Simulation Experiments (OSSE) studies and demonstrations would be required, but it would offer solutions that otherwise may be totally elusive to Earth observations.
- **Technology for GSO observations:** Table 5.3 lists measurements at GSO that are new and difficult, such as precipitation or cloud properties. Through OSSEs, the observational and sensor requirements can be developed so that the technological challenges can be laid out. If large antennae cannot be deployed at GSO, sensor carpets might be an alternative. The spatial-temporal trade-off for precipitation measurements at GSO versus LEO will determine how this technology can

contribute, although it is often stated that these measurements at GSO cannot be done, but this must be re-examined continuously.

- **Spaceborne lasers:** We must continue technology solutions to improve spaceborne lasers. This is critical for a number of science needs, including improved ranging, altimetry, and measuring chemical constituents, including CO_2. For example, the development of accurate laser ranging and signal processing could enhance future GRACE-like missions, as well as provide interferometric observations.

New solutions for old problems: There are a number of yet-unresolved technology problems that require additional effort. Lasers have already been cited. Others include the development of large-scale antennae, either inflatable or ultra-large, lightweight mesh antenna to provide improved resolution for low-frequency radiometry. Some years ago there was a test deployment of an inflatable antenna of $\sim$100 m, but any continued work on this important technology does not seem to take place. Additionally, there has been little progress in developing cheaper, light weight, low power space radars. There must be a technological solution to this (old) problem as well. Another mundane, but important area is the development of improved correlators, especially if stick antenna or sensor carpets are implemented.

Advanced radiative transfer methods: Satellite remote sensing requires inverting the radiative transfer equation to retrieve geophysical quantities of interest. In view of the sensitivity of retrieval products to even small changes in outgoing radiation, the inversion (retrieval) process must be based on the best possible radiative transfer model(s) that cover all wavelengths from Ultraviolet (UV) to microwave, discriminate polarizations in incoherent and coherent radiation, and accommodate both passive and active sensing techniques. To support the development of advanced satellite retrievals, radiation transfer models must have physically consistent representations of atmospheric composition, cloud ice and water particles, hydrometeors and precipitation, surface roughness, and vegetation or soil properties across the whole electromagnetic spectrum. Models should specify each component in terms of physical variables (even if values have to be assumed) instead of empirical relationships, and aggregate detailed scattering and emission parameters on satellite footprint scales. Further advances in fundamental quantum physics and spectroscopy may be required to accurately model continuum absorption and emission of gases.

Radiation transfer codes that fully satisfy these requirements do not currently exist. The development of more accurate radiation transfer models is the fundamental underpinning of any major new advance in satellite remote sensing. Therefore, we call for the development of a new generation of radiation transfer codes for water cycle remote sensing applications. The radiative transfer processes for wavelengths from UV to microwave for irregularly shaped objects and uncommon size distributions, and in complex inhomogeneous media, such as ice and snow particles and snow packs, grass and tree leaves and vegetation canopies, and soils, are major difficulties of radiation calculations, and must be investigated in detail.

Most current retrieval techniques exploit only a few selected wavelengths from a single satellite instrument, and therefore do not provide either the best analysis of the available satellite data nor total physical consistency across data products. New, faster retrieval techniques (e.g., adjoined model equations or statistical-inverse models) must be developed, that are general enough to allow the simultaneous retrieval of a range of geophysical quantities from multiple wavelength, multiple sensor and multiple platform measurements. Such methods must be based on rigorous forward models of the measured radiation. We must develop advanced multi-variate retrieval methods that can exploit the totality of the spectral information acquired by future satellite constellations, and eventually analyze data from the whole energy and water cycle observing system. The development of more powerful radiative transfer codes and multi-variate retrieval methods is a prerequisite for acquiring crucial information for the success of water cycle research.

5.6 Weather: Improving weather information, forecasting, and warning

Predicting the weather is of utmost importance for many routine human activities today, ranging for example from agriculture to energy distribution, including transportation of passengers and freight by air and sea routes, and many other activities. Daily weather prediction also aims at the early prediction of extreme weather events which can have significant human and financial impacts. This is exemplified for example in the coastal regions of the Gulf of Mexico where population has increased in the past twenty years and so has the overall population vulnerability to hurricanes. Issuing weather prediction today requires numerical models which ingest millions of atmospheric observations every day.

Much time has passed since scarce observations of the weather were made by passionate individuals with limited instrument technology. Since the 1960s, the Earth has been observed from space by weather satellites. Observations of the atmosphere today have increased to large numbers, increasing tenfold in the last five years and approaching now 10^7 individual observations per day. This wealth of information is ingested at first and within a few hours after collection into atmospheric models to issue weather forecasts. As for the observations collected a few years ago, they are also useful in the framework of reanalysis initiatives that recreate a best estimate of the past weather using all available information at the time. This means that the atmospheric measurements of today serve two purposes: today, the prediction of weather, and, tomorrow, the reconstruction of today's climate via reanalyses. However, many aspects of our atmosphere such as the global wind field and the water cycle remain poorly observed.

An increasing number of observations of the atmosphere are collected every day. These observations are useful for weather forecasting and climate monitoring. In order to be useful to everybody, atmospheric observations would need to be referenced in a consistent reference frame. Weather prediction would benefit from such a consistent reference frame to locate various platforms, the number and the variety of which is bound to increase. Also, weather prediction lacks good water vapor observations.

Like many other technology areas, weather observing and data dissemination systems move toward more distributed systems, rather than centralized systems. Such distributed observing networks will eventually communicate intelligently with each other and some of them be located optimally. The timing and location of observations will be targeted to areas of particular forecast error sensitivity. Complementing the existing and developing weather space missions, UAV will collect measurements in the lowermost part of the atmosphere where weather affects the most human activities and violent events such as tornadoes unleash their destructive force. UAVs present the advantage of bringing to atmospheric scientists and meteorologists the power of *in situ* observations with an on-demand location and timing possibility. Furthermore, the easily upgradeable technology onboard UAVs requires less technological advances than space-compatible solutions at a fraction of the cost. Finally, the vicinity of the measurements to the phenomena of interest puts less constraints on optics and detector technology that do not need to reach the fast integration times required by fast-moving space platforms located several hundred of kilometers above the Earth atmosphere. However, unlike their satellite counterparts, UAVs will measure phenomena of very fine horizontal and vertical scale (about ten meters) and location errors acceptable today for satellite location determination will be unacceptable for UAV measurements, should one want to assimilate the data. With the modernization of the GPS constellation and the addition of a more precise L2C signal, and the advent of the GALILEO system, positioning UAVs in real-time with high accuracy in ITRF will become feasible at low cost.

Year-round, manifestations of violent weather remind us that improvements in forecast accuracy and lead time need to be achieved in order to increase preparedness of affected populations. Among the various basic meteorological observables, water vapor is both important in regard to the phenomena in which it is involved (e.g., hurricanes, flash floods, etc.) and difficult to observe, for it presents a very high space- and time-variability. Water vapor plays a crucial role in the dynamics and thermodynamics of many atmospheric processes that act over a wide range of temporal and spatial scales, covering the global hydrological and energy cycles that effectively define the local and global climate change, contributing largely to the greenhouse effect, and playing a critical role in the vertical stability of the atmosphere and in the structure of the evolution of atmospheric storm systems. The scarcity of traditional meteorological observations, especially over the Southern Ocean and Polar Regions, as well as the shortcomings of the traditional methods over the land, have contributed greatly to uncertainties in a global and regional weather analysis. Given that we currently do not have the means to observe on a global scale the sources and sinks of water vapor (evaporation at the surface and precipitation over land and ocean), observing the water vapor field is a another way of studying the water cycle. This also bears on water resources management. Improving the quality of weather forecasts will require observations of water vapor with more geographical coverage and higher temporal frequency than what is available today from *in situ* measurements or passive infra-red and micro-wave sounders. GNSS offers a new, more economical, and in principle, real-time method of atmospheric water vapor recovery (see Section 2.9.1). Therefore, GGOS can provide atmospheric scientists with near-real-time observations of atmospheric delays encountered by GNSS signals. If observed along the vertical, these delays can bring information on the lower tropospheric water vapor content. If observed between a GNSS satellite and a receiver in low-Earth orbit, these delays can help gather information on the stratospheric mass field.

5.7 Ecosystems: Improving the management and protection of terrestrial, coastal, and marine ecosystems

Management and protection of ecosystems depends on our ability to observe certain key environmental parameters that control their behavior, and upon which they depend. All ecosystems are closely linked to the global carbon cycle, and terrestrial wetlands are particularly important, as they are both highly productive yet vulnerable to land use, and readily monitored remotely.

5.7.1 Measurements of CO_2 spatial and temporal distribution to better understand the Earth's carbon cycle

There is very clear scientific evidence that the increasing concentration of greenhouse gases is seriously affecting the Earth's climate. Of greatest concern is the increasing concentration of CO_2 which is resulting from human activity such as the burning fossil fuels and the tropical biomass. Current CO_2 mixing ratios in the atmosphere are 30% above those at the beginning at the industrial revolution and are increasing at about 1.5 parts per million per year.

The processes that generate CO_2 are relatively well understood; the substantial sinks that absorb CO_2 are not. In order to understand how the CO_2 cycle operates and how we are affecting this cycle over the long term, the nature and location of these sinks must be understood because this influences the spatial and temporal distribution of CO_2 within the atmosphere's boundary layer. Mapping and monitoring the CO_2 distributions, combined with transport models, can be used to better understand the CO_2 cycle and thereby better determine the processes and spatial location of sinks.

Presently, CO_2 is measured with *in situ* sensors near the surface and on towers. Occasional measurements are made via sampling on research aircraft. Several activities are underway to extend the observation of CO_2 from the surface of the Earth and from space. Differential absorption LIDAR techniques are being developed at Goddard Space Flight Center and other institutions to monitor the vertical profile on and off a CO_2 absorption line in the near infrared (see also Section 2.4.5). Initial activities are focused on measurements in the boundary layer, but as more powerful lasers become available this work will be extended to higher altitudes. This program is being developed in response to requirements from the North American Carbon Program to make measurements of CO_2 within the boundary layer of 1 pppm volume (1 ppmv) with a vertical resolution of $\sim$10 meters. The first instruments have been built during the last few years and are being tested.

Laser-based sensing techniques are also being developed at Goddard Space Flight Center, other NASA centers, and by ESA for measuring CO_2 column densities from airborne and eventually from space space The first step is to demonstrate a LIDAR that will measure the CO_2 column density from an aircraft. The ultimate goal is a space based LIDAR sensor that can measure the CO_2 mixing ratio to 1 ppmv to provide a continuous synoptic measurement of its spatial distribution in the lower troposphere. These will measure at all times of day and continuously over the ocean. A flight mission approach is being developed for an anticipated flight opportunity in the 2013 - 2016 time frame.

5.7.2 Monitoring wetlands

Flows of water in wetlands, especially in extensive lowland systems such as the Pantanal, Amazon, Everglades, Niger inland delta and Okavango, are forced largely by subtle slopes or differences in head, e.g., mm to a few cm. These slopes are generated by spatial variations in the inputs and losses of water across the wetlands (Lesack & Melack, 1995). Since stage gauging of rivers or lakes very seldom includes measurements within adjacent wetlands and because of the spatial heterogeneity of the subtle differences in head within wetlands, it is quite difficult to determine water, solute or particulate fluxes in wetlands. Recent applications of interferometric synthetic aperture radar have demonstrated the possibility of quantifying the differences in elevation across extensive wetlands (Alsdorf et al., 2000, 2001b,a). However, validation of these results and implementation of hydraulic models within wetlands requires accurately and precisely measuring the levels of multiple stage gauges. Current capabilities with high end GPS units permit making the required measurements, in principle, but doing so in large remote systems is quite difficult. Establishing a network of base stations or repeatedly moving and repositioning a base station is necessary and lengthy recording at each site from a floating platform compounds the difficulty of the process.

5.8 Agriculture: Supporting sustainable agriculture and combating desertification

Agriculture, including forestry, has had a profound impact on the composition of soil, land cover, and changes in topography (e.g., Turner II et al., 1990). Monitoring the impact of agriculture and society at large on land cover therefore is of crucial importance in order to understand the interactions of human activities and the environment. The advent of remote sensing has been a boon for monitoring human modifications of the Earth's land surface so that the impacts of land use on global systems, including climate, can be better assessed. With remote sensing, continuous and consistent characterizations of the Earth's land surface became possible. Landsat satellite data has especially become the standard for mapping land cover changes due to human activities. Landsat 1 was launched by NASA in 1972, and measurements continue today with Landsat 7, making it the longest running remote sensing program. Monitoring the changes in land use and cover caused by deforestation and agriculture are considered in Sections 5.8.1 and 5.8.2, respectively.

Considering the great impact of agriculture on the environment, efficiency in terms of nutrients and chemicals introduced into the environment is beneficial. Based on geodetic techniques, Precision Farming (PF) is developed and increasingly used to reduce the resources required in food production, thus not only keeping prices for food at a low level but also reducing the environmental impact of farming (Section 5.8.3).

5.8.1 Monitoring deforestation and logging

Landsat data, with 30 m resolution, is ideal for mapping land cover changes over large areas. In particular, it has become valuable for monitoring deforestation around the world. The first large-scale deforestation mapping using satellite imagery was applied to the Brazilian Amazon. More recently, through the NASA Pathfinder Humid Tropical Deforestation project, repeat assessments have been made by various studies for the Amazon (Tardin & da Cunha, 1990; Skole & Tucker, 1993), and for much of the tropics (Skole et al., 1994).

However, while Landsat data are valuable for mapping land cover change over large areas, they have still been too expensive in terms of effort, time, and labor costs, for mapping land cover change continuously over the entire tropics or the entire planet. Therefore, several attempts have been made to map deforestation using a sampling approach. The FAO Remote Sensing Survey (see, e.g., FAO, 2001; Mayaux et al., 2005) used a 10% sampling of Landsat scenes to map tropical deforestation for the 1980s and 1990s. However, FAO's 10% sampling approach has been deemed insufficient to map forest change because deforestation is clustered, and not randomly distributed (Townshend & Justice, 1995). Some studies have suggested that complete wall-to-wall mapping is necessary (Townshend & Justice, 1995), while others have suggested that a 10% sample is sufficient for large-area estimates. A sampling of Landsat scenes was also used by the recent TREES II project of the Joint Research Center of the European Commission to map deforestation rates for the entire humid tropics (Achard et al., 2002, 2004). However, to address the concern about the clustered nature of deforestation, they used a stratified random sampling approach, which focused the samples within *a priori* delineated deforestation hotspots in the humid tropics. Another approach to map deforestation over large areas has used the coarse-resolution AVHRR Pathfinder data (about 8 km resolution) over the 1982-1999 period (Hansen & DeFries, 2004), calibrated to regions with known estimates of deforestation mapped using higher-resolution remote sensing. Both the TREES II and AVHRR studies indicated that deforestation rates were much lower than reported in the Forest Resources Assessment of Food and Agriculture Organization (FAO) (Hansen & DeFries, 2004).

Two other recent studies using remote sensing to examine deforestation dynamics are worth mentioning. While Landsat data have been used to map large-scale deforestation around the world, good estimates of selective logging have not been available. Asner et al. (2005, 2006) developed a method to estimate selective logging over the Amazon Basin using Landsat data, and found that forest area damage from selective logging matched or exceeded reported rates of deforestation. This has been a remarkable advance in our ability to use remote sensing data to map fine-scale patterns of land use practices such as logging. In another recent study, Pongratz et al. (2006) examined the dynamics of land cover change following deforestation in Mato Grosso, Brazil. Mato Grosso state has seen enormous expansion of soybeans in recent years, and it has been debated whether the soybean expansion has occurred on abandoned pasture land, or whether it is resulting in new deforestation. Pongratz et al. (2006) estimated that total conversion of forest to cropland exceeded 540,000 ha over the 2001-2004 period, peaking at 23% of the total deforestation in 2003.

5.8.2 Agricultural land cover and land use

Mapping agricultural land cover and land use over large areas has proved surprisingly more challenging. Global satellite data such as the 1 km AVHRR data or the more recent 250 m-1 km MODIS data have been used to derive global land cover maps (e.g., IGBP DISCover product (Loveland et al., 2000), University of Maryland land-cover maps (Hansen et al., 2000), or Boston University land cover), but have paid scant attention to detailed characterization of the world's agricultural lands (McCabe & Wood, 2006). While deforestation registers a clear signal in global satellite data, it has been difficult to characterize the heterogeneous agricultural land use practices around the world. While large-scale mechanized intensive agriculture in regions such as Iowa and Kansas are clearly visible from satellites, it is more difficult to characterize the heterogeneous landscapes in places like West Africa (Ramankutty, 2004).

5.8.3 Precision farming

There is a trend to PF, which in common with many engineering activities (see Section 4.3.3) requires precise geopositioning of farm machinery, and reliable spatial information (e.g., maps of the property, terrain, soil type, growing patterns, etc.). Farmers in Europe, North America and Australia are adopting PF practices very rapidly. Benefits of PF include sustainable farming practices, 24 hour-operations, increased automation, which all together result into reduced environmental impact of farming and contribute to lower food prices.

A particularly challenging form of PF is Control track farming (CTF). Currently, GNSS-RTK is used to guide the farm machinery during all stages of the grain growing cycle, so that the wheels of the machinery always travel along the same "ruts" of compacted soil. This leads to less breakup of the soil, decreases soil erosion, and makes possible innovative practices such as growing a second crop interlaced with the main grain crop. The ability of the farmer to guide his/her machinery along the same ruts, over and over again, is possible because the GNSS datum can be defined to the centimeter level. However, currently PF and CTF depend on the availability of local augmentation system providing the accuracy at the few centimeter level. Improved real-time access to ITRF and compatibility of databases with this reference frame would support these applications globally without local expertise in establishing and maintaining augmentation systems.

Forestry practices also benefit from geodetic innovations such as directly georeferenced airborne laser scanners. These instruments known as LIDAR, can determine the DEM and DSM, as well as scan understorey, etc. In effect the difference between the two surfaces can be used to estimate biomass, and help forest management practices, such as when to harvest, where tree growth is sub-optimal, etc.

The increased availability of GNSS and improved access to ITRF also provides the means to change livestock practices. Examples could be livestock equipped with GNSS sensors, which would make them trackable. Particularly in open ranching, the combination of such sensor on free ranging livestock with spatial information could also be used for warning systems indicating animals approaching roads through open ranching country.

Chapter 6
Geodesy: Foundation for exploring the planets, the solar system and beyond

J. F. Zumberge, J. S. Border, V. Dehant, W. M. Folkner, D. L. Jones,
T. Martin-Mur, J. Oberst, J. G. Williams, X. Wu

The utility of a Global Geodetic Observing System is not limited to our home planet, but also extends to scientific studies of the planets and their moons (including our own Moon), observations beyond the solar system, and the exploration of space in general. Examples of the first two of these include Planetary Geodesy and Radio Science, where the GGOS infrastructure is a requirement for making and interpreting measurements for these sciences; we include a section on each.

The exploration of space in general involves spacecraft in Earth orbit and beyond. Tracking these spacecraft from Earth depends critically on the GGOS infrastructure. In the section on inter-planetary navigation we describe the current and future requirements of GGOS for this application.

6.1 Planetary geodesy

The most accurate estimates of the time-dependent orientation of the planets and their satellites are based on radio-frequency range and Doppler measurements between spacecraft in orbit or landed on those bodies and Earth tracking stations. Planetary geodesy provides invaluable information about the distribution and state of the matter of which they are comprised. Future measurements are expected to help in determining whether liquid water exists under the surface of bodies such as Europa. Rotation variations also provide means to measure the interaction between the planetary surface and the atmosphere. The most accurate planetary orientation measurements are for Mars, where the large number of past and present landers and orbiters provides a large data set.

The accuracy of the determination of the Mars orientation is approaching that of the Earth not very long ago. The accuracy is expected to improve over the next decade with a new generation of landers with improved radio capabilities.

The reference frame for planetary geodesy is established by measurement of the Earth's orientation. Thus maintaining and even increasing the accuracy of knowl-

H.-P. Plag, M. Pearlman (eds.), *Global Geodetic Observing System*,
DOI 10.1007/978-3-642-02687-4_6, © Springer-Verlag Berlin Heidelberg 2009

edge of the orientation of the Earth is needed to improve our understanding of the other bodies in the solar system.

In the following sections, we first discuss the relevance of geodetic quantities to rotation and interior properties of the planets. Sections 6.1.2 to 6.1.4 illustrate the challenges in planetary geodesy using examples of Mars, the Moon, and Europa. Finally, Section 6.1.5 discusses future geodetic infrastructure on or near to the planets and their satellites.

6.1.1 Planetary rotation and interior properties

The principle of using rotation parameters, as discussed in Section 3.11 for obtaining information on the interior properties of Earth, may be applied to other terrestrial planets. Similarly, the interiors of other terrestrial planets will affect their nutations and wobbles. Librations are also influenced by interior properties, and as such, observations of librations will lead to further knowledge on the interior of slowly rotating planets such as Mercury.

Precession and nutation are induced by the tidal gravitational torque on an oblate planet; this is the case for Mars and for the Earth. This motion is very helpful for studying the deep interior of Mars, mainly because it is different for a planet with a solid core than for a planet with a liquid core. The dimension of the core (or equivalently the core moment of inertia) can also be determined from these observations. Figure 6.1 shows the relative influence of the dimension of the core on the nutation transfer function of Mars.

Similarly for the planet Mercury in a spin-orbit resonance 3:2, librations of amplitude 500 m are expected if the core is liquid and half as much if the core is solid. The dimension of the core also influence these values. Generally speaking the global parameters such as the mass, the moment of inertia, the rotation, the orientation and their changes may be used to better constrain the interior properties of terrestrial planets.

Length-of-day variations are deviations from the uniform rotation speed of a planet. They are mostly related to geophysical fluids (core, ocean, atmosphere, hydrosphere, etc.) in the system, if they exist. For Mars the seasonal condensation/sublimation of the icecaps induce polar motion as well as large changes in the length-of-day at the seasonal periods. This is easy to understand from the moment of inertia changes induced by the mass repartition and from angular momentum conservation between the solid part of the planet and the fluid surrounding it. The seasonal length-of-day changes correspond to a change with respect to the uniform rotation at the level of 15 meters at the equator, which can be computed from general circulation models and constraints from the observations.

In parallel, information on planetary interiors can also be obtained from measurements of global parameters from the gravity field. The determination in parallel of Love numbers representing the tidal effects are also important global parameters that can be used to determine the internal structures of the terrestrial planets and their satellites in the solar system.

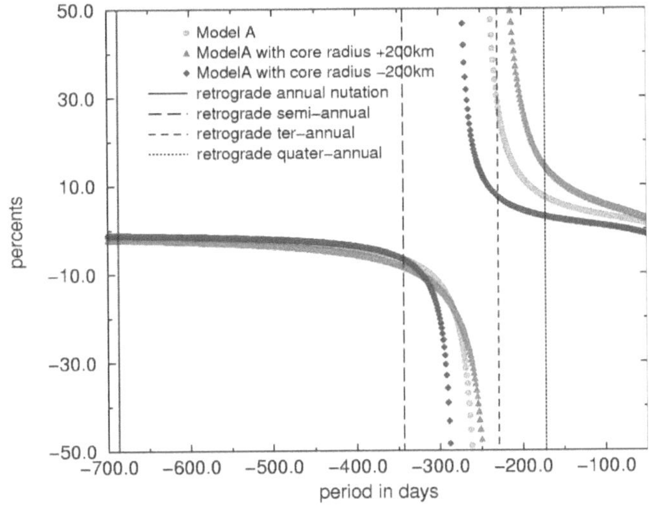

Fig. 6.1. Resonance induced when different dimensions of the core are considered. The vertical lines show where nutation could be observed.

6.1.2 Example: Mars

The only observations of Mars which provide insight into its interior structure are based on geodesy. (Other clues to the interior structure are provided by meteorites found on Earth which originated on Mars, though these may be more indicative of the surface than of the interior.) A series of spacecraft, from the Viking orbiters in the 1970's to the current Mars Reconnaissance Orbiter and Mars Express spacecraft have provided increasingly accurate determinations of the Martian gravity field (Smith et al., 1998; Konopliv et al., 2006). The martian precession constant was originally determined by measurements of the Viking landers and the Mars Pathfinder lander radio signals (Folkner et al., 1997). The combination of the precession constant and the low-order gravity field determine the polar moment of inertia, which gives the first indication of the size and mass of the martian core. The more recent gravity field determinations are increasingly sensitive to the martian orientation and can provide estimates of the precession constant with accuracy comparable to the Viking and Pathfinder landers (Konopliv et al., 2006).

The martian gravity field is also sensitive to the distortion of the planet as its distance from the sun varies due to Mars' eccentric orbit. Estimates of the tidal deformation Love number k_2 suggest that the core is liquid rather than solid (Yoder et al., 2003; Konopliv et al., 2006). Comparison of the gravity field to topography allows estimation of the crustal thickness and possible explanation for the dichotomy between the southern lowlands and the northern highlands (e.g., Smrekar et al., 2004).

The martian rotation rate varies due to seasonal evaporation and condensation of the polar CO_2 icecaps. The lander and orbiter radio measurements provide estimates of the variation in the distribution of the mass of the icecaps, providing insight into the climate and formation of polar layered deposits.

Future landers with improved radio instrumentation are expected to improve the accuracy of the martian precession constant by a factor of ten or more, and also be accurate enough to measure nutation and polar motion with sufficient accuracy to determine the size, density, and state of the martian core. These will complement expected seismometry to provide a much more thorough understanding of the structure and formation of Mars. The future martian geodesy measurements will approach milli-arcsecond accuracy, comparable to the current accuracy of Earth orientation measurements. Continued monitoring of Earth orientation to at least this accuracy will be required to accurately interpret the martian measurements.

6.1.3 Example: Earth's Moon

Missions to the Moon include orbiters, impactors, landers and surface rovers. There have also been flybys of spacecraft heading for other destinations. Orbiting missions may have experiments to image the surface at different wavelengths and also to measure surface topography by altimetry, gravity field by spacecraft tracking, magnetic field, remote sensing of surface composition, and near Moon particles and fields. Landers can deliver instruments (fixed or roving) and astronauts to the surface for exploration. The future LCROSS experiment will impact the Moon throwing up material to be analyzed. Also in the future, there will be a need to follow the movements of rovers and astronauts on the lunar surface.

Orbiting, landing, roving and impacting missions must know how the position of the spacecraft changes with time. Orbiters need the direction and distance to surface targets under study. Tracking of the highest accuracy is needed for gravity field determination. For an impactor, the location and time of impact are needed. Landers need to arrive at their intended destination.

The use of tracking data for navigation of spacecraft requires Earth orientation information. For operations and maneuvers this information must be extrapolated forward from recent accurate measurements, so quick analysis is important. The fitting of tracking data for determining the gravity field can take advantage of the more accurate past Earth orientation parameters available from analyzing measurements after the tracking data is taken.

Radio range and Doppler as well as laser ranging techniques have been used for tracking at the lunar surface. LLR requires no power source on the Moon which has allowed ranges to be acquired for more than three decades. LLR data analysis has contributed to lunar science, gravitational physics, ephemerides, and Earth science.

LLR is both a contributor to and a user of Earth orientation information. The laser range accuracy is presently making a transition from centimeter to millimeter levels. With current renewed interest in lunar exploration, studies on future lunar geodesy experiments are currently under way involving active lasers on the lunar surface pointed at Earth. Current lunar ranging requires sophisticated ground equipment and is carried out by few Earth stations only. On the other hand, signals from strong (50 mJ) lasers on the Moon could be within the sensitivity of all stations in the global SLR network. Such a laser station should be combined with a co-located small radio transmitter, which would be observable to VLBI stations and which would enable us to firmly tie lunar coordinates to the celestial reference frame of radio sources. Earth orientation parameters of comparable accuracy are needed.

6.1.4 Example: Europa

A proposed Europa orbiter to detect a possible underground liquid water ocean remains a high science priority for the planetary science community and ESA/NASA (Wu et al., 2001). The primary scientific goal of a systematic geodetic and geophysical experiment is to measure tidal gravity, displacements and libration on Europa to determine the existence and dimensions of the ocean and other interior structures. Doppler tracking from Deep Space Network (DSN) stations is essential to achieve these measurement objectives by determining the Love number k_2 and libration amplitude through orbit dynamics, and by providing accurate orbital positions for surface geodetic measurements from altimeter or InSAR instruments. To assess effects of uncertainties in geocentric positions of DSN stations, tropospheric delay, and EOP, orbit determination simulations were carried out and covariance analyses for a proposed nominal Europa orbiter were considered. Even with conservative uncertainties in these ground geodetic parameters, their effects on dynamic parameters such as k_2 and libration are negligible compared with their achievable accuracies. The effects on orbital positions are generally at the level of several centimeters along the unfavorable (along- and cross-track) directions. This level of uncertainty is fairly small compared with the anticipated peak tidal displacement amplitude of 30 m with the liquid ocean. Current levels of accuracies in geocentric DSN station coordinates, polar motion and UT1 are also adequate for the planned ESA BepiColombo mission to Mercury with very precise multi-frequency (including Ka band) tracking from the DSN sites equipped with water vapor radiometers for tests of general relativity and the internal structures.

6.1.5 Planetary mapping

Thanks to the radio experiments on the US Mars Pathfinder and to the radio and laser ranging experiments Mars Orbiting Laser Altimeter (MOLA) on the Mars Global

Surveyor spacecraft, Mars possesses a well-defined coordinate system. MOLA has mapped the planet at high accuracy and with globally consistent quality within this coordinate system. Consequently, morphologic features anywhere on the surface of Mars can be located within the Mars coordinate system to within 100 m. The High Resolution Stereo Camera (HRSC), a camera on the Mars Express mission, dedicated to accurate topographic mapping, is currently refining the resolution of these maps.

For the Moon, in spite of the fact that it is closer to Earth than Mars, the currently available mapping control is less accurate. The Apollo Laser retroreflectors provide excellent control points and practically define the coordinate system of the Moon. Hence, coordinates of features near the Apollo sites (where the Apollo spacecraft in their near-equatorial orbits provided image coverage at high resolution) are well established. Unfortunately, geodetic control rapidly decreases towards higher latitudes and towards the Lunar farside. Coordinate uncertainties may amount to several kilometers. Such "map-tie" errors are annoying to spacecraft engineers who would like to target a spacecraft to a specific surface location seen in images. The situation is expected to improve with release of data from the Japanese Kaguya (SELENE) mission or the US Lunar Reconnaissance Orbiter (scheduled for launch in 2009), which are both equipped with laser altimeters and sophisticated camera systems.

6.2 Radio science and interferometry

Radio science with tracking data: Radio science encompasses a wide range of techniques, with a correspondingly diverse set of geodesy requirements. Studies of planetary atmospheres, rings, and the interplanetary medium through occultation and scattering measurements do not often depend on extremely accurate Earth orientation or antenna location knowledge; however, the study of local gravitational fields through accurate spacecraft orbit determination does require a very good terrestrial reference frame. Errors in UT1, polar motion, or antenna positions translate directly into errors in the angular position of a spacecraft.

State-of-the-art radio science observations rely on ground system performance that often exceeds the formal requirements. The geodetic aspect of this is after-the-fact knowledge of UT1 and pole offsets to 0.5 cm, and antenna positions to 1 cm. The current (2005) requirements are 5 cm for after-the-fact Earth orientation and 3 cm for antenna positions.

Imaging Very-Long-Baseline Interferometry: VLBI imaging of radio sources generally relies on self-calibration to remove the effects of antenna-dependent errors. This technique can dramatically improve the dynamic range and fidelity of radio images, but absolute position information is lost. Consequently, VLBI imaging is relatively immune to errors in antenna positions or Earth orientation. Requirements on the terrestrial geodetic system imposed by other uses will be more than adequate for high-resolution radio source imaging.

Astrometric Very-Long-Baseline Interferometry: Precise position measurements of radio sources are directly dependent on the accuracy of the terrestrial geodetic system. For narrow-angle differential (phase-referenced) astrometry most terrestrial error sources are reduced by the angular separation of the radio sources in radians. For wide-angle (absolute) astrometry, there is no reduction in terrestrial error sources. Wide-angle astrometric measurements are the basis of the celestial reference frame.

The best current wide-angle VLBI position measurement accuracies are $\sim$1 nrad. This implies knowledge of UT1 and polar motion to 0.5-1.0 cm, exceeding the current Earth orientation requirement by up to an order of magnitude. The requirement on antenna positions is less straightforward because the effect of a baseline error on a position measurement depends on the relative orientation of the baseline and the line of sight to the radio source. Knowledge of antenna positions to 2 cm would be adequate, with a long-term goal of 1 cm.

Earth-space interferometry: The use of baselines between ground antennas and antennas on Earth-orbiting spacecraft (e.g., the Japanese VSOP and VSOP-2 missions) requires an accurate spacecraft orbit as well as geodetic parameters. Normally errors in orbit determination exceed those of the global geodetic system. Consequently this type of observation does not impose any new or more stringent requirements on geodetic parameters.

6.3 Interplanetary navigation

Geodesists and students of Geodesy do not always realize the critical role that geodesy plays in the exploration of worlds beyond our own. Although this application is often overlooked, national space agencies must provide geodesy-related "calibrations" to support their lunar and planetary missions. These calibrations are necessary to chart the course of spacecraft en route to other bodies in the Solar System. The radio link from spacecraft to Earth not only serves for communications but also provides tracking observables that are the primary source of data about spacecraft position and velocity for most missions. While spacecraft move in a celestial reference frame, observers on Earth have positions known in a terrestrial reference frame. Definitions of reference frames, catalogs of objects, relations between reference frames, and modeling of radio signal propagation delays are all important aspects of spacecraft navigation.

6.3.1 Current and future tracking data types

Thornton & Border (2000) give an excellent treatise on radiometric tracking techniques for deep-space navigation. Consider their equation 3.2-1 to illustrate the importance of geodetic quantities:

$$\dot{\rho}(t) = \dot{r}(t) + \omega_e r_s \cos\delta \sin(\omega_e t + \phi + \lambda_s - \alpha). \tag{6.1}$$

Here ρ is the range from a tracking station to a distant spacecraft and $\dot{\rho}$ its time derivative. On the right-hand side $\dot{r}$ is the geocentric range rate, ω_e the Earth's rotation rate, r_s the distance of the tracking station from the axis of rotation of the Earth, λ_s the longitude of the tracking station, α the right ascension of the spacecraft, and δ its declination. If t is in universal time then ϕ is the instantaneous right ascension of the mean Sun. The radiometric observables transmitted from a spacecraft and received by an antenna on Earth are modeled with this equation or some variant, making it immediately clear that knowledge of station locations and Earth orientation are critical in using Doppler (the ratio of received frequency to transmitted frequency is $1 - \dot{\rho}/c$), for example, to determine spacecraft coordinates.

As an example, the time-of-day component of Earth orientation is UT1 $\approx R_e \omega_e t$ (in dimensions of length; $R_e \approx 6378$ km is the Earth equatorial radius). Thus a 10-cm error in UT1 corresponds to an error in spacecraft right ascension of about 10 cm/6378 km $\approx$ 16 nrad. For a spacecraft approaching Mars, assuming an Earth-to-Mars distance near its minimum of $\sim$80 million km, this translates to about 1.3 km. Details on the sensitivity of deep-space navigation to EOP errors can be found in Estefan & Folkner (1995), while the determination of station locations of NASA's deep-space tracking stations is described in Folkner (1996).

In addition to Doppler measurements, range and interferometric observables are also used for navigation. Range measurements are based on the round-trip light time for a code uplinked from a tracking station, transponded by a spacecraft, and received back at a tracking station. Normally the same station is used for the uplink and downlink functions, though this isn't strictly necessary given good clock synchronization. Interferometric observables are based on the difference in arrival time for a spacecraft signal that is received at two separate stations. Observations of natural radio sources are usually included as part of an interferometric measurement session in order to calibrate the instrument, and this data type is referred to as Delta-VLBI. While Doppler and range provide line-of-sight determinations of spacecraft state, Delta-VLBI determines spacecraft position and velocity on the plane-of-sky. The different data types have differing sensitivities to geodetic parameters, but they share one thing in common: the precision of the radio link at the microwave frequencies of the Space Research bands is quite high. Given adequate signal-to-noise ratios, Doppler observables provide precision of 0.01 mm/sec, range precision is 10 cm, and VLBI precision is 1 cm. Any unmodeled effects in the propagation equation at this level or worse will affect interpretation of the data. Transition to higher frequencies, driven by the need for more communications bandwidth, will result in even better metric observable precision and hence interest in improved geodetic calibrations.

Spacecraft equations of motion are integrated in a celestial reference frame. Today, the celestial reference frame is defined by the positions of quasars in the ICRF as described by Ma et al. (1998). The origin of right ascension is a certain linear combination of catalog coordinates. The Earth's equator and the equinox of the Earth's orbit are measured quantities in this frame. Planetary ephemerides, based largely on radar range measurements of the planets, and range and interferometric measurements of spacecraft landed or in orbit about planets, are constrained to be consistent with the ICRF. By the same token, accurate geodetic models are needed to interpret measurements of planetary positions. Sources from the ICRF catalog are used for Delta-VLBI observations. Enhancements to the catalog, including densification and refinement of the coordinates of individual sources, would directly benefit the Delta-VLBI technique.

Table 6.1. Current and future requirements for radiometric observables, geodetic coordinates and related calibration parameters. The future columns are speculative, and will ultimately depend on specific mission requirements.

Tracking Error Source (1 sigma Accuracy)	units	current capability	2005 reqt	2010 reqt	2020 reqt	2030 reqt
Doppler/random (60s)	mm/s	0.03	0.05	0.03	0.03	0.02
Doppler/systematic (60s)	mm/s	0.001	0.05	0.003	0.003	0.002
Range/random	m	0.3	0.8	0.5	0.3	0.1
Range/systematic	m	1.1	0.6	2	2	1
Delta-VLBI	nrad	2.5	5	2	1	0.5
Troposphere zenith delay	cm	0.8	1	0.5	0.5	0.3
Ionosphere	TECU	5	5	5	3	2
Earth orientation (real-time)	cm	7	30	5	3	2
Earth orientation (after update)	cm	5	5	3	2	0.5
Station locations (geocentric)	cm	3	3	2	2	1
Quasar coordinates	nrad	1	1	1	1	0.5
Mars ephemeris	nrad	2	-	3	2	1

Tracking stations have coordinates given in the ITRF. Tracking networks used for deep-space applications tend to have just a few stations at a few widely separated sites. Building a global reference frame to include these stations has greatly benefited from ties and modeling consistency between stations employing different techniques including VLBI, GPS, DORIS, SLR, and LLR. With spacecraft data arcs spanning days to years, and measurement precision at the cm level, it is important to model effects including plate motion, solid Earth tide, pole tide, and loading from ocean, atmosphere, hydrology, and ice.

Space agencies understand the importance of determining the state of the Earth platform and are active contributors to cooperative efforts to improve the knowledge of reference frames and the interrelation between the terrestrial and celestial reference frames. Data reduction employs transformations between reference frames and models for precession, nutation, and EOP. Rapid delivery of current estimates of the parameters that are more difficult to predict, especially UT1, is necessary to support targeting for an encounter.

Calibration of media delays is also important to the interpretation of radiometric observations. Data from GPS receivers co-located with tracking antennas, and data from receivers at surrounding locales, are used to develop calibrations for zenith troposphere delay and for line-of-sight ionospheric delay. Even though GPS data may be very precise, the modeling used to transform GPS measurements to the spacecraft line-of-sight introduces errors at the level of precision of the spacecraft tracking data or worse. The availability of additional satellites and receivers, with more geometric coverage, could lead to useful improvements in media calibrations.

Spacecraft *in situ* measurements are required in some scenarios, such as landing on an asteroid with a poorly known ephemeris, and may be efficient for use in other scenarios, such as relative navigation between networks of landed and orbiting assets at another body. But Earth-based observations are likely to remain an important input to the navigation process for at least some phases of most missions. Details of measurement techniques may change. Large arrays may be used for spacecraft communications, enabling more powerful interferometric techniques. Higher frequencies and eventually optical links may be used. But in all cases geodesy-related calibrations are needed for interpretation of Earth-based measurements.

6.3.2 Interplanetary trajectory determination

Many future interplanetary missions will require higher navigational accuracy than that needed by past and current missions. Some future mission scenarios, such as pinpoint landing, will require very precise delivery to either an atmospheric interface, or to the surface for bodies without atmosphere. The main navigational tracking methods can be split into ground-based methods, such as range, Doppler and VLBI, and spacecraft-based methods, such as optical navigation and spacecraft-to-spacecraft tracking. Ground-based tracking methods are used, at least for some part of the missions, by all missions, and most times ground-based and spacecraft-based methods complement each other more than replace each other. All of the ground-based tracking methods rely on the precise knowledge of the position of the Earth station in the celestial frame that is used for trajectory integration. This requires precise knowledge of the position of the station in the terrestrial reference frame, including tidal effects and plate motion, and precise transformations between the terrestrial and the celestial reference frame, including up-to-date knowledge of polar motion and Earth rotation (UT1), as well as accurate models for nutation and precession. Also very important for spacecraft navigation is the determination of the ephemeris of the natural bodies that the spacecraft target to, and the determination of the position of celestial sources used for VLBI and optical navigation. These ephemerides and positions are mostly derived from data from ground observatories, and the same kind of precise geodetic data is needed in order to reduce the observations to estimate the location of the celestial bodies.

6.3.3 Current and future requirements of GGOS for interplanetary navigation

Indicated in Table 6.1 are ESA's and NASA's current and anticipated deep-space mission requirements for Earth orientation, station coordinates, and related calibration parameters (Martin-Mur et al., 2006). The EOP components PMx, PMy, UT1 change from day to day, depending primarily on atmospheric effects. The tightening requirements shown in the table would likely require near-real time VLBI measurements of UT1.

The future columns in Table 6.1 are speculative, and will depend on specific mission requirements. Martin-Mur et al. (2006) indicate that there will be an "Increased need for higher accuracy in guidance, navigation, and control, in order to perform pinpoint landing, and to take advantage of higher resolution instruments." An example of a demanding future requirement could come from a Mars sample return mission or a manned mission to Mars.

To achieve a pinpoint landing, very high accuracy would be needed to deliver a spacecraft to the atmospheric interface. Given this, the requirements on the complexity of the guidance system to lower the spacecraft through the atmosphere to the desired landing location could be kept to a minimum. Improvements would be needed in several areas, as indicated in Table 6.1, to substantially improve overall navigation performance. In addition to improvements in measurement precision and geodetic and Earth orientation models, there would also have to be reductions in errors due to transmission media, improvement in celestial reference frame models, and an improved ephemeris for Mars.

Chapter 7
Integrated scientific and societal user requirements and functional specifications for the GGOS

R. Gross, G. Beutler, H.-P. Plag

7.1 Introduction

As discussed in the previous chapters, the terrestrial reference frame is the foundation for virtually all space-based and ground-based Earth observations. Through its tie to the celestial reference frame by the time-dependent Earth rotation parameters it is also fundamentally important for interplanetary spacecraft tracking and navigation. Providing an accurate, stable, homogeneous, and maintainable terrestrial reference frame, celestial reference frame, and the Earth rotation parameters linking them together is one of the essential goals of GGOS.

In recent decades, the geodetic techniques also contribute to the database of Earth observations in particular related to mass transport, dynamics, and ionosphere and troposphere parameters. Observations of changes in the Earth's geometry (solid Earth surface, sea surface, lake surfaces, and ice surfaces) are an important contribution to the Earth observation database serving a wide range of applications.

In this chapter, the requirements of the diverse set of scientific and societal users concerning the terrestrial and celestial reference frames, the associated Earth rotation parameters, and the complementary gravity measurements are first summarized. Subsequently, the requirements in terms of a number of quantities observed by geodetic techniques or determined in geodetic analysis are compiled. The tasks, products, and specifications of the GGOS that are needed in order to meet the most demanding requirements of the users are then presented.

H.-P. Plag, M. Pearlman (eds.), *Global Geodetic Observing System*,
DOI 10.1007/978-3-642-02687-4_7, © Springer-Verlag Berlin Heidelberg 2009

7.2 Summary of user requirements

7.2.1 Societal applications

Most societal applications are concerned with determining the position of some object, whether it be the fixed position of survey markers on the ground, buildings, bridges, dams, and oil platforms or the mobile position of cell phones, farm equipment, automobiles, trucks, trains, airplanes, and ships. While most users may not be aware of it, geodesy and GGOS play a fundamental role in societal applications by providing the infrastructure, including the underlying terrestrial reference frame, that enables the position of the object to be determined. The position of the object is determined within some underlying reference frame and the accuracy of the position determination will ultimately depend on the accuracy of that frame. As a general rule-of-thumb, the reference frame should be an order of magnitude more accurate than the required accuracy of the position determination. Moreover, the accuracy of the position depends on the latency with which the position is required. Positions determined in near-real time will in general be less accurate than those determined with longer measuring and processing time. Thus, the mode in which access to positions in the reference frame are determined is crucial for the achievable accuracy.

GNSS is increasingly being used to determine the vertical position of objects. Since GNSS determines the height of the object above the ellipsoid, an accurate geoid is required in order to convert the ellipsoidal height to a geopotential-related height above the geoid (often equated to the "mean sea level"). For the most part, societal users do not need accurate Earth rotation parameters except to the extent that they are needed when determining positions with GNSS. However, for highly accurate positions, the requirements in terms of Earth rotation are rather demanding.

Geodesy and GGOS also provide the infrastructure that allows different spatial information, such as imagery from different space and airborne platforms, to be georeferenced and aligned with each other. The importance of accurately georeferencing spatial information is being recognized by many national governments (see Section 4.1).

An important aspect of georeferencing is the use of a global reference frame to allow spatial information to be converted into a common system that is consistent for all users. Global aviation, marine traffic, and cross-boundary land traffic are examples benefiting from compatible spatial databases.

7.2.2 Earth observations

Recognizing the need for coordinated and comprehensive monitoring of the Earth and its interacting systems to support sustainable development, the intergovernmental GEO was formed with the task of implementing a GEOSS (see Section 5.1). All of the nine SBAs of GEO (see Table 5.1 on page 155) require an accurate, stable, and homogeneous terrestrial reference frame as the foundation for the observations, and they rely on geodetic measurements to provide accurate trajectories of the aircraft and satellite platforms housing their instruments. In addition, the disaster, climate, water, and weather SBAs depend directly on geodetic measurement techniques to provide some of their observations. The requirements in terms of geodetic quantities for the nine SBAs as extracted from GEO (2005b) are given in Table 5.2 on page 156.

7.2.3 Natural hazards

Mitigating the impact on human life and property of natural hazards such as earthquakes, volcanic eruptions, debris flows, landslides, land subsidence, sea level change, tsunamis, floods, storm surges, hurricanes and extreme weather is an important scientific task to which GGOS can make fundamental contributions. GNSS and InSAR can be used to monitor the pre-eruptive deformation of volcanoes and the preseismic deformation of earthquake fault zones, aiding in the issuance of volcanic eruption and earthquake warnings. GNSS can also be used to rapidly estimate earthquake fault motion, aiding in the modeling of tsunami genesis and the issuance of tsunami warnings. Gravity measurements can be used to track mass motion within volcanic conduits. Gravity and altimetric measurements can be used to track floodwaters in river basins. Altimetric and tide gauge measurements can be used to monitor sea level change. Essential to all such measurements is the underlying terrestrial reference frame in which the measurements are made.

In 1990, 23% of the world's population lived both less than 100 km from the coast and less than 100 meters above sea level. Nearly a fourth of the world's population is therefore vulnerable to the effects of a rising sea level combined with expected changes in extreme events. Although the long-term average rate of sea level rise is only a few mm/yr, mitigation and adaptation efforts need to be planned well in advance in order to be prepared for rare extreme events. But great demands are placed on GGOS because the sea level rise signal is so small. For example, the terrestrial reference frame, which should be at least an order of magnitude more accurate than the amplitude of the signal being measured, needs to be accurate and stable to within about 0.1 mm/yr to support studies of sea level change (e.g., Plag, 2006a; Blewitt et al., 2006a). This makes sea level change studies one of the most demanding users of GGOS.

7.2.4 Earth science

The solid Earth is surrounded by a fluid, mobile atmosphere and oceans and upon its land surface lies a continually changing distribution of ice, snow, and ground water.

The mantle is both thermally convecting and rebounding from the glacial loading of the last ice age, and the fluid core is undergoing some type of hydromagnetic motion to generate the Earth's magnetic field. The changing distribution of mass associated with the migration of the surficial fluids and the motion of the mantle and core changes the Earth's gravitational field, changes the Earth's rotation by changing its inertia tensor, and changes the Earth's shape by changing the load on the solid Earth. GGOS measurements of the Earth's changing gravity, rotation, and shape can therefore be used to study these and other dynamical Earth processes. In general, the science requirements translate into the best possible accuracy for the observations, which will then foster the best scientific advance in these studies.

7.2.5 Lunar and planetary science

Estimates of the shape, gravity, and rotation of the planets and other celestial bodies is obtained by accurately tracking spacecraft that are in orbit about or have landed on those bodies. Spacecraft tracking measurements are based upon radio range and Doppler measurements taken between the spacecraft and the Earth-based tracking stations. Errors in the positions of the tracking stations, including errors in Earth orientation, translate directly into errors in the angular position of the spacecraft and hence in the derived estimates of the shape, gravity, and rotation of the body. In addition, laser ranging has been used to track retroreflectors on the lunar surface. The accuracy of the laser ranges is currently improving from centimeter to millimeter levels. Making full use of this improving accuracy requires knowledge at the same millimeter level of the position of the laser tracking stations on Earth and of the Earth orientation parameters.

The interior structure of the planets and other celestial bodies can be inferred from the estimates of their shape, gravity, and rotation that have been determined from spacecraft tracking measurements. For example, the size and mass of the Martian core was first derived from estimates of Martian gravity and precession obtained from spacecraft tracking measurements. Estimates of the tidal deformation of Mars derived from determinations of Martian gravity suggest that its core is liquid rather than solid. The thickness of the Martian crust can be estimated by comparing the topography of Mars to its gravity field. The presence of an underground ocean of liquid water on Jupiter's moon Europa could be detected by tidal gravity, deformation, and libration estimates obtained from tracking measurements of a spacecraft orbiting Europa. All such inferences of the interior structure of the planets and other celestial bodies rely on the fundamental spacecraft tracking measurements, the accuracy of which is limited by the errors in the position and orientation of the Earth-based tracking stations.

Tracking and navigating interplanetary spacecraft requires accurate terrestrial and celestial reference frames, station positions, and Earth orientation parameters. While spacecraft move in the celestial frame, tracking stations are located in the terrestrial frame. Errors in the positions of the tracking stations and errors in the Earth orientation parameters used to transform the coordinates of the tracking stations from the terrestrial to the celestial frame translate directly into errors in the angular position of the spacecraft. For example, an error of 10 cm in Earth UT1 translates to a position error on Mars of about 2.6 km assuming an Earth-Mars distance of $150 \cdot 10^6$ km. Reducing errors in Earth orientation and tracking station position enables more accurate tracking and navigation of interplanetary spacecraft and hence more precise targeting of the spacecraft for pinpoint landing on bodies like Mars.

Table 7.1. URs for access to position. Fr. stands for Frame, where we distinguish L: local frames, N: national frames, G: global frame. Repro. stands for Reproducibility and gives the time window over which positions are expected to be reproducible with the stated accuracy. Note that navigation has been excluded since it has complex requirements depending on the particular application. From Plag (2006a).

Application	Parameter	Accuracy	Latency	Fr.	Repro.
Surveying with precise point positioning	3-d coor.	10 to 50 mm	days	N	decades
	velocity	1 mm/yr	n/a		
Monitoring	3-d coor	< 10 mm	days	L	decades
	velocity	< 10 mm/yr	weeks	L	decades
Control of processes	horizontal	10 to 100 mm	seconds to minutes	L	decades
Construction	3-d	< 10 mm	seconds to minutes	L	months to years
Early warning	3-d	10 mm	seconds to minutes	G	days
Hazards and risk assessments	3-d	< 10 mm	days to months	G	decades
Numerical weather prediction	IPWV	1-5 kg/m^2	5-30 minutes	G	decades
Climate variations	IPWV	1 kg/m^2	1-2 months	G	decades
Scientific studies	3-d coor.	< 10 mm	n/a	G	decades
	velocity	< 1 mm/yr	n/a	G	decades
Earth observations	3-d coor.	< 10 mm	days	G	decades
	velocity	< 1 mm/yr	n/a	G	decades

National space agencies recognize the importance of accurate terrestrial and celestial reference frames, station positions, and Earth orientation parameters and levy requirements on the accuracy of these and related radiometric observables (see Table 6.1 on page 205). However, the requirements levied are typically just the accuracy with which the observable can be measured. If the observable could be measured more accurately, then the requirement would be changed to reflect the more accurate measurement. This means that, in effect, tracking and navigating interplanetary spacecrafts requires as accurate a determination as can be made of the terrestrial and celestial reference frames, station positions, and Earth orientation parameters. This makes tracking and navigating interplanetary spacecrafts one of the most demanding users of these GGOS products.

In summary, it can be stated that planetary geodesy, radio science, interferometry (including imaging VLBI, astrometric VLBI, and Earth-space VLBI), and interplanetary navigation in order to carry out and interpret their measurements all require accurate terrestrial and celestial reference frames and that these frames are linked to each other through accurate Earth rotation observations. The performance of the GGOS is not a limiting factor in all of these applications. However, to meet demanding future requirements, especially those imposed by inter-planetary navigation, GGOS will have to be enhanced to meet the requirements identified in Table 6.1 (on page 205). In particular, GGOS needs to allow the real-time determination of Earth orientation accurate to ≤ 2 cm, which will most likely require near-real-time VLBI measurements of UT1. GGOS also should enable calibrations of troposphere delay and ionosphere accurate to ≤ 0.3 cm and ≤ 2 TEC units, respectively.

7.3 Quantitative requirements

Table 7.2. Overview of latency and accuracy requirements of main user categories. Modified from Plag (2006a).

Requ.	Latency	time scales	accuracy
UR1	real time	sec. to min.	< 10 cm
UR2	hours to days	up to diurnal	< 5 mm
UR3	weeks to months	monthly to seasonal	2-3 mm
UR4	$>$ months	interannual to secular	< 1 mm/yr

Quantitative requirements make more sense if they show a clear connection to the specific application from which they originate. Table 7.1 summarizes the current and likely future requirements for access to positions in a terrestrial reference frame for main applications in terms of accuracy, spatial and temporal resolution, reference frame, and reproducability. The accuracy requirements depend on both the time scales and the latency with low-latency applications in general considering shorter time scales. Table 7.2 summarizes the most demanding accuracy requirements as a function of latency and time scales.

For users requiring real-time positioning, the most extreme accuracy requirements are expected to be considerably less than 10 cm and down to 1 cm (e.g., sensor positioning, hydrographic measurements, automated snow-plowing), and in some cases even less than 1 cm (e.g., control of large mining and construction equipment). Some real-time applications will require high integrity (e.g., process control) and high update rates. For near-real time positioning and other near-real time applications with latencies of hours to days, accuracy requirements will be close to 1 cm in most of these applications (monitoring of infrastructure, meteorological applications) while other applications will require less accuracy (e.g., of the order of 5 cm) but higher integrity (e.g., land surveying). Many applications can accept considerable latency but will require accuracy at the 1 cm level or better for daily coordinates and a few millimeters or better on intraannual time scales. For long-term monitoring tasks, 1 mm/yr or better in stability seems to be a critical boundary both for scientific and nonscientific tasks. This number also applies to collection of geo-databases, which are to be maintained over time scales of several decades.

Presently, GGOS contributes significantly to meeting the requirements UR3 and UR4 (Table 7.2), although the stability requirement of < 1 mm/yr may not be meet. Based on GGOS products alone, the requirement UR1 cannot be met due to properties of the GPS-alone system combined with the large latency for required IGS products. For this UR, local and regional augmentations are currently required. Some but not all needs of the UR2 are met by GPS&IGS but the large latency of the precise IGS products and the insufficient accuracy of the rapid IGS products leave a considerable share of this UR in need of local or regional augmentation systems. While UR1 and partly UR2 in Table 7.2 can be met by local to wide-area augmentation systems, UR3 and UR4 depend crucially on the quality of ITRF and the available products. Moreover, achieving UR1 and UR2 through a system based on Signal in Space (SiS) only (that is the signal received from the GNSS satellites), would considerably increase the areas of applications and provide significant economic advantages.

Table 7.3 gives a more detailed overview of the quantitative requirements for typical science applications. For most of these applications requiring knowledge of the kinematic of the Earth's surface, the accuracy requirement in terms of motion is of the order of 1 mm/yr or less. Similarly, using precise point positioning for the determination of coordinates in a national reference frame, also requires knowledge of the velocity field with an accuracy of 1 mm/yr in all three components. Monitoring of infrastructure and hazardous areas have the same requirement on the accuracy of the velocity field.

The accuracy requirements for the geoid for the full utilization of satellite altimetry are of the order of 1 cm for wavelengths down to a few tens of kilometers, translating into an accuracy of 10^{-9} or better (Table 7.4).In order to monitor the mass movements in the Earth system and particular the global water cycle, accuracy requirements are on the order of < 10 mm of equivalent water column for spatial wavelengths of < 500 km, which translates into < 0.2 mm in geoid height and < 3 nms^{-2} for gravity. Temporal resolution is on the order of 1 month or, even better, 10 days.

For practical applications, the requirements for Earth rotation are dominated by the effect on positioning and the operation of satellite systems. For precise point positioning, errors in Earth rotation map directly into position errors. For example, an error of 1 mas (milliarcseconds) in polar motion corresponds to errors in horizontal displacements of the Earth's surface of about 30 mm, while an error of 1 ms (millisecond) in time corresponds at the equator to an error of about 460 mm in displacement. These numbers illustrate the high consistency between the terrestrial reference frame and Earth rotation, which is required to links the satellite frame to the terrestrial frame. For a low-latency access to precise point positions with an accuracy of 10 mm, the corresponding instantaneous accuracy for Earth rotation (being a factor of ten better than the position requirement) would be 0.03 mas and 0.002 ms in polar motion and rotation, respectively. Rothacher et al. (2001) report discrepancies between Earth rotation parameters determined with high temporal resolution

Table 7.3. User requirements for scientific applications. S.R. stands for spatial resolution, T.R. for Temporal resolution, Fr. stands for Frame, where we distinguish L: local frames, N: national frames, G: global frame. R. stands for Reproducibility and gives the time window over which the parameters are expected to be reproducible with the stated accuracy. 1 μGal is equal to 10^{-8} ms^{-2}. From Plag (2006a)

Application	Parameter	Accuracy	S.R.	T.R.	Fr.	R.
Mantle convection and plate tectonics	3-D velocities	< 1 mm/yr	n/a	n/a	G	several
	static geoid	$< 10^{-9}$	n/a	n/a	G	decades
	secular strain rate	10^{-19} s^{-1}	10^3 km	n/a	G	and longer
Postglacial rebound	3-D velocities	< 1 mm/yr	10^2 km	n/a	G	several
	geoid	$< 10^{-9}$	n/a	n/a	G	decades
	strain rates	10^{-15} s^{-1}	10^2 km	n/a	G	and longer
	Earth rotation	0.1 mas/yr	n/a	n/a	G	
	local sea level	< 1 mm/yr	0.2 to $1 \cdot 10^3$ km	n/a	G	
Climate change, including present changes in ice sheets and sea level	3-D displacements	1 mm	10^2 km	months	G	decades
	3-D velocities	< 1 mm/yr	$< 10^2$ km	n/a	G	decades
	local gravity	< 0.3 μGal	$< 10^2$ km	n/a	L	decades
	geoid	< 10 mm	200 km	n/a	G	decades
	Earth rotation	0.1 mas/yr				
	local sea level	< 1 mm/yr	10^2 km	months	n/a	decades
Ocean circulation	gravity field	$< 10^{-9}$	10^2 km	months	G	decades
Hydrological cycle	gravity field	$< 10^{-9}$	10^2 km	months	G	decades
	3-D displacements	< 1 mm	10^2 km	months	G	decades
Seasonal variations	gravity field	$< 10^{-9}$	10^2 km	months	G	decades
	local gravity	< 1 μGal	n/a	months	L	decades
	3-D displacements	< 1 mm	10^2 km	months	G	decades
	Earth rotation	1 mas				
Atmospheric circulation	Earth rotation	1 mas		days		decades
Earth tides	gravity	0.01 μGal	10^3 km	hours	G	years
	3-D displacements	1 mm	10^3 km	hours	G	years
	strain rates	10^{-15} s^{-1}	1 to 10^3 km	< 1 day	G	years
Surface loading	3-D displacements	< 1 mm	$< 10^2$ km	< 1 day	G	years
	local gravity	0.1 μGal	$<< 10^2$ km	< 1 day	G	years
Seismotectonics	3-D displacements	1 mm	$< 10^2$ km	days	G	hours to years
	strain rates	10^{-19} s^{-1}	< 10 km	years to secular	L	decades
Volcanoes	3-D displacements	1 mm	1 to 10^2 km	< 1 day	L	years
	gravity	1 μgal	1 to 10^2 km	days	L	years
Earthquakes, tsunamis	3-D displacements	1 mm to 1 cm	$<< 10^2$ km	sec to days	L	weeks to to decades
	strain	10^{-8}	≈ 10 km	offsets	L	n.a.
	strain rates	10^{-15} s	≈ 10 km	< 1 yr	L	years
	local gravity	0.3 μGal	$<< 10^2$ km	sec to days	L	weeks to decades

from GPS and those determined from VLBI and SLR to be orders of magnitude better than these requirements. However, at sub-daily temporal resolutions, the present

Table 7.4. Measurement requirements in terms of geoid height and gravity anomaly accuracy. Taken from Drinkwater et al. (2003). Note that the requirements for both scientific and nonscientific applications are given. 1 mGal is equal to 10^{-5} ms^{-2}.

Application	Accuracy Geoid (cm)	Accuracy Gravity (mGal)	Spatial Resolution half wavelength (km)
Oceanography:			
Short scale	1-2		100 km
	0.2		200 km
Basin scale	~0.1		1000 km
Solid Earth:			
Lithosphere and upper mantle density structure		1-2	100 km
Continental lithosphere			
– Sedimentary basins		1-2	50-100 km
– Rifts		1-2	20-100 km
– Tectonic motions		1-2	100-500 km
– Seismic hazards		1.0	100 km
Ocean lithosphere and interactions with asthenosphere		0.5 - 1.0	100-200 km
Geodesy:			
Leveling by GPS	1.0		100-1000 km
Unification of worldwide height systems	1.0		100-20000 km
Inertial navigation system		~1-5	100-1000 km
Orbits (1 cm radial orbit error for altimetric satellites)		~1-3	100-1000 km
Ice sheets:			
Rock basement		1-5	50-100 km
Ice vertical movements	2.0		100-1000 km
Sea-level change:	Many of the above applications, with their specific requirements, are relevant to sea-level studies		

low-latency or near-real time accuracy of Earth rotation observations and predictions is most likely not meeting these requirements. de Viron et al. (2005) point out that for the determination of gravity field changes with missions like GRACE, diurnal and sub-diurnal effects of the atmosphere on Earth rotation are of importance for the orbit determination. Their model study show that atmospheric angular momentum variations at diurnal time scale can produce polar motion near 0.2 mas. On time scales of several days, atmospheric effects can reach several milliarcseconds (Lambert et al., 2006), corresponding to 10 cm or more in displacement.

Other requirements on Earth rotation result mainly from scientific applications, and for these applications, an increasing accuracy of the observations normally leads to new applications. Examples are questions related to the effect of earthquakes (e.g., Chao & Gross, 2005), volcanic eruptions, and seasonal mass motion on the Earth's surface (e.g., Chen & Wilson, 2003; Gross et al., 2004) on Earth rotation, where the current accuracy of the observations as well as the sophistication of models (see Salstein et al., 2001) are limiting the scientific understanding of the processes on a rotating Earth. Likewise, the current accuracy is at the margin of what is required to achieve improvements in understanding and modeling of Earth rotation changes induced by interactions of the solid Earth with its fluid envelope (e.g., Plag et al., 2005). For studies of the interaction between fluid core and solid mantle, the length of the space-geodetic Earth rotation record with high accuracy appears to be the main limitation. Secular rates are compromised by the lower accuracy and potential instabilities of the older parts of the record, which limits its application to studies of, for example, postglacial rebound effects on Earth rotation (see, e.g., Mitrovica & Milne, 1998).

As illustrated in Section 2.9, geodetic observations increasingly are used in non-geodetic applications such as numerical weather predictions, climate studies, and space weather monitoring. Table 7.5 summarizes the requirements in terms of geodetic infrastructure and parameters for numerical weather prediction and climate applications. For numerical weather prediction, the low latency combined with high accuracy in station coordinates constitute demanding requirements. For climate studies, the requirement for long-term stability appears to exceed the present capabilities.

Table 7.5. Requirements for meteorological applications of GPS. Accuracy requirements are for IPWV in kg/m^2 or path delay in mm. Values are from Elgered et al. (2005).

Nowcasting		
Requirement	Generic	GPS Meteorology network
Horizontal domain	Sub-regional	Europe to national
Horizontal sampling	5-50 km	10-100 km
Repetition cycle	0.25 - 1 h	5 min - 1 h
Absolute accuracy	1-5 kg/m^2	1-5 kg/m^2
Timeliness	0.25-0.5 h	5 - 30 min

Numerical Weather Prediction				
Requirement	Generic		GPS Meteorology network	
Horizontal domain	Global	Regional	Global	Regional
Horizontal sampling	50-500 km	10-250 km	50-300 km	30-100 km
Repetition cycle	1-12 h	0.5-12 h	0.5-2.0 h	0.25-1.0 h
Integration time			MIN(0.5 h, rep. cycle)	MIN(0.25 h, rep. cycle)
Absolute accuracy	1-5 kg/m^2	1-5 kg/m^2	3-10 mm	3-10 mm
Timeliness	1-4 h	0.5-2 h	1-2 h	0.5-1.5 h

Climate Monitoring		
Requirement	Generic	GPS Meteorology network
Horizontal domain	regional-global	All
Horizontal sampling	10-100 km	10-250 km; indiv. stat.
Time domain	>> 10 years	Weeks to many years
Repetition cycle	1 h	1 h
Absolute accuracy	0.25-2.5 kg/m^2	1 kg/m^2
Long-term stability	0.02-0.06 kg/m^2/decade	0.04-0.06 kg/m^2/decade
Timeliness	3-12 h	1-2 months

7.4 Tasks of GGOS

The tasks of GGOS are to encourage, facilitate and promote the following activities, based mainly on the combined work of the IAG Services:

- define a unique celestial reference system;
- define a unique terrestrial reference system, including geodetic datum;
- define a unique geodetic reference system;
- define a unique gravity reference system;
- define all the physical and mathematical models needed to analyze GGOS observations;
- provide and maintain an accurate, stable, and homogeneous celestial reference frame;
- provide and maintain an accurate, stable, and homogeneous terrestrial reference frame including its origin;
- provide and maintain the time-dependent Earth orientation parameters that are used to transform coordinates between the terrestrial and celestial reference frames;
- provide and maintain definitions, constants, models, etc. of the geodetic reference systems;
- provide and maintain parameters describing the static and time-dependent components of the Earth's gravity field;
- provide and maintain parameters describing the static and time-dependent components of the shape of the land, ice, and ocean surfaces;
- provide and maintain parameters describing the total electron content of the ionosphere;
- provide and maintain parameters describing the water vapor content of the troposphere;
- provide and maintain parameters describing the transport of mass within and between the atmosphere, oceans, and land.

7.5 Products available through GGOS

As a result of the above tasks, the principal products that are determined and provided by the IAG Services and made available through GGOS include:

- a catalog of celestial radio sources including their coordinates that provides the celestial reference frame;
- a catalog of terrestrial sites defining the reference polyhedron associated with the terrestrial reference frame, including their reference coordinates at a common epoch and time series describing the temporal evolution of the coordinates;
- time series of coordinates of additional terrestrial sites or points that are needed to densify the terrestrial reference frame in order to provide access to the frame anywhere on the Earth's surface;

- a model predicting the motion of the Earth's surface caused by loading effects of atmospheric surface pressure, ocean-bottom pressure, and continental water storage including snow and ice;
- precise orbits and clocks for GNSS satellites that allow access to the terrestrial reference frame;
- time series of Earth rotation parameters (UT1, polar motion, nutation/precession) including their time rates-of-change that provides the link between the celestial and terrestrial reference frames;
- values of the defining constants and derived physical and geometrical parameters of the geodetic reference system;
- values of parameters describing the static component of the Earth's gravity field;
- time series of parameters describing the time-dependent component of the Earth's gravity field;
- time-dependent maps of the total electron content of the ionosphere;
- time series of zenith path delays that provides the water vapor content of the troposphere;
- time series of angular momenta of the atmosphere, oceans, continental water storage including ice and snow, mantle, and core that provide estimates of the mass transport within the Earth system;
- time series of sea surface height and sea level measurements that provide estimates of the changing shape of the ocean surface;
- time series of ice sheet and glacier elevations that provide estimates of the changing shape of the ice surface;
- similar time-dependent, body-fixed site coordinates, orientation parameters, and gravity parameters for other planets and celestial bodies in the solar system such as the Moon and Mars.

7.6 Accuracy of GGOS products

The GGOS products listed in the previous section, which are produced by the IAG Services, must have sufficient accuracy, temporal and spatial resolution, and latency to meet the requirements of all users. This can be done if the requirements of the most demanding users are met. The most demanding user of the terrestrial reference frame is likely to be scientific studies of sea level change caused by climate change. Since global sea level is rising at a rate of a few millimeters per year, and since the frame in which the rise is being measured should be at least on order of magnitude more accurate than this, the terrestrial reference frame should be accurate at a level of 1 mm and be stable at a level of 0.1 mm/yr (Blewitt et al., 2006a).

The most demanding users of the geoid models are likely to be: (1) the use of the geoid in oceanic general circulation models to define the mean sea surface topography, and (2) the GNSS determination of the height of objects at the millimeter level. These applications require the static geoid to be accurate at a level of 1 mm and to be stable at a level of 0.1 mm/yr, consistent with the accuracy and stability of the terrestrial reference frame.

The most demanding user of the Earth orientation parameters is likely to be the tracking and navigation of interplanetary spacecraft. This user is capability-driven and requires the most accurate EOPs that can be determined, realizing that those determined in near real-time are somewhat less accurate than those determined with a delay of a couple of weeks. This user also requires that the EOPs be consistent with the celestial and terrestrial reference frames. So if the terrestrial reference frame is accurate at a level of 1 mm, consistency demands that the EOPs also be accurate at the same level of 1 mm.

Specifically, the accuracies of the GGOS products required by the most demanding users are:

- celestial reference frame: accurate to 25 μas, stable to 3 μas/yr;
- terrestrial reference frame: accurate to 1 mm, stable to 0.1 mm/yr, including geocenter; scale accurate to 0.1 ppb, stable to 0.01 ppb/yr;
- Earth orientation parameters: accurate to 1 mm with a latency of 2 weeks, 3 mm in near real-time, with daily resolution;
- static geoid: accurate to 1 mm, stable to 0.1 mm/yr, with a spatial resolution of 10 km;
- time varying geoid: accurate to 1 mm, stable to 0.1 mm/yr, with a spatial resolution of 50 km and a time resolution of 10 days.

7.7 Functional specification for GGOS

The second part of this chapter focuses on the functional specifications of the geodetic observing system in 2020. The goal here is to have a clear picture of what GGOS should be able to deliver in 2020, based on the known or expected user requirements as summarized above. For the functional specifications, the most demanding of these user requirements have been considered as described in the previous sections.

7.7.1 Determination, maintenance, and access to the global terrestrial reference frame

The following functional specifications define the inherent accuracy of the ITRF at the time of determination, as well as later epochs. Currently, reference coordinates can only be predicted for the reference points of the ITRF, and the underlying model for the prediction of the reference frame coordinates is a linear model. In the future, a requirement on the terrestrial reference frame is that reference positions can be predicted for any point on the Earth surface. Therefore, the basis for access to the reference frame is a dynamic Earth reference model, which assimilates observations of variations in Earth's geometry, gravity field, and rotation. Through data assimilation, the model will be forced to closely reproduce the observed changes in the reference polyhedron (the present-day ITRF) as well as observed variations in the gravity field and rotation in a self-consistent way.

ITRF-001-DER: *Provision of the reference frame through a dynamic Earth reference model* — The terrestrial reference frame will be provided by an operational dynamical Earth reference model which will assimilate observations of variations in the Earth's geometry (in particular, for a reference polyhedron), the shape of the ice and ocean surfaces, the gravity field, and Earth rotation. Moreover, the reference model will also assimilate auxiliary observations, in particular meteorological observations. This dynamic Earth reference model will allow the prediction of reference trajectories for any point on Earth with temporal resolution of 1 hour and a sub-kilometer spatial resolution. The stability of the model in terms of geokinematic will be that of the reference polyhedron, i.e., sub-millimeter per year.

ITRF-002-ORI: *Tie between RFO and CM* — The deviation between the reference frame origin of the terrestrial reference frame and the center of mass of the Earth system will be smaller than 1 mm at any time.

The specification ITRF-002-ORI implicitly limits the secular trend between RFO and CM to 0.1 mm/yr over 10 years and even less if we look at longer time intervals. With this specification, GGOS meets the most important requirement for global sea level studies.

ITRF-003-PRE: *Precision of reference coordinates* — The precision of coordinates of the points of the reference polyhedron of the ITRF will be better than 1 mm in the horizontal and 3 mm in the vertical component at any time.

Since the terrestrial reference frame will not utilize a reference polyhedron based on positions and secular velocities, it does not make sense to specify the accuracy of velocities or a secular stability. However, ITRF-003-PRE implicitly determines also the accuracy of any secular motion.

ITRF-004-SCA: *Scale of the reference frame* — The scale of the reference frame will be accurate to 0.1 ppb and stable to 0.01 ppb/yr.

ITRF-005-ACC: *Access to the ITRF* — Standard access to the ITRF will be through precise orbits and clocks of the GNSS satellites and Earth rotation parameters. Low latency (down to real-time) orbits and clocks will have an accuracy equivalent to a range error of less than 5 cm for an availability of 99.999%. Post-processed orbits and clocks will allow the determination of single daily point coordinates with an accuracy of 1 cm in the horizontal and 2 cm in the vertical anywhere on the Earth surface with high availability (99.999%) and high integrity.

The specification ITRF-005-ACC is the basis for accurate precise point positioning facilitating many of the applications discussed in the previous chapters. For point motion observed with stationary, continuously observing GNSS sites, the observed trajectory can be compared to the trajectory predicted for that point with the dynamic Earth reference model in order to determine anomalous motion as specified in ITRF-001-DER. At the same time, this specification implicitly provides the basis for monitoring surface kinematics.

7.7.2 Earth rotation

ERP-001-EOP: *Earth Orientation Parameter* — Earth Orientation Parameters will be determined with an accuracy of 1 mm, a temporal resolution of 1 hour, and a latency of 1 week; near real-time determinations of the Earth Orientation Parameters will be determined with an accuracy of 3 mm.

7.7.3 Earth's gravity field

GRAV-001-GEOID: *Accuracy of the static geoid* — The static geoid will be provided with an accuracy of 1 mm, a long-term stability of 0.1 mm/yr and a spatial resolution of 10 km.

GRAV-002-VAR: *Accuracy of the time variable gravity field* — The time variable geoid will be provided with an accuracy of 1 mm, a long-term stability of 0.1 mm/yr, a spatial resolution of 50 km, a temporal resolution of 10 days, and a latency of 0.5 months.

7.7.4 Earth system monitoring: mass transport and mass redistribution

ESM-001-SSH: *Sea surface height variations* — The sea surface height variations will be determined globally with an instantaneous local accuracy of 10 mm, a temporal resolution of 10 days, a spatial resolution of 10 km, a latency of 5 days, and a local secular accuracy of 0.5 mm/yr.

ESM-001-GSL: *Global sea surface changes* — The globally average sea surface height changes will be determined with an instantaneous accuracy of 1 mm, a temporal resolution of 10 days, a latency of 10 days, and a secular accuracy of 0.1 mm/yr.

ESM-002-CRY: *Cryosphere mass balance* — The variations in the surface elevation for the large ice sheets will be determined with an instantaneous local accuracy of 20 mm, a temporal resolution of 10 days, a spatial resolution of 10 km, a latency of 20 days, and a local secular accuracy of 0.2 mm/yr.

ESM-003-WCY: *Mass transport in the global water cycle* — The mass transport in atmosphere, ocean, cryosphere, and terrestrial hydrosphere will be determined to an accuracy of an equivalent of 10 mm water cover with a temporal resolution of 10 days, a spatial resolution of 400 km, and a latency of 0.5 months.

7.7.5 Determination, maintenance, and access to the celestial reference frame

ICRF-001-DET: *Determination of the Celestial Reference Frame* — The ICRF will be defined by the coordinates of a number of extragalactic radio sources distributed throughout the sky. The coordinates of the ICRF sources will be accurate to 25 microarcseconds and stable to 3 microarcseconds/yr..

7.8 Operational specifications for GGOS

GGOS must ensure performance of certain operations in order to generate products of sufficient accuracy to meet the requirements of the users, as expressed in the functional specification given in the previous section. The high level activities that GGOS together with the IAG Services must ensure are to:

- operate global networks of geodetic reference stations, gravimeters, and tide gauges;
- operate a global sub-network of core reference stations at which the techniques are co-located;
- determine the survey ties between the co-located techniques;
- process all observations with an accuracy and consistency of at least 1 ppb;
- operate a dynamic Earth reference model assimilating comprehensive observations of variations in Earth's geometry, gravity field and rotation;
- document the procedures, standards, and conventions used to generate the products;
- maintain databases of observations and products;
- ensure continuity, accuracy, and consistency of observations and products as the networks and data reduction procedures evolve.

Chapter 8
The future geodetic reference frame

T. A. Herring, Z. Altamimi, H.-P. Plag, P. Poli

8.1 Introduction

The ingredients for future geodetic reference frames can be separated into at least two categories, namely (1) theoretical, conventional, and definitional aspects, and (2) practical implementation aspects and physical components. In this Chapter, we focus on category (1).

The present theoretical basis for the current ITRF is fairly simple at the moment (see, for example, McCarthy & Petit, 2004), which is appropriate as long as most of the infrastructure for the determination of the ITRF is ground-based. As more components of its realization and maintenance move away from the Earth's surface into near-Earth space (and maybe beyond) and as accuracy "requirements" become more strict it is likely that closer attention will be paid to relativistic aspects. While the IAU has elaborated its recommended metric, some work with respect to ITRF determination and maintenance currently still uses Newtonian reductions with some general relativity "corrections" applied. This may no longer be an adequate approach in 2020. But also the Newtonian part of the theory is not fully at the targeted accuracy level. In particular, the theory of Earth rotation is especially weak with respect to increased accuracies. Moreover, motions of the RFO with respect to the CM are not explicitly accounted for, and higher degrees of surface deformation are neglected. Any coupling of the angular and linear momentum balance is not considered.

As pointed out in Section 2.7, it is conceivable that by 2020 the international timescale will be maintained using ultra-stable frequency standards in orbit. Earth-based clocks, which are subject to large environmental "noise", might only be used as local standards slaved to the orbital clock system. However, it is mentioned here that synchronization poses a severe problem which is not solved so far.

Below, we do not consider relativistic effects. However, the concept and theory described below strives to provide a fully self-consistent system, which can then be the basis for a conventional system and frame. Currently, this is not the case in a strict sense and most likely also not at a perceptible level. There seems to be a

H.-P. Plag, M. Pearlman (eds.), *Global Geodetic Observing System*,
DOI 10.1007/978-3-642-02687-4_8, © Springer-Verlag Berlin Heidelberg 2009

degree of arbitrariness in assigning some components or features to one feature or another (like the partition between nutation and polar motion). In general, if the decision is truly arbitrary, then the critical issue is to be clear and unambiguous in the definitions and conventions.

The current concept of the terrestrial reference frame is that of a polyhedron with secular motion of the individual points. This assumption is a severe limitation as it imposes a global filter of all geophysical signals and hampers the comparison of models and observations. The central question addressed here is how the non-secular deformation should be included in the mathematical models that define the future reference system. In principle, there are two alternatives to achieve this:

- (A) Maintain the secular frame similar to the current ITRF and associate to it with a deformation model that allows the computation of non-secular motion at any location; or
- (B) consider forming a complete model that describes both the secular and non-secular parts.

Here, we consider alternative (B). This alternative poses considerable challenges in its implementation.

8.2 Concept of reference system and reference frame

A geodetic reference system is a set of definitions and mathematical models that allow geodetic measurements to be related to each other in a systematic fashion. Two basic classes of geometric systems are required in modern geodesy: one related to a non-rotating inertial reference frame; and the other attached in a prescribed fashion to the rotating Earth(see Section 2.2). In addition, a reference system for potential based measurements is also required. Often this latter system is related to the geometric reference frame through the height of an equipotential surface above or below a geometric reference surface defined through the geometric reference system. For geometric measurements, such as ranges and angles, the reference system is most commonly based on a set of Cartesian axes. Part of the reference system definition specifies how the origin and the direction of the axes in this reference system are specified, but equally important are the mathematical models that relate measurements to coordinates in the realization of reference frame. On a deformable Earth all reference system definitions are complicated by the need to account for the deformation.

The evolution of reference system definitions has progressed with the accuracy of geodetic methods with refinements in the system definitions often leading to improved accuracy of geodetic methods. Conceptually, a reference system may be defined with its origin (denoted here as RFO) at the CM, its Z-axis along a mean direction of the rotation axis during a specified interval of time and its X-axis passing through a specific location. Such definition would appear to be useful when positional measurements (to define axes directions) and gravitational measurements are available to define the CM and to contribute to the axes directions through the inferred moments of inertia. However, in an era of millimeter level position determinations and gravity measurements accurate to 1 ppb of the main field, this type of definition poses problems primarily because at these levels mass motions in the system are easily detected. In modern definitions of the reference system and its realization, which is the reference frame, non-secular motions of the surface of the Earth and mass re-distributions must be explicitly accounted for. The advances in computing power also allow the parameters of reference frame to be simultaneously estimated from potentially a wide variety of data types. The ability to simultaneously and rigorously realize an Earth reference frame is one of the major benefits of GGOS.

In this context, we introduce two Cartesian XYZ coordinate systems. Precisely how one of these systems is attached to the Earth and how the other will define a non-rotating inertial reference frame will be defined later. For the Earth attached system, we can give the time-dependent coordinates of every mass element in the Earth. A geodetic measurement site is simply one of the mass elements at the surface of the Earth. The mass elements undergo a variety of motions: Tidal displacements due to both solid Earth tides and ocean loading, loading signals from atmospheric and hydrology mass movements, tectonic motions, and other secular motions from internal mass movements. Certain sites will also under go seismic displacements in the forms of coseismic and postseismic displacements, or be affected by anthropogenic subsidence.

In the terrestrial system, the external potential will be given by

$$\begin{aligned} V(x_p,y_p,z_p,t) = {} & G\int_{\text{SolidEarth}} \frac{\rho_s(x,y,z,t)}{d_{sp}}\mathrm{d}V + \\ & G\int_{\text{FluidCore}} \frac{\rho_c(x,y,z,t)}{d_{cp}}\mathrm{d}V + \\ & G\int_{\text{FluidEarth}} \frac{\rho_c(x,y,z,t)}{d_{fp}}\mathrm{d}V \end{aligned} \tag{8.1}$$

where V is the gravitational potential at position x_p, y_p, z_p, ρ_s, ρ_c and ρ_f are the density of the solid earth, fluid core, and fluid earth (ocean and atmosphere), d_{sp}, d_{cp}, and d_{fp} are the distances from the integration points to x_p, y_p, z_p, and G is the gravitational constant. The reference system will determine how the densities change with time. In each of the regions of the Earth, the motion of the mass points is different. In the solid Earth, except at times of earthquakes, much of the motion is secular (although at seasonal and interannual time scales, significant motion takes place). In the mantle, typically motion rates are similar to plate tectonic rates of 10 cm/yr. In the fluid core, the motions are much more uncertain but could be as high as km/yr and vary on decadal time scales (as evidenced by changes in the magnetic field and decadal changes in length of day). The motions in the fluid Earth are much more rapid and have large spatial and temporal variations. When the potential is sensed from space, the fluid Earth component can be treated as a surface density layer.

A consequence of the moving fluids in this system is that the loading displaces masses and consequentially the solid Earth mass motions have a contribution from the loading phenomena. If the changes in the mass loading are known, then this effect in the solid Earth integration can be accounted for and the temporal variations in density can be treated as secular motions.

In defining a reference system that fully exploits the accuracy of today's measurements care must be taken. The motions of the system need to be accounted for in a consistent fashion so that results in the reference frame can be related at different times. The geodetic signals that reveal new things about the Earth are those that show significant deviations from the motions predicted in the reference system. The system definition must treat consistently the deformations in the system and here in lies the current limitations in current reference system definition and realization. Ideally, all sources of mass motion and deformation would be known but this currently is not the case. Some components of mass motion are reasonably known such as the solid Earth tides while others such as hydrological loading are not well known. In a coherently constructed reference system for GGOS, addressing the inconsistencies in the current models should be possible.

The largest gains in non-secular deformation modeling are likely to be in the area of modeling hydrological variations but there could be some potential problems. Missions that measure time variable gravity such as GRACE yield estimates of mass movements, which are treated as thin-shell surface density layers on the surface of the Earth. With current gravity missions, temporal and spatial resolution of hydrological mass estimates is limited, but increased spatial and temporal resolution is expected for future missions. These mass estimates could be used to compute elastic deformations of the Earth surface provided that the mass loads the surface and is not supported by dynamical effects in the atmosphere. Atmospheric dynamics typically considers two definitions for surface pressure, namely (1) the dynamic surface pressure (denoted as P_s) that is the physical quantity measured by a barometer, and (2) the hydrostatic surface pressure (denoted by π_s) that would be measured by an instrument only sensitive to the downward force induced solely by the weight of the column of air without its acceleration (if any). The first quantity is in fact the one collected by *in situ* barometers, and assimilated into atmospheric weather models, while the second quantity is the one predicted by global weather models, which assume that vertical acceleration is zero at their spatial and temporal scales (hydrostatic assumption). A general assumption is that the vertical atmospheric acceleration is negligible if averaged over synoptic or larger scales (i.e., a few hours and a few tens of km in the horizontal). However, at smaller smaller scales (for example, < 1 hour, < 1 km), (1) there may be some sources and sinks to the atmospheric mass load (sources: volcanoes, evaporation; sinks: e.g., rain, pouring down at a rate that can make the local air column with an approximate mass of 104 kg per square meter lose up to 30 kg of water over the same area in an hour of very strong rainfall events), and (2) there may be some local acceleration upward or downward (strong updrafts or downdrafts associated with moist processes, and sustained downhill (so-called "katabatic") winds (e.g. over Antarctica and Greenland) which would induce vertical accelerations at the beginning and end of the slopes on which these winds run. But these two effects are usually considered to start to become significant only at the small scales indicated above. Thus, for atmospheric contribution to the mass and loading, the surface pressure output of current (hydrostatic) global weather models appears to be sufficient.

As pointed out by Blewitt (2003), the appropriate Love numbers need to be used when the loading signals are computed to ensure the origin of the reference frame for the load is known. The most common choices here are Center of Figure of the solid Earth (CF), the Center of Mass of the Solid Earth (CE), or the CM of the whole Earth system (Blewitt, 2003). If CF is used, the gravity field associated with the loading will have degree-1 terms. If CM or CE are used, then the transformation to CF position estimates will need translations added.

In a dynamic reference system realization, the expected motions of all geodetic sites are computed. For those sites that form the reference frame, these time dependent motions should be well known. Such sites might be those away from coastal area and that have small or well computed hydrological signals. The well monitored frame realization sites, the rotation and translation that align the coordinates to the reference frame values would be used to realize the reference system. Unlike current reference frames, such as ITRF2005, these new reference frames need to be continuously maintained with non-secular motions being measured and computed. For the reference frame to be useful to all users, the non-secular motions need to be computed at all geographic locations (gridded at the appropriate resolution), or a service is needed that evaluates the deformation at given location and time specified by users. The scientifically interesting product from a reference system realization is displacement, which is the difference between measured coordinates and those determined from the reference system.

8.3 Future reference frame formulations

Current global reference frames basically consist of a set of point coordinates at a reference epoch t_R and constant velocity vectors for each point. This set of coordinates describes the secular evolution of a polyhedron over time, and the points implicitly determine the axes, the RFO, and the scale of the underlying reference system. In addition to the secular polyhedron, the frame also includes a set of models that describe deviations of the actual motion of the Earth surface from the secular polyhedron.

The mathematical model for the description of point position $\mathbf{X}$ of the secular polyhedron as function of time t is that of regularized coordinates

$$\mathbf{X}^{(i)}(t) = \mathbf{X}_0^{(i)} + \mathbf{V}_0^{(i)} \cdot (t - t_0) \tag{8.2}$$

(McCarthy & Petit, 2004). In order to be able to assign predicted reference coordinates to any point on the Earth surface, knowledge of the global velocity field $\mathbf{V}_0 = f(\mathbf{X})$ would be required. This is currently not available. Therefore, precise satellite orbits and clocks are used to make a reference positions relative to the vertices of the polyhedron available at any point and anytime.

The simple mathematical model of regularized coordinates has two major problem, one being the fact that the motion of the mass elements is not linear in general. The second problem is that the velocity vectors have errors, which over time can deform the polyhedron considerably. Therefore, frequent adjustments of the coordinates and velocities of the vertices of the reference polyhedron are necessary, leading to significant temporal inhomogeneities (if complete recomputations of past reference frame-related quantities are not performed). However, if initial analysis of the geodetic data is preformed with minimal constraints applied to the system, reference frames can be updated from covariance matrix and solutions or from normal equations very rapidly.

Having more elaborated models for the prediction of the Earth's surface motion available, the secular model can be replaced by

$$\mathbf{X}(t) = \mathbf{X}_0 + \delta\mathbf{X}(t) \tag{8.3}$$

where $\delta\mathbf{X}(t)$ is predicted by a reference Earth model. In principle, this reference Earth model predicts the global displacement field $\delta\mathbf{X}(t)$ for any point on the Earth surface and for any time t. This also implies that the station motion model used in space-geodetic analyses to describe the point motion as function of time could fully be based on the predictions of the reference Earth model. Conventionally, only those parts of the point motion changing significantly over the analysis interval are accounted for.

Over the next few years, the displacement field in practice will be a composite of different contributions, that is,

$$\delta\mathbf{X}(t) = \mathbf{V}_0 \cdot (t - t_0) + \sum_{k=1}^{K} g_k(t, \mathbf{X}), \tag{8.4}$$

where g_k; $k = 1, K$ are the displacement fields determined from geophysical models of Earth tides, surface loading, earthquake processes, and other surface displacements.

Considering the current and expected accuracy, the reference Earth model will have to be based to some degree on assimilation of observations. Alternatively, a completely empirical approach would be based on the observed polyhedron, which would preserve the internal consistency of the observation technique and represent the time-dependent coordinates of the polyhedron points as

$$\mathbf{X}^{(i)}(t) = \mathbf{O}^{(i)}(t), \tag{8.5}$$

where $\mathbf{O}$ are observed time series. This references frame would be aligned to the secular model described by (8.2) on average with no internal deformation of the observed polyhedron. This approach would required means for the interpolation of $\mathbf{O}^{(i)}$. For that, again the orbits and clocks of the GNSS could be utilized. However, this approach would benefit strongly from improve station motion model (i.e., predictions of the $\mathbf{O}^{(i)}$) for the analysis.

8.4 Origin and orientation of the TRS

The origin of the future reference frame will remain to be defined as the CM, which is the only point in the Earth system that physically is special. Concerning the orientation of the axes, one could consider to have these axes aligned to the principal moments of inertia. As a consequence, polar drift would be zero in such a system provided the linear motions of the reference sites include the effects of motion of the moments of inertia with respect to the crust of the Earth.

8.5 Scientific challenge of the future reference frame: the need for an Earth system model

As stated above, motion of points at the Earth's surface results from internal and external forcings, which do not only affect the shape of the Earth but also the gravity field and the rotation of the Earth. The internal forcing is mainly associated with geodynamic and tectonic processes, and integrated models that would predict surface motion at a wide range of spatial and temporal scales are not yet available. Locally and regionally, pre-, co- and postseismic deformations can be modeled to a certain extent (e.g., Okada, 1992; Pollitz, 1996, 1997; Kreemer et al., 2006b; Sun et al., 2006) with the accuracy of the predictions depending on the local and regional processes and tectonic setting. Models for the effect of earthquakes on the gravity field and Earth rotation have also been developed (e.g., Chao & Gross, 1987; Sun & Okubo, 1998; Chao & Gross, 2005). However, secular plate motion models that agree well with the observed present-day secular plate motion are generally empirical models (e.g., DeMets et al., 1994; Bird, 2003; Kreemer et al., 2003; Sella et al., 2002), and some of these models use the secular motion determined from space-geodetic observations as constraints.

The external forcing can be separate into body and surface forcing with the former acting as a volume force on the whole solid Earth and the latter acting as a surface force on the surface of the solid Earth. While the effects of the body forces on the solid Earth, that is, the tidal forcing due to the tidal potential, on the shape, rotation and gravity field of the Earth is well understood and model predictions are at an accuracy level comparable to the accuracy of observations (e.g., McCarthy & Petit, 2004), this does not seem to be the case for the surface forces resulting from surface mass loads. Mass motion in the fluid envelope of the solid Earth and the interior of the solid Earth. Rearrangements of mass within the fluid of the solid Earth, including the atmosphere, oceans, and terrestrial hydrosphere, cause changes of the Earth's gravitational field, force Earth's rotation changes by changing the solid

Earth's inertia tensor and angular momentum, and induce changes in Earth's shape by changing the load acting on the deformable solid Earth. The main uncertainties in modeling these effect appears to be in the surface mass loads (e.g., Van Dam et al., 2003).

Because of the main uncertainties being in the surface mass, geodetic observations have increasingly been used to invert for mass change. However, despite the coupling of the effects in shape, rotation, and gravity field, changes in the Earth's shape have been used independently to infer global scale mass motions (e.g., Blewitt et al., 2001; Blewitt & Clarke, 2003; Wu et al., 2002, 2003, 2006) without utilizing the concept of consistency across all geodetic observations. Others have used changes in the Earth gravity field to infer ice load changes (e.g., Velicogna & Wahr, 2006) and changes in the global and regional hydrology (e.g., Tapley et al., 2004a) without rigorously testing the consistency of the inferred mass transports with changes in Earth rotation and shape. Only recently, the importance of consistency has been emphasized (e.g., Clarke et al., 2005). Thus, Gross et al. (2004) inverted Earth's rotation and shape changes for mass loads; Gross (2006) inferred mass loads from observations of changes in the gravitational field and rotation; and Kusche & Schrama (2005) combined changes in Earth shape and gravitational field.

All of these attempts are hampered by the fact that the reference frame is biased because of the assumption of linear secular motion of the reference points, which in fact constitutes a global filtering of the geophysical signals, which results in a reduction of these signals in the geodetic time series. Moreover, the incomplete modeling of the reference point motion may also affect the scale of certain techniques because of different station networks (in particular, different ratios of hemispheric station numbers), and different observation times. As an extreme example, with the current practices, if one system only observed in winter and another only in summer, there would be an apparent scale difference due to the annual vertical signal caused by surface loads.

Therefore, more elaborated models for the point motion need to be integrated in the process of determining the reference frame motion. However, integrated model development is in an initial state, and particularly models that couple a rotating, deformable solid Earth fully with advanced models of atmosphere, ocean, and terrestrial hydrosphere, are just beginning to emerge. It is clear that such model development and validation will depend crucially on the availability of consistent observations of the geodetic quantities in a well-defined reference frame that does not by definition suppress the signals predicted by the model.

8.6 Towards an Earth system model

The Earth is a dynamic system. It has a fluid, mobile atmosphere and oceans, a continually changing global distribution of ice, snow, and water, a fluid core that is undergoing hydromagnetic motion, a mantle both thermally convecting and rebounding from the glacial loading of the last ice age, and mobile tectonic plates. In

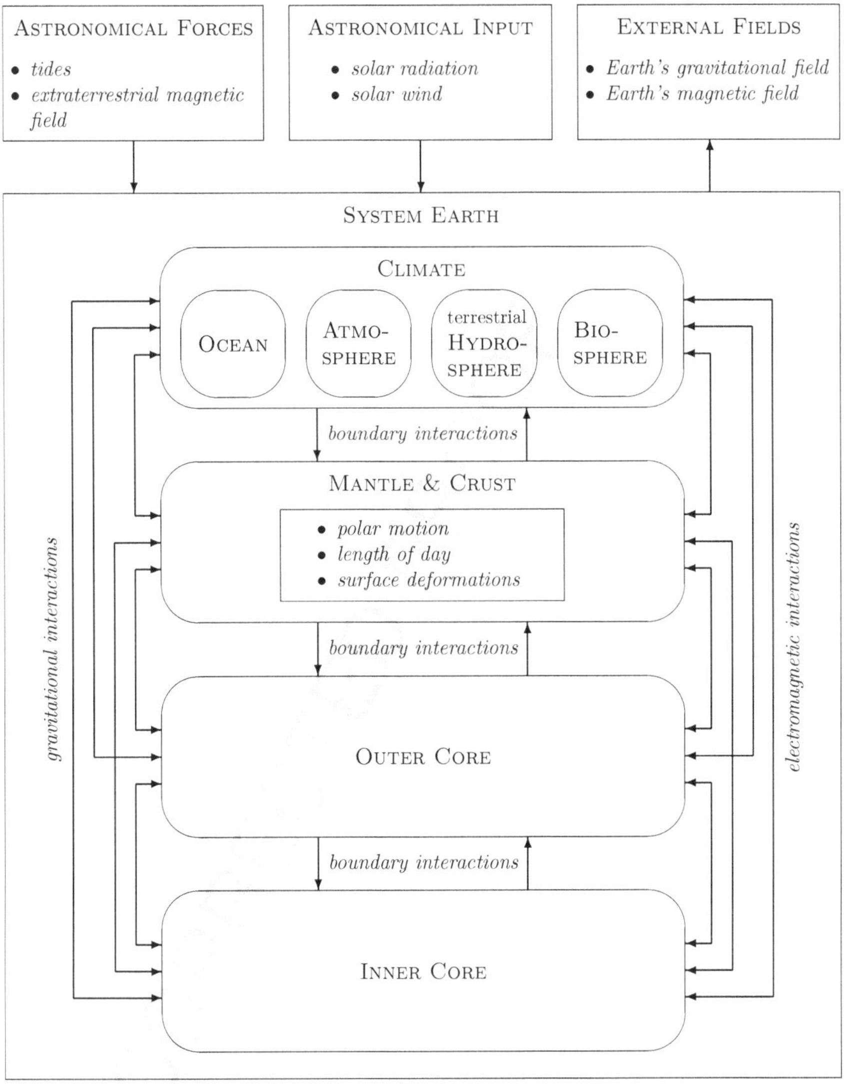

Fig. 8.1. Components of the Earth system and their mechanical interactions. From Plag (2006a).

addition, external forces due to the gravitational attraction of the Sun, Moon, and planets also act upon the Earth. These internal dynamical processes and external gravitational forces exert torques on the solid Earth, or displace its mass, thereby causing the Earth's rotation, gravitational field, and shape to change. Only if all these processes can be modeled and predicted in a consistent Earth system model can we expect to make progress towards the dynamic Earth reference model discussed above.

For the modeling of the mechanical processes in the Earth system, the system can be viewed as composed of subsystems such as crust and mantle, outer and inner cores, and the fluid envelope of the solid Earth (Figure 8.1). The latter consists of ocean, atmosphere, and terrestrial hydrosphere, which are the prominent components of the climate system. The biosphere is also interacting with the components of the climate system, and, considering the anthroposphere as part of the biosphere, also the solid Earth. We have chosen not to separate the cryosphere from the ocean and terrestrial hydrosphere but rather consider the ice load on land as part of the terrestrial hydrosphere and sea ice as part of the ocean.

The subsystems depicted in Figure 8.1 interact through surface forces at the joining boundaries and through volume forces due to gravity or electromagnetic fields. The overall system is subject to external forces including tides and the extraterrestrial magnetic field. Additionally, radiation absorbed in the system and interaction with solar wind change the dynamical state in the system and thus these external forcings have to be considered as input to the mechanical system.

The Earth's rotation is an integral quantity affected, in principle, by all processes changing the mass distribution and the dynamics of the system. Thus, the rotation is ultimately coupled to deformations and variations of the gravity field of the Earth.

In this mechanical view, the geometry of the solid Earth as well as the mass distribution in its interior are determined by the forces acting on the solid Earth, such as tidal forces, surface loading, and variation in the Earth's rotation and gravitational field, as well as forces inside the solid Earth, such as slow redistribution of mass due to convection, or rapid redistributions during earthquakes. With respect to surface loading due to mass redistribution in the ocean, the atmosphere, and the terrestrial hydrosphere, it is important to note that any of these mass movements changes the Earth's gravitational field primarily due to the mass movements, and, secondarily, due to deformations of the solid Earth. Any of these changes will affect the mass distribution in the ocean and thus create additional loads and variations in the three geodetic quantities.

In addition to the mechanical forces, on longer time scales we also have to consider thermodynamical forcing driving the convection in the Earth's mantle and core and creating phenomena such as volcanism and plate tectonics. However, for a description of the main characteristics of the geodetic variables, the mechanical view provides a valid basis.

Modeling of the Earth focusing on mechanical properties of the geosphere traditionally attempts to describe the whole Earth by a single system of equations specialized for specific phenomena (see e.g. Lambeck, 1988; Wahr, 1981, for rotation and loading deformations, respectively). In order to be feasible, this approach requires a high degree of simplification and many interactions and feedbacks have to be neglected. Consequently, even the most advanced geophysical models presently available are highly simplified and, moreover, specialized for the description of specific phenomena (such as nutation, Earth tides, surface deformations, geoid anomalies, glacial loading).

Over the last decades, several studies have demonstrated that complex systems can be modeled using a modular approach, with the individual modules representing subsystems or components that interacting through boundary conditions (surface forces, energy transfer, and particle transfer), and far-field interactions (gravimetric and electromagnetic volume forces). Complex climate models are built in this way, with separate submodels for, e.g., the ocean, atmosphere, cryosphere, clouds, and land surface.

In a modular approach to the dynamics of a rotating planet, the planet is represented by a number of physically defined subsystems coupled to each other both by boundary conditions and far-field interactions. Thereby, different subsystems are described each on its own by dynamical equations. Couplings between different subsystems in this approach have to be defined independently of the structure of dynamics of the subsystems as physically meaningful quantities, e.g. forces, moments or fields. There might be, moreover, external excitations acting on one or several particular subsystems such as for instance a tidal potential.

With respect to geodetic variables, integrated systems have been studied mainly for Earth rotation. Jüttner & Plag (1999) used a simple modular model, i.e., Dynamical Integrated Modular Earth Rotation System (DIMERS), with submodels for the Earth's mantle, fluid core, and solid core, as well as the atmosphere and ocean to study system characteristics and to model polar motion forced by atmospheric loading. Based on a system model similar to DIMERS, i.e., Dynamic Model for the Earth Rotation and Gravity (DyMEG), Seitz et al. (2005) studied the noise characteristics of polar motion, while Thomas et al. (2005) investigated the contribution of the ocean to polar motion excitation.

These model studies demonstrate that a modular approach to a Earth system model serving geodetic applications is feasible. In particular, these model studies show that so-called emerging system properties (e.g. the model period of Chandler wobble and nearly diurnal wobble) are sensitive to sub-model properties and coupling between the submodels, and these models allow the studying of the emerging properties as function of model parameters.

However, as pointed out by Jüttner & Plag (1999), there are a number of challenges in developing a consistent theory for an advanced model with more realistic representation of all submodels. For most subsystems, individual reference frames will be needed, and in order to exchange of body forces and boundary conditions between the system, the relation of the individual frames to a common model frame will have to be determined. For some of the submodels, such as the ocean circulation models, feedback from a deforming solid Earth with variable rotation is not sufficiently considered.

Challenges are also in the solid Earth processes themselves. Unlike a uniform description of the rotating planet as a whole, a modular theory of planetary dynamics requires direct time domain integration as an initial value problem for basically two reasons. First, the transformation into the frequency or Laplace domain and back into the time domain by a Greens function formalism requires linearity of the mathematical description of the whole system dynamics. That would restrict any subsystem model to a linear theory. The modular approach, however, should not restrict the internal structure of the subsystems in any way except for the match of couplings. Secondly, the definition of a boundary value problem refers to boundaries of particular subsystems but is characterized by eigenmodes of the system as a whole. Thereby additional links between all the distinct subsystems are introduced which are not physically determined as interactions of subsystems but mathematically as matches of certain kinematic patterns. In the treatment as boundary value problem a complete classification of possible types of normal or eigenmodes has to be achieved beforehand. This classification has to be done anew after any substantial change in any subsystem thus contradicting the spirit of the modular approach. Contrarily, in the modular approach to an initial value problem new properties of the whole system emerge in course of integration.

The modular approach allows for successive sophistications individually inside each of the subsystems without requiring any changes in the other subsystem. The only demands on the mathematical description of dynamics of a subsystem are that it has to supply the other subsystems with time-dependent values of the prescribed physical coupling parameters and that it has to work with such coupling parameters supplied to it by other subsystems at each time step anew. Of course, special attention has to be paid to the definition of the interactions right at the outset. Indeed, the isolation of subsystems of the planet and convention on the kind of their mutual interactions defines the structure of the modular theory. Unlike changes within any subsystem, the mere addition of a new interaction of two subsystems requires changing both of them. Moreover, the introduction of a new subsystem even requires changing all other subsystems interacting with the new one.

The choice of subsystems already characterizes a certain structure of the planetary interior and of the circumstances at the planet's surface. For the dynamic Earth reference model, the appropriate choice of modules will be pivotal for the accuracy of the model predictions. It can be expected that submoduls will have to be introduced for different time scales. For example, the modeling of co- and postseismic processes may require a module separate from the one used for surface loading. In the end the dynamic Earth reference model may turn out to be a combination of a large number of models for various processes, which run in parallel and interact through boundary conditions and far-field forces, with the overall model development "guided" by geodetic observations in real-time.

Chapter 9
The future Global Geodetic Observing System

M. Rothacher, G. Beutler, D. Behrend, A. Donnellan, J. Hinderer, C. Ma, C. Noll, J. Oberst, M. Pearlman, H.-P. Plag, B. Richter, T. Schöne, G. Tavernier, P. L. Woodworth

In this Chapter, we focus on the design of the geodetic observing system that will meet the specifications summarized in Section 7.7 and be able to sustain the products listed in Section 7.5. Thus, this Chapter treats GGOS as an observing system (see Section 1.3 for a discussion of the two different meanings of "GGOS"). In Chapter 10, the main focus will be on GGOS as an organization and the integration of GGOS in the global context of Earth observation.

GGOS has been organized by the IAG to work with the established IAG Services in order to provide the geodetic contribution to global Earth monitoring, including the metrological and reference system basis for many other Earth observing systems. GGOS is therefore one of the basic observing systems comprising GEOSS.

GGOS is complex, addressing relevant geodetic, geodynamic and geophysical problems, which have deep impact on vital issues for humankind, such as global change, sea level rise, global water circulation, water supply, natural disasters, risk reduction, etc.(see Chapter 5 for details). It is a visionary concept based on the requirements and specifications given in Chapter 7 and on the assessment of what components are needed to meet the very demanding goals.

In order to address the ambitious GGOS goals, we will integrate a multitude of sensors into one global observing system. In the following sections the focus will be on the technical design and rationale for the proposed GGOS. The individual components of the system will be discussed and the interaction between the components will be outlined, from the geodetic observations and the interfaces to the products for the users.

9.1 The overall system design

The overall GGOS is designed in such a way that it meets the requirements and needs of the scientific and the societal users (see Chapter 7). The tasks listed in Section 7.4 have been identified as the most important high-level tasks for GGOS,

H.-P. Plag, M. Pearlman (eds.), *Global Geodetic Observing System*,
DOI 10.1007/978-3-642-02687-4_9, © Springer-Verlag Berlin Heidelberg 2009

however, the tasks that will actually be performed individually or collectively by the IAG Services, not GGOS itself. This list implies a very complex system with many different sensors and instruments, on the Earth, in the air and in space, that are integrated to form a global observing system appearing to the outside world as one large, comprehensive "geodetic instrument" for monitoring the Earth system. In order to function as a large Earth observatory for the benefit of science and society, GGOS has to encompass not only global terrestrial networks of observatories and space missions devoted to geodetic Earth observation and planetary exploration, but also the communication infrastructure, analysis centers, coordinating centers, and Internet portals. GGOS will eventually generate the well-defined products that will provide the metrological basis for Earth sciences, geo-information science, and terrestrial and planetary navigation. GGOS, therefore, consists of the following four crucial components:

- **Instrumentation:** global terrestrial networks of observatories, Earth observing satellites and planetary missions;
- **Data infrastructure:** data transfer, communication links, data management and archiving systems, data and product dissemination centers, web portals, etc.;
- **The GGOS Portal:** a unique access point for all GGOS products with a database of relevant metadata compliant with international standards; and
- **Data analysis, combination, modeling:** complete and consistent data processing chains ranging from the acquisition and processing of vast amounts of observational data to the consistent integration and assimilation of these observations into complex numerical models of the Earth system.

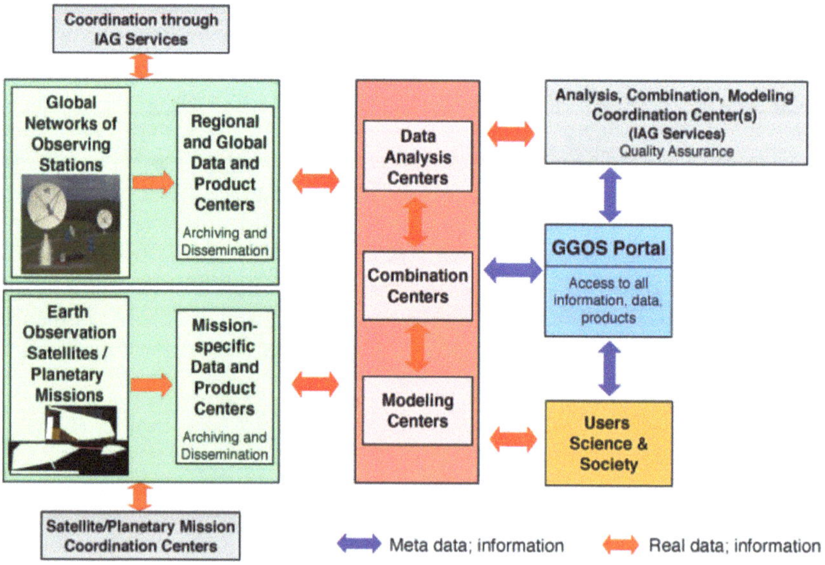

Fig. 9.1. The overall system design of the future GGOS including global observing networks, satellite missions, data centers, analysis centers and coordination centers, etc.

These four components are shown in Figure 9.1 and will be described in more details in the sections below. Figure 3.1 on page 92 shows how GGOS is designed to connect the space and terrestrial geodetic observations (left-hand side) to the Earth system components and their interactions (right-hand side) by way of the "three pillars of geodesy" (the Earth geometry and deformation, the Earth rotation and its variations, and the Earth gravity field with its temporal changes) in the center of Figure 3.1. The principal products of GGOS are summarized in Section 7.5 and the general accuracy requirement is provided in Section 7.6. From the accuracy requirement for GGOS of 1 ppb (including consistency between all GGOS products), it follows that consistent permanent (as opposed to sporadic) ground and space observations are required to meet the needs of science and society. In order to serve the purposes mentioned in Section 7.5, long-term stability and consistency among all GGOS products at a level better than 1 ppb is required, and the products must be available in due time (e.g., in real time for some of the applications) in order to meet user requirements.

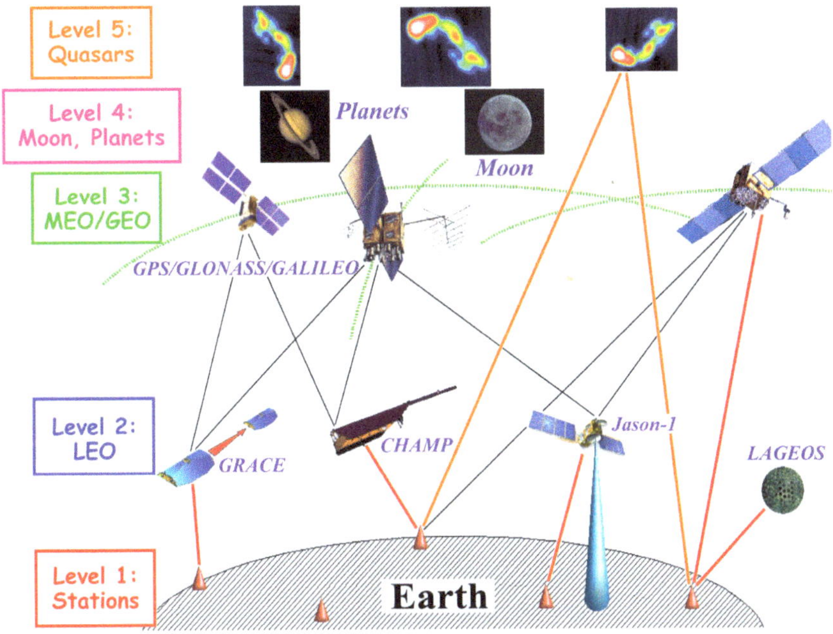

Fig. 9.2. The five levels of GGOS and their interactions with observations of various types. The combined infrastructure allows the determination and maintenance of the global geodetic reference frames, and the determination of Earth's gravity field and rotation. The ground networks and GNSS are crucial foir positioning. In particular, they allow the monitoring of volcanoes, earthquakes, tectonically active regions and landslide-prone areas. The LEO satellites monitor sea level, ice sheets, water storage on land, atmospheric water content, high-resolution surface motion, and variations in the Earth's gravity field. The latter are caused, to a large extent, by regional and global transport in the hydrological cycle.

9.2 The overall observing system design: the five levels

The GGOS will have five major levels of instrumentation and objects, which actively perform observations, or which are passively observed, or both, namely:

- Level 1: the terrestrial geodetic infrastructure;
- Level 2: the LEO (Low Earth Orbiter) satellite missions;
- Level 3: the GNSS and the Lageos-type SLR satellites;
- Level 4: the planetary missions and geodetic infrastructure on planets; and
- Level 5: the extragalactic objects.

These five levels, independent of whether they are active or passive, receivers or emitters or both, are connected (see Figure 9.2) by many types of observations in a complex way to form the integrated geodetic observing system. The major observation types at present are:

- Microwave observations of the GNSS satellites from the ground and from LEO satellites;
- Laser ranging to LEOs, dedicated laser ranging satellites, GNSS satellites, and the Moon;
- Microwave observation of extragalactical objects (quasars) by VLBI;
- Instrumentation onboard the LEO satellites measuring accelerations, gravity gradients, satellite orientation, etc.;
- Radar and optical observations of the Earth's surface (land, ice, glaciers, sea level, etc.) from remote sensing satellites;
- Distance measurements between satellites (K-band, optical, interferometry, etc.);
- absolute and relative gravity measurements; and
- tide gauge measurements.

In the future, new measurement techniques will evolve and will be included into the system. The individual parts (observation types) of the overall system are connected by the co-location of different instruments at the same site on the Earth, or on the same satellite or object. This co-location of instruments and sensors is extremely important for the consistency and accuracy of the system, so that it will act as one large "instrument" (see Section 9.3.8). Each of the techniques has its own strengths and weaknesses, and through co-locations, it is possible to exploit the strengths and mitigate the weaknesses so as to build the strongest possible observing system.

GGOS is not the first global geodetic observing system. Such systems existed for a long time to monitor seasons, to produce maps and to navigate reliably and accurately on the Earth. Prior to the space age, "predecessors" of GGOS consisted of only three levels, namely, globally distributed observatories (Levels 1), the Moon, the Sun and the planets (Level 4) and "fixed" stars and quasars (Level 5). Level 4 (the Sun, Moon and planets) of the historic systems was, so to speak, the predecessor of the GGOS Level 3 (the GNSS). Cross staffs, and then later optical telescopes and watches (first mechanical, then atomic) were the hardware components in Level 1 of the historical systems. Level 5 traditionally was the system of "fixed" stars. The star catalogues were realizations of the celestial reference frame.

9.3 Level 1: Ground-based infrastructure

The first level of GGOS consists of all terrestrial networks of geodetic ground stations contributing to the terrestrial reference frame or to Earth monitoring:

1) The global network of radio telescopes coordinated by the IVS;
2) The global network of SLR and LLR stations of the ILRS;
3) The global network of GNSS stations of the IGS;
4) The global network of DORIS stations coordinated by the IDS;
5) The global network of superconducting gravimeters comprised in the GGP and the global network of sites occupied episodically with absolute gravimeters;
6) The global network of tide gauge stations coordinated by the IOC; and
7) Global networks of geodetic timing stations.

Most of these observing stations are equipped with additional, complementary sensors and instruments (e.g., meteorological sensors, water vapor radiometers, etc.) and at many of the stations more than one instrument are co-located. The design of these networks as fundamental and integral parts of the GGOS is described in the following subsections (see also the respective Sections in Chapter 2).

9.3.1 Core network of co-located stations

The core of the terrestrial global GGOS network, the part realizing the integration of the various instruments on a global scale, will be a set of about 40 globally well-distributed core sites. These stations co-locate the major geodetic observation techniques and a variety of additional sensors. The co-location of the different techniques allows not only the integration of the individual technique-specific networks into a unique terrestrial reference frame (ITRF) but also the assessment of the observation quality and accuracy and the mutual validation of the results. A network of such core sites is mandatory in order to monitor the global reference frame at an accuracy of 1 mm or better over decadal time scales.

These core sites will be equipped with the following instruments, which are based on most recent sensor technologies, connected to real-time communications (data streaming), collecting data at the highest possible observation rates, operated automatically, and are highly reliable:

- at least two geodetic VLBI telescopes to ensure continuous VLBI observation (24 hours per day, 7 days a week), allowing for maintenance periods for individual telescopes;
- an SLR/LLR telescope to track all major satellites equipped with laser retro-reflectors and, for some core sites, the Moon;
- at least three GNSS receivers and antennas to guarantee that individual antennas (and receivers) can be upgraded (e.g., for the tracking of new GNSS, such as GALILEO) without losing the precise local ties to the other antennas, thus ensuring long-term millimeter-level stability;

- a DORIS beacon of the most recent generation;
- terrestrial geodetic survey instruments to permanently and automatically monitor the local ties between the reference points of the space-geodetic technologies with 1 mm accuracy;
- ultra-stable oscillators for time and frequency keeping and transfer (with VLBI, GNSS, laser links, etc.);
- a superconducting and an absolute gravimeter to support gravity satellite missions and geocenter determination;
- meteorological sensors for measuring pressure, temperature and humidity;
- seismometer for earthquake detection, epicenter localization and the determination of rupture parameters in combination with deformation from the space-geodetic techniques and GNSS seismology; and
- a variety of additional sensors (water vapor radiometers, tiltmeters, large gyroscopes, groundwater sensors, etc.).

If major new observation technologies are developed in the future, which will supply complementary information, these sensors must be added to the instrument ensemble of a core site.

9.3.2 VLBI station network

The VLBI station network for 2020 is foreseen to have a size of about 40 globally-distributed sites with one or, even better, two telescopes at each site. These telescopes should be of the VLBI2010-type. Most of the currently used VLBI equipment was developed in the 1970s and 1980s, and the equipment is being pushed to the limits of performance and is costly to maintain. The existing antennas at many sites move slowly, which makes it difficult to provide the rapid whole sky coverage needed for the highest accuracy. Therefore, a rejuvenation of the VLBI network is crucial. In view of the requirements of GGOS, IVS Working Group 3 (WG3) on VLBI2010 was charged with examining the current and future requirements for geodetic VLBI systems. The group compiled their findings in the so-called VLBI2010 Vision Paper (Niell et al., 2006) and made recommendations for the next generation of the VLBI system. Recognizing the need for a standing body within IVS that would ensure the realization and implementation of the new system, the VLBI2010 Committee was set up.

The VLBI2010 system is envisioned to meet the following criteria: low cost of construction, low cost of operation, and rapid analysis and delivery of final results. To accomplish this, the center piece of the new system will be a small-antenna observing system (dish diameter of 12 m or larger) in concert with global high-speed network links. The lower sensitivity of a smaller antenna, as opposed to the present $\sim$20 m antennas, will be more than compensated for by high slew rates of at least 5 degree/sec and higher observational data rates (8-16 Gbps and higher), which will allow many more observations to be taken. The observing will be done over a broad, continuous frequency range (broadband delay approach) of 2-18 GHz allowing mitigation of any radio frequency interference.

The rapid advance of both magnetic-disk technology and global high-speed network technology will be utilized in VLBI2010. All data collection and transmission interfaces and formats will adhere to the set of internationally agreed VLBI Standard Interface (VSI) specifications. An array of antennas directly connected to the correlator via high-speed networks provides the possibility for real-time and near real-time processing to produce geodetic results within hours, which is particularly important to the rapid turnaround of Earth orientation parameter results.

The GNSS community has demonstrated the value of increasing the number of receiving sites and improving the geographic distribution. The present geodetic VLBI network has a very irregular distribution of antennas over the surface of the Earth. Africa, South America, and Asia are particularly under-represented compared to the other continents. Thus, important considerations for the planning of a new network are the number and the locations of the sites needed to satisfy the 1 mm goal. Although the detailed choices for deployment of new stations will be driven by a combination of science, economics, and politics, a quantitative estimate can serve to specify the lower limit for the number of sites.

The goal of combining GNSS, VLBI, SLR, and DORIS geodetic networks provides a guideline for the number of VLBI sites. The current uncertainty in GNSS daily horizontal measurements for a global network is approximately 3 to 5 mm and is unlikely to improve significantly. In contrast, the repeatability in regional GNSS networks of ~1000 km is down to approximately 1 to 2 mm. For VLBI the horizontal repeatability of the Very Long Baseline Array (VLBA) antennas has been 1.5 to 3 mm over the past decade, while for the new VLBI system specified by the VLBI2010 Committee the horizontal accuracy is expected to be better than 1 mm. In order to take advantage of the best attributes of both GNSS and VLBI, the spacing of combined VLBI/GNSS sites should be of the order of 2000 km. Such spacing would require approximately forty sites (Eurasia (14), Africa (7), Australia (3), Antarctica (2), Greenland (1), North America (6), South America (6), Southern Pacific (2)) equipped with one (or preferably two or more) telescope(s) to allow continuous operation.

9.3.3 SLR/LLR station network

The estimated size of the GGOS SLR/LLR network is based on meeting 1 mm/decade stability in the origin and scale requirements for the reference frame. This stability is presumed achievable under realistic weather conditions and local network operational strategies. The same network is also expected to address the tracking needs of the large set of satellites anticipated in the GGOS 2020 time frame. In addition to the current distribution of satellite categories being supported by the ILRS, it is anticipated that there will be a significant increase in the number of GNSS satellites (GPS, GLONASS, GALILEO, COMPASS, etc.) that will be tracked in campaign mode. GGOS requires a globally distributed network of 30-40 SLR stations co-located with GNSS and VLBI, where a high percentage of these stations must also be co-located with either gravity instruments or DORIS beacons. These core observatories should be globally distributed, at sites with good weather conditions and stable geology. "Good weather sites" should permit ranging at least 60% of the time, and have weather patterns lacking strong seasonal signatures. Sites with stable geology do not show local motion, which would otherwise corrupt the reference frame stability. They should be several hundred kilometers away from plate boundaries, faults, and ridges. Bedrock would be ideal, but may not be practical at every site. Four stations (on four different continents) should have lunar ranging capability to continue the long time series of LLR since 1969. A number of the current SLR/LLR stations would likely be sites for the GGOS network. All SLR/LLR sites must be co-located with GNSS, and several should, in addition, be co-located with other instruments.

The network should be equipped with fourth generation systems with high repetition rates (10^2-10^3 Hertz), higher quantum efficiency detectors (either avalanche photodiodes or PMT quadrant or pixilated detectors), shorter dead-time between events, increased automated or autonomous operations, real-time communications for data flow and centralized operations monitoring, and improved calibration and diagnostic monitoring. The higher data rate will allow more rapid satellite acquisition and improved pass interleaving for satellite conflict resolution. Real-time data flow will improve upon the current 1 - 2 hour availability cycle.

Many of the fourth generation capabilities are now being demonstrated in current stations. 2 KHz operation is presently operational at the Graz SLR station. Others are actively pursuing it. Semi-autonomous and automated operations are currently routine at the Zimmerwald and Mt. Stromlo stations. The NASA SLR2000 prototype is being developed with many of the fourth generation capabilities. The stations at Zimmerwald, Matera and Concepcion have initiated studies of the two-wavelength concept for a more accurate refraction correction.

Earth ground-based laser experiments in 2005 to the Mercury Laser Altimeter (MLA) onboard the MESSENGER spacecraft enroute to Mercury, and to the MOLA onboard the Mars Global Surveyor (MGS) orbiting Mars, demonstrated that there is sufficient signal strength for interplanetary ranging measurements and that laser transponder experiments are capable of providing accurate spacecraft ranging and timing information. With the deployment of an optical receiver and accurate on-board timing system on the upcoming Lunar Reconnaissance Orbiter (LRO), one-way ranging to the moon should be operational in late 2008. Several groups are now working on two-way transponders for use in lunar and planetary ranging for studies of lunar and planetary dynamics and gravity field. As a result, many lunar range measurements with higher accuracy and much better temporal coverage will become available.

9.3.4 GNSS station network

The future global GNSS network (maintained by the IGS) will be a multi-purpose observation network. It will be of vital importance for:

- the reference frame realization, monitoring and maintenance;
- the densification of the network of core sites, and the basis for regional densifications of the global reference frame;
- time and frequency transfer between time laboratories equipped with GNSS receivers;
- the monitoring of global plate tectonics and deformation phenomena (loading, etc.);
- the monitoring of the displacements after and during an earthquake (GNSS seismology, i.e., observing the seismic waves with 20-50 Hz sampling rates) to give additional information on earthquake magnitudes and rupture processes;
- the connection of tide gauges to the global reference frame through co-location; and
- for ground-based atmospheric sounding (troposphere and ionosphere).

To meet these goals the IGS station network of the proposed GGOS shall consist of a few thousand GNSS stations with the following characteristics:

- State-of-the-art receivers tracking all GNSS satellites, i.e., GPS, GLONASS, GALILEO, COMPASS, and similar navigation satellite systems yet to be developed. To achieve utmost accuracy and reliability all available GNSS satellites have to be tracked. The collection of data from more than one system makes GGOS independent of the individual systems. The receivers will record all major measurements of codes and carrier phases on all frequencies relevant for Earth observation.
- Homogeneous global distribution of sites, densely covering all major tectonic plates. In the case of a large earthquake, the effects on the global reference frame should be known and available in near real-time.
- Core sites (i.e., sites co-located with other space-geodetic instruments) shall be equipped with more than one receiver and antenna to allow for equipment upgrades without loss of accuracy and time-series continuity.
- All VLBI, SLR and the majority of DORIS sites shall be equipped with a co-located GNSS receiver.
- Sites shall be equipped with real-time data communication links and the possibility to collect data at a sampling rate of a few tens of Hz.
- GNSS receivers shall be connected to (and ideally driven by) ultra-stable oscillators, especially from time laboratories.

This network will be fundamental to connect, through co-location of instruments, all the other networks and to ensure that the positions of all sensors of the global geodetic observing system will be known in a unique global reference frame.

9.3.5 DORIS station network

The quality, density and homogeneity of the DORIS network have been continuously improving for 20 years (Fagard, 2006). With 56 well-distributed stations around the globe, it guarantees an excellent orbit coverage for the DORIS-equipped satellites. The density and homogeneous distribution of the DORIS network an important contribution, on the one hand, to the realisation of the ITRF, both by densifying the IERS network and through the co-locations available for a majority of the DORIS stations, and, on the other hand, to sea level monitoring, through co-locations with tide gauges at one third of the stations. Thanks to the general renovation process carried out over the past six years, almost all antenna monuments have now an excellent long-term stability of the antenna reference point. Moreover, the massive deployment of third generation beacons will ensure a higher performance and reliability of the network.

The IDS station network of the proposed GGOS shall consist of 60 to 70 stations with the following characteristics:

- A homogeneous global distribution of sites, covering all major tectonic plates. The IDS network shall permit the determination of the motion of all major plates, and shall provide a global and robust coverage for all DORIS-equipped satellites.
- The current network will be maintained, and a few new stations will be installed to fill gaps or improve robustness.
- The network provides support to current missions (SPOT, JASON-1, ENVISAT), and planned future missions (JASON-2, CryoSat-2, ALTIKA, etc.), at least until 2020.
- Third generation, or new state-of-the-art, beacons will improve the measurement quality and reliability.
- More beacons shall be connected to atomic clocks to provide a better connection to the International Atomic Time.
- All DORIS sites shall be equipped with accurate meteorological sensors to permit precise atmospheric corrections of the measurements.
- Long-term stability of the antenna reference points shall be at the cm-level over time frames of a few decades.
- DORIS equipment will contribute to core sites, i.e., stations co-located with other space-geodetic instruments (SLR, VLBI, GNSS). The DORIS/VLBI interference issue will be investigated and resolved to the extent possible.
- A homogeneous global distribution of sites co-located with tide gauges will help to accurately calibrate sea level change.

The network monitoring will be enhanced to ensure that the DORIS stations consistently provide reliable and precise measurements:

- Periodic site visits for equipment inspection, antenna stability checking, local geodetic survey;
- Daily monitoring of parameters such as status, failures, voltage, transmitted power on both frequencies, time synchronization, meteorological sensors status, USO warming time, etc.;

- Remote management and control of the DORIS beacons; and
- Increase the level of the operating rate to 90%, with daily monitoring of performance indicators.

9.3.6 Networks of gravimeters

For the proposed GGOS it is extremely important to couple the space-geodetic techniques delivering information about the geometry of the Earth (shape, deformation, orientation, and rotation) with gravimetric measurements not only from gravity satellite missions, but also from sensors on the Earth's surface. Most of the processes in the Earth system have an impact on all geodetic observations and the complementarity of gravimetric sensors is crucial for the separation of various processes involving mass transport.

To obtain time series of gravimetric measurements that improve the monitoring of the Earth system on a global level, a network of about 30 gravimetric stations (identical to the extent possible with core sites, see Section 9.4.1) should be established. Each of these stations should consist of a superconducting as well as an absolute gravimeter, both continuously measuring the gravitational acceleration and its time variations.

9.3.7 Network of tide gauge stations and ocean bottom geodesy

As reported in Section 2.9.3, tide gauge sea level measurements are coordinated internationally through GLOSS, which coordinates a network of about 300 tide gauge stations (see Figure 2.45 on page 83). By 2020, it is expected that all of the core tide gauge network sites, the majority of all other sites with long sea-level records, the stations which provide comparison data for altimeter calibration, and indeed many other tide gauge stations, will be equipped with GNSS receivers. These receivers have two functions: to enable the tide gauge measurements to be located in the same reference frame as the altimeter data, and to determine the rates of vertical land movement (see Section 2.9.3).

The historical tide gauge record has been derived primarily from float and stilling well technology. However, nowadays one can deploy acoustic, radar and pressure tide gauges, as well as digital float systems, each of which has its advantages and disadvantages (see Section 2.9.3 and IOC, 2006). Although GLOSS standards simply require tide gauge stations to provide measurements to better than 1 cm accuracy in all weather conditions, one would expect that any new GLOSS installation would learn from the experience of the Sumatra tsunami of December 2004, and therefore include dual gauges (e.g., a "sea level" gauge based on radar, and a "tsunami" gauge based on pressure measurements) and dual telemetry. Data flow should be both near real time (especially for tsunami and storm surge applications) and delayed-mode for scientific applications.

Currently, much experience is available from float, acoustic and pressure systems, while radar devices are relatively new. However, their low cost and ease of installation and maintenance, means that they may be widely used in future. By 2020, one would expect such devices to be both accurate and affordable. However, one would expect there to be an ongoing need for capacity building in their use in developing countries.

BPRs are also of importance for geodetic applications (see Section 2.9.3). Data from deep ocean bottom pressure recorders are particularly relevant for comparison to temporal space gravity data from missions such as GRACE. However, only a few BPRs have been deployed so far explicitly for such comparison purposes; the POL BPRs in the South-West Atlantic being one example (Hughes et al., 2007). By 2020, the installation of a permanent global ocean network of 50-100 BPRs would be technically feasible. However, there are currently no firm plans for such a network, and the challenge of data transmission would need to be addressed seriously, unless the community wished to work only with delayed-mode information.

9.3.8 Co-location of instruments and auxiliary sensors

The co-location of different and complementary instruments is crucial for several reasons:

- Without co-location sites and highly accurate local tie information, it is impossible to establish a unique and common global reference frame for all major space-geodetic techniques.
- Co-location sites allow the comparison, validation and combination of estimated parameters common to more than one technique. The comparison is crucial for the detection of technique-specific biases, and furthermore the combination of common parameters strengthens the solutions.
- Complementary observation techniques may be the only way to separate the signals of different processes taking place in the Earth system.

Co-location should therefore not only be limited to the space geodetic techniques but include additional sensors that aid integration and combination. A list of such instruments is given in Section 9.3.1.

The measurement and monitoring of the local ties between different instruments should have a similar status and accuracy in the future as the observations of the space-geodetic techniques themselves. The local tie measurements should be performed with 0.1 mm accuracy, in a fully automated way and on an almost continuous basis, since local ties may change over time. These measurements have to account for any deflection of the vertical when relating the local ties to the geometric frame. Because of discrepancies in the results from co-located techniques, it is extremely important to be able to fully rely on the measured local ties. This will help to identify (and eventually correct) the considerable remaining systematic effects in the results of the individual observing techniques.

At core sites, local ties do not only have to be established between the reference points of the major space-geodetic observing technologies, but also to other sensors. As an example, the height differences between the reference points of the space-geodetic technologies and atmospheric sensors have to be known with decimeter accuracy for comparison, validation and combination purposes (e.g., the tropospheric delay difference between the GNSS and VLBI antenna reference point has to be taken into account when combining the tropospheric delay estimates from GPS and VLBI). The location of a water vapor radiometer with respect to the other observation techniques has to be known as well, and the same is true for gravimeters, tiltmeters, large gyroscopes, etc.

The core network (see Section 9.3.1) will be fundamental for the co-location of instruments. However, due to environmental conditions at the individual sites (geological and geodynamic stability, weather conditions for SLR/LLR, multipath environment, change in ground water table, etc.), it will not always be possible (or reasonable) to co-locate all instrument types at one location. "Isolated" instruments should then at least be tied to a unique global reference frame by setting up a GNSS receiver at the same location. All instruments must be co-located with GNSS.

9.4 Level 2: Low Earth Orbiter satellite missions and their applications

Satellites observing the Earth from space will be an indispensable component of GGOS in 2020. Satellites have the big advantage that they collect data homogeneously and consistently over large parts of the Earth surface. They also allow the collection of data that cannot be recorded at the Earth's surface. These satellites are nowadays equipped with a multitude of sensors and instruments, monitoring the land, ocean and ice surfaces as well as the Earth's gravity field and its temporal variations.

The potential and impact of satellite missions on Earth observation will increase considerably due to the fact that: (1) more and more satellite constellations instead of individual satellites will be launched increasing the temporal and spatial resolution of the data, and (2) satellites will be flown in "formations," forming large observing instruments composed of sensors on more than one satellite.

Due to the importance of the satellite component for the GGOS design and products, the observation of certain geodetic/geophysical parameters of the Earth (e.g., the gravity field and its temporal variations) by a satellite mission should not end with this dedicated mission, but has to be continued with follow-on missions establishing eventually a chain of missions (as in the case of the altimetry missions TOPEX, JASON-1, JASON-2, and ERS-1, ERS-2, ENVISAT, etc., see Table 2.2). Such "chains" of satellite missions are crucial for monitoring the Earth system over long time periods and for the detection of long-term trends and changes in the Earth system. Therefore, they should be viewed as a strategic element of the GGOS.

9.4.1 Gravity satellite missions

The gravity field missions CHAMP and GRACE (see Section 2.6.5) have made to a huge improvement to our knowledge of the Earth's static and time-variable gravity field. The missions have improved the accuracy of the static gravity field models by a factor of at least 100 compared to pre-CHAMP models, which were mainly determined from satellite laser ranging data. Based on monthly gravity fields determined from CHAMP and, in particular, GRACE data, seasonal variations and trends in the Earth's gravity field can be monitored, providing unique information about relevant mass transport phenomena like the water cycle in large river basins, the melting of ice sheets in Antarctica and Greenland and the associated sea level change, as well as in the ocean current systems. ESA's GOCE mission will lead to another significant improvement in the resolution and accuracy of the Earth's static gravity field and of our knowledge of the ocean current systems. GOCE will also mark an important step toward a more accurate unified global vertical reference frame.

In view of these developments, it is clear that present and future satellite gravity field missions will play a crucial role in GGOS. An uninterrupted monitoring of the temporal variations of the gravity field is of utmost importance for global change studies, i.e., the reliable detection of small trends in the gravity field due to sea level rise, the melting of ice sheets and changes in the ocean current systems.

To avoid any gaps in the time series – GRACE may last till 2013 – a GRACE follow-on mission with only minor design changes is crucial, because the development of new technologies may require several years and might not be ready before the decommissioning of the GRACE pair of satellites. GGOS will have to work with the space agencies to ensure this follow-on mission.

For mission concepts beyond 2013, new scientific challenges, for example, global ocean circulation, hydrological cycle, secular trends of geoid, ice sheet and glacier evolution, crust and lithospheric structure and dynamics, big earthquakes, and vertical datum improvement, require higher temporal resolution, higher spatial resolution and higher accuracy (see Figure 3.5). Accordingly, different sensor designs such as quantum-gradiometers, low-low and high-low Satellite-to-Satellite Tracking (SST) or ranging, optical clocks, etc., might be required.

One obvious concept to improve the accuracy of inter-satellite measurements (low-low SST) is the replacement of the K-band link of the GRACE mission concept by an optical link (i.e., a laser interferometer). A gain of a factor of 100-1000 in the accuracy of the inter-satellite measurements can be expected from such a development. Initial studies of such a concept have been performed by NASA and are presently being conducted by ESA as well. With such accuracies, the de-aliasing of the gravity field determination (removal of effects from high-frequency signals from the atmosphere, ocean tides, etc.) will become a major challenge. The same is true for the separation of the gravity signals resulting from different Earth processes. Different orbit constellations and different types of satellite formations will have to be considered for this purpose. In addition, complementary sensor systems (surface deformation monitoring with GNSS, ocean bottom pressure sensors, airborne gravimetry, superconducting gravimeters, etc.) will be crucial to allow for the separation of different processes. Sensor integration is therefore at the very heart of the GGOS concept.

It is possible that optical clocks will reach stabilities of 10^{-18} in about 10 years (see Section 2.7.3). Using the theory of General Relativity such clocks will allow the direct determination of potential differences between clocks corresponding to height differences on the level of 1 cm (geoid). With frequency comparisons between clocks in space and on ground, a consistent global vertical reference frame can then be established with very high accuracy.

9.4.2 Ocean and ice altimetry satellite missions

Radar altimetry proved to be a reliable and efficient technique for monitoring the global sea level and its changes. With currently four active radar altimetry missions (ERS-2, JASON-1, ENVISAT, GFO-1) and one launched recently (JASON-2), the global ocean can be observed with a reasonable accuracy. However, it must be ensured that the current constellation is maintained also for the future.

In January 2008, a CEOS Ocean Surface Topography Constellation Strategic Workshop held in Assmannshausen (Germany), discussed and outlined implementation plans for the next 15 years (Figure 9.3). Among others it was recommended to:

- maintain continuity of high-accuracy JASON-type altimetry;
- maintain continuity with altimeters on at least two complementary, high-inclination satellites; and
- extend the capability of altimetry to denser observational coverage through the use of the swath altimetry technique.

With JASON-2, high-accuracy missions will continue until at least 2013. Because JASON-2 for the first time is an EUMETSAT-operated mission to provide critical weather forecast data, continuation is likely. JASON-3 is now under approval.

For medium-accuracy high-inclination orbits, also covering much of the polar oceans, HY-2A (China) and Saral/ALTIKA (India/France) missions are to be launch in 2009. The HY-2A mission may be followed by similar missions. In addition, the Sentinel-3A (Europe) mission, to continue on the current ENVISAT orbit, may have follow-on missions.

With the launch of CryoSat-2 in 2009, also a dedicated mission for measuring the polar caps and the ice-covered oceans is planned. However, with no follow-

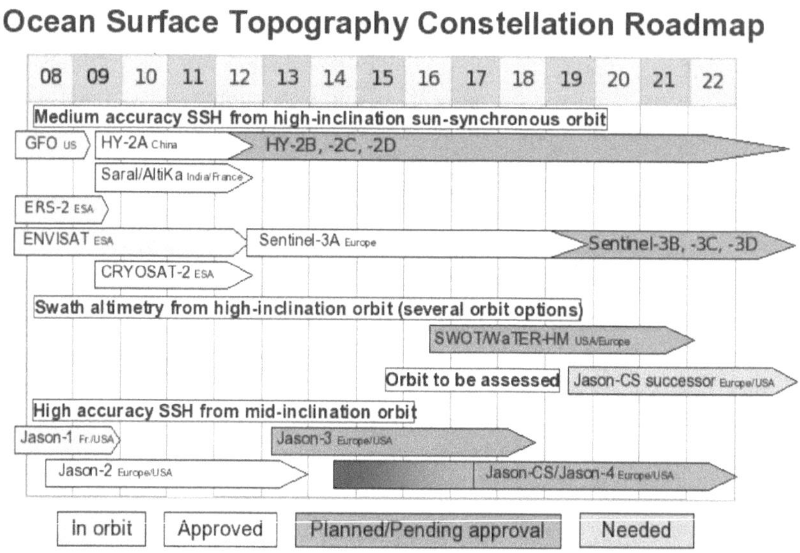

Fig. 9.3. Ocean Surface Topography Constellation, Strategic Workshop on Ocean Surface Topography Constellation, Assmannshausen (Germany, January 2008).

on missions there will be a critical gap in the observations of the climate-sensitive polar regions. GGOS efforts are needed to ensure the continuation of dedicated ice missions.

In the case of swath altimetry, allowing a more dense and flexible coverage, no plans exist so far for a mission before 2016 (SWOT/WaTER-HM mission). Previous attempts to carry a wide-swath altimeter on JASON-2 were cancelled in 2005 by NASA.

In summary it can be noted that, while in the past all missions have been operated by either the USA or ESA and France, future satellite constellations will benefit from contributions by other nations. This change makes it critical that the current open data policy be maintained, including the near real-time data distribution for operational applications as well as to ensure accompanying scientific studies.

9.4.3 InSAR and optical satellite missions

InSAR observations produce spatially continuous images of the deformation of the Earth's surface (see Section 2.4.5 for examples). These images are complementary to other space-based geodetic observations, which produce temporally smooth, but spatially discontinuous point measurements of surface motions. The need for improved coverage of the Earth's surface is obvious, particularly for geohazards and Earth sciences (see Chapter 5).

The recent National Academy of Sciences report "*Earth Science and Applications from Space: National Imperatives for the Next Decade and Beyond*" (National Research Council, 2007) represents U.S. scientists' consensus on critical Earth observations from space that are required to address issues of climate change, water resources, ecosystem health, human health, solid-earth natural hazards, and weather. The Report recommends that the planned DESDynI mission, an L-band InSAR and laser altimetry mission, be launched in the 2010-2013 time frame. DESDynI would measure surface and ice sheet deformation for understanding natural hazards and climate, and vegetation structure for ecosystem health. DESDynI would help scientists understand the effects of changing climate and land use on species habitats and atmospheric carbon dioxide, the response of ice sheets to climate change and the impact on sea level, and would be used to improve forecasts of the likelihood of earthquakes, volcanic eruptions, and landslides.

Geodetic networks support InSAR by providing geodetic control for the observations. The geodetic networks also provide tropospheric and ionospheric maps for improving the quality of the interferograms. The geodetic data will be used to calibrate and validate the InSAR observations and, as mentioned above, will complement the InSAR observations by providing temporal continuity to the images.

NASA has proposed that a International SAR Information System (ISIS) be established. A group concerned with the ISIS has not yet been formed. This group would set data policies, would establish the ISIS as the vehicle for delivering InSAR data to the general science community, and would coordinate acquisition and processing of data. GGOS could be an important advocate or umbrella organization for this group.

By 2020 it is anticipated that a constellation of InSAR satellites with contributions from the USA, Europe, Brazil, Taiwan, China, and Japan. A coordinated constellation of InSAR satellites would provide multi-baseline observations for detailed topographic mapping and vegetation structure studies. The constellation would also allow for more frequent observations at particular locations, enabling more rapid response to events such as earthquakes, volcanoes, and landslides, as well as a better determination of time-dependent phenomena.

A combined treatment of imaging and point techniques will be crucial in order to calibrate the dynamic Earth reference model proposed in Chapter 8. For that purpose, SAR images need to be available at least for the globally distributed fundamental stations.

9.4.4 Future satellite mission concepts

Over the last few years satellite technology developments have been extremely rapid and have resulted in new concepts for future satellite missions. The most important new concepts are:

- Design of micro- or even nano-satellites;
- Constellations with large numbers of satellites;

- Formation flying; and
- New instrumentation.

Initially, the concept of micro- or nano-satellites has mainly been realized by university projects for students. Nowadays these developments are also pursued by national space agencies as an interesting alternative to small or large satellites to achieve certain mission goals. Due to the miniaturization of satellite components and sensors, micro-satellites can nowadays be used for challenging mission tasks at a fraction of the costs of satellites such as CHAMP or GRACE. Especially in connection with formation flying, or satellite constellations with a large number of satellites, this alternative concept becomes very attractive.

A constellation of a large number of satellites, possibly in different orbital planes and configurations, has the big advantage that the temporal and spatial resolution of Earth observations can be drastically improved. This can be seen from the number of daily radio occultations generated from the constellation of six COSMIC satellites, compared to individual satellites like CHAMP. Near real-time Earth monitoring based on satellite observations (e.g., for early warning systems) will require a constellation approach and inter-satellite communication to allow for a near real-time analysis of the data on the ground. Together with the micro- and nano-satellite concept mentioned above, constellations of 10-100 satellites will become feasible and affordable in the future.

Formation flying is a very interesting new aspect of satellite missions. Compared to conventional missions it adds two new "dimensions": (1) it allows for inter-satellite measurements (e.g., the K-band link between the GRACE satellite pair), and (2) it opens the door to build a virtual, more efficient/accurate instrument by integrating the instruments on several satellites into one large observing system (e.g., the integration of the Terra_SAR-X and TanDEM-X satellites for the generation of high-resolution DEMs). An example of a mixed concept of constellation and formation flying is the ESA SWARM mission, where two of the three different satellites fly in a formation to measure the East-West gradients of the magnetic field, and the third satellite orbits the Earth at a higher altitude to allow separation of different parts of the magnetic field.

In addition to the developments mentioned above, there will also be considerable progress in the instrumentation for satellite missions. Optical clocks may reach a stability of about 10^{-18} in the next decade. This will allow the direct measurement of the gravitational potential based on the effects of General Relativity on clocks (clocks in a strong gravitational field run slower than clocks in a weak gravitational field), and thus enable the unification of physical height reference frames at the centimetre level. Microwave links between satellites will be replaced by optical links (optical interferometers; e.g., for GRACE-type measurements) that will increase the precision of the inter-satellite measurements by at least a factor 100-1000. Reflectometry and scatterometry antennas for GNSS altimetry applications may become important add-ons to Earth observation satellites. In addition, inter-satellite communication technologies will make possible (near) real-time transfer to ground stations, as required for early warning systems.

9.4.5 Co-location onboard satellites

The co-location of different sensors and observation types onboard a satellite is extremely important to establish connections between the different observation techniques. These connections, and their complementarity, may be crucial to correctly modeling certain aspects of the observations (e.g., correction for non-gravitational forces with accelerometers in gravity field determination) and to separate effects stemming from different processes or components of the Earth system. In addition, the availability of complementary instruments on a satellite (e.g., different tracking systems like GNSS, SLR, and DORIS for precise orbit determination) allows for the connection of techniques at the satellite, which is complementary to those at co-location sites on the Earth's surface, and in addition allows for the detection of technique-specific biases.

A good example for this development is the rapid progress achieved in orbit determination with the tracking data of the TOPEX/Poseidon satellite using DORIS, GPS, SLR and altimetry (crossovers). For future satellite missions, emphasis should therefore be placed on satellites that establish links between different observation and tracking techniques. It is of particular importance that all GNSS satellites be equipped with laser retro-reflector arrays (Figure 9.4). Future VLBI observations of GNSS satellites should also be performed, establishing another link between techniques, which would directly connect the satellite frames to the ICRF.

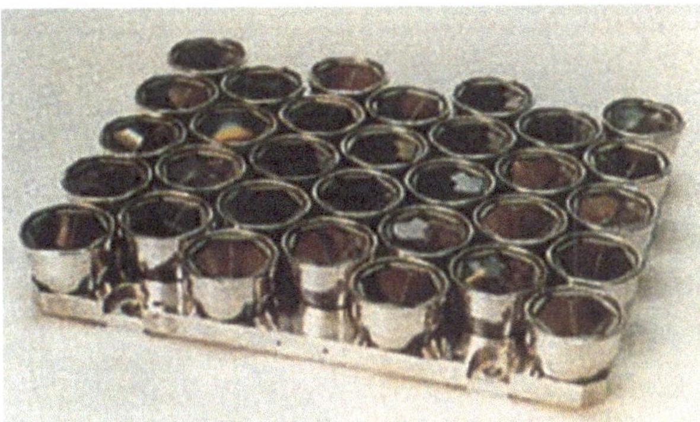

Fig. 9.4. Retro-reflector arrays on GPS-35 and GPS-36 satellites.

9.4.6 Airborne and shipborne sensors

Data from terrestrial and spaceborne instruments should be supplemented with data obtained from airborne platforms and ships. Typically, the data stemming from aircrafts and ships are rather local or regional in nature compared to the data collected by satellite missions. However, airborne and sea surface data with high spatial resolution are very important for assessing the quality and accuracy of satellite or ground-based data. They provide more detailed information about the processes being studied. Although the main focus of GGOS is on global aspects of Earth monitoring, most of the natural hazards are rather regional or local in extent. To understand them in detail, GGOS will strive, starting with the global perspective, for higher and higher resolution of the Earth monitoring in space and time.

Airborne and shipborne gravimetry illustrate how our knowledge of the global Earth gravity field from satellite missions can be densified and improved. The regional gravity data is combined with the global gravity field models from satellites to obtain the high-frequency part of the field.

9.5 Level 3: GNSS and laser ranging satellites

9.5.1 Global Navigation Satellite Systems

The GNSSs are evolving rapidly and a Global Navigation Satellite System of Systems (GNSSS) becomes realistic (Hein et al., 2007). The GLONASS is being replenished with a new generation of satellites to be completed by 2010. The first two GALILEO engineering satellites have been launched, with the full constellation to be completed by 2013. China is also working on a civil satellite navigation system (COMPASS). GPS will be upgraded as well: the first new generation satellites with a second civil signal (L2C) have already been launched. New GPS IIF satellites with three civil frequencies will be launched from 2009. A new GPS III constallation is planned for the end of the 2020 time frame.

Both Japan and India plan to launch smaller regional systems. The Japanese Quasi-Zenith Satellite System (QZSS) is planned to have three satellites in highly inclined geostationary orbits. The Indian Regional Navigation Satellite System (IRNSS) will consist of a seven-satellite constellation.

In addition, Satellite-Based Augmentation System (SBAS) for the GNSSs are developing and adding relevant infrastructure. The European Geostationary Navigation Overlay Service (EGNOS), the U.S. Wide Area Augmentation System (WAAS), the Japanese Multifunctional Transport Satellite Space-based Augmentation System (MSAS), the Indian GPS Aided GEO Augmented Navigation (GAGAN), and the Nigerian Communication Satellite (NIGCOMSAT). The SBAS are transmitting or will transmit additional signals to those of the GNSS signals, and all SBAS are planned to be interoperable with the GNSS on two frequencies.

GNSS are also crucial for the reference frame realization and for many applications in Earth science and Earth observation (see Chapters 2 to 5). After $\sim$2013 approximately 100 GNSS satellites will be available, promising a new level of positioning quality and accuracy. This will have a fundamental impact on most GGOS products, from the reference frame to GNSS atmospheric sounding, to reflectometry and scatterometry.

It is therefore essential for GGOS to make the best possible use of a combination of the GNSS systems available for civil applications. GGOS, through the IGS, will have the goal to generate consistent products of the highest accuracy for all GNSS systems. The ground network of GNSS stations should support this by the installation of receiver technology able to track all relevant GNSS signals at the same time.

In order to link the GNSS to SLR, laser retro-reflectors should be installed on all new GNSS satellites (see Section 9.4.5). All GLONASS and GALILEO satellites are (or will be) equipped with laser retro-reflectors.

9.5.2 Laser ranging satellites

Stations in the ILRS network range to a constellation of both, passive and active satellites including the Moon (see Section 2.4.2). The SLR network will track the set of passive, spherical geodetic satellites such as LAGEOS-1 and -2 (see Figure 2.8), Etalon-1 and -2, Starlette and Stella for reference frame maintenance and measurements of time-varying components of the gravity field. SLR measurements will continue on the GPS-35 and GPS-36 satellites, the GALILEO satellites and selected satellites in the GLONASS series. Efforts are underway to include retroreflectors on the upcoming GPS-III constellation. Tracking these GNSS satellites is crucial for the assurance of positioning quality, long-term stability, verifying orbit and timing accuracy, and aligning other reference frames (e.g., WGS 84) with the ITRF. This tracking will also guarantee the interoperability of the different GNSSs.

The retro-reflector arrays flown in space to date have been made from solid quartz cubes, (either back-coated or uncoated). Engineering studies indicate that hollow cubes made from either aluminum or glass may provide considerably higher return signal strength for similar weight and area conditions on arrays at GNSS altitudes.

9.6 Level 4: planetary missions

In the coming years, there will be significant advances in studies of reference frames, gravity fields and rotation of Solar System planets and their satellites. With the current interest of space agencies worldwide in lunar exploration, much progress is expected in the geodesy and cartography of the Moon. The Japanese spacecraft Kaguya (former name: SELENE), the Chinese Chang'e 1 spacecraft, and the Indian spacecraft Chandrayaan-1 are currently in lunar orbit. Kaguya is in a circular polar orbit, and its powerful laser altimeter has completed the first global topographic map. Coverage includes the polar areas for the first time, which were beyond the reach of the altimeter on the previous Clementine spacecraft. Using an elaborate radio tracking scheme, Kaguya and its two sub-satellites will also improve the lunar gravity field, in particular on the far side, where reliable data are hitherto lacking.

The LRO, scheduled for launch in May 2009, will provide further significant contributions to lunar geodesy. The orbiter will carry the Lunar Orbiter Laser Altimeter (LOLA) to further densify the topographic grid (25 m along-track shot spacing, 1.2 km across-track spacing at the equator, after one year). LRO will also carry a laser receiver, which can be targeted from ILRS stations for precise range measurements to the orbiter at the 10-cm level. Images obtained by LRO will also form the basis for a new generation of accurate lunar standard maps which will essentially realize the Moon-fixed reference frame. The GRAIL mission in the NASA discovery program scheduled for launch in 2011 will focus on lunar gravity. Using techniques pioneered by the joint U.S.-German Earth GRACE mission, the GRAIL twin spacecraft will aim at improving the current knowledge of the lunar gravity field to higher-resolution (30 x 30 km) and higher accuracy (< 10 mGal).

A robust international program of lunar robotic lander missions is expected in the coming decade. The strawman payloads that are currently being discussed will include geodesy packages, involving active lasers for range measurements at mm-level and radio transmitters suited for observations by terrestrial VLBI stations. These new techniques will firmly tie the Moon into the ICRF and should improve the knowledge of the tracking of lunar orbital and rotational dynamics as well as tidal deformation.

Likewise, future missions are expected to further our knowledge of the dynamics of the terrestrial planets. The Exomars spacecraft, scheduled for launch in 2013, would deploy a geophysics package on the surface of Mars. The package would include a radio experiment for monitoring of variations in the Mars rotation, caused by atmospheric dynamics and the condensation/sublimation cycles of ice in the polar caps (see Section 6.1.2). Accuracies are expected to approach the milli-arcsecond level. The MESSENGER spacecraft, the first ever to enter Mercury orbit (orbit entry scheduled for 2011) is expected to produce the data base for the definition of a new coordinate system and for the production of a standard global map. MESSENGER will also determine the Mercury gravity field (parameters of expected degree and order 16) and study the planet's rotation including the complex librational motion (see Section 6.1.1).

The MESSENGER spacecraft on its way to Mercury recently demonstrated a 2-way laser link experiment over a distance record of 24 Mio km. By operating the onboard laser altimeter transmitter and receiver in combination with the terrestrial SLR station at the Goddard Space Flight Center, the spacecraft position was determined to 20 cm formal standard deviation, along with parameters of onboard clock offsets and drift (Smith et al., 2006). Laser link techniques (highly collimated laser beams not affected by the Earth's troposphere and ionosphere) are expected to become the tracking method of choice, which will establish firm ties of distant planets into the Solar System reference frame.

9.7 Level 5: extragalactic objects

The quasars and other compact radio sources included in the ICRF have point-like optical images. Their red shifts indicate great distances, hence their emissions must be powered by processes different from stars and galaxies, probably mass flows into massive black holes. At the resolution of geodetic/astrometric VLBI using S-band (2 GHz) and X-band (8 GHz), the objects are generally not point-like but have a structure that may change with time. Such structure changes can be seen as angular position changes of up to 1 milliarcsecond. The brightest extragalactic radio sources have in fact too much detectable structure to be good astrometric objects. By balancing the competing criteria of source strength, compactness and constancy of structure and position, a set of about 100 geodetic sources has been selected for routine geodetic VLBI observations, while the rest of the ICRF sources improve the distribution and density over the sky (Figure 9.5). It should be noted that the small number of VLBI stations in the southern hemisphere causes the ICRF to be weaker in all aspects in the southern sky.

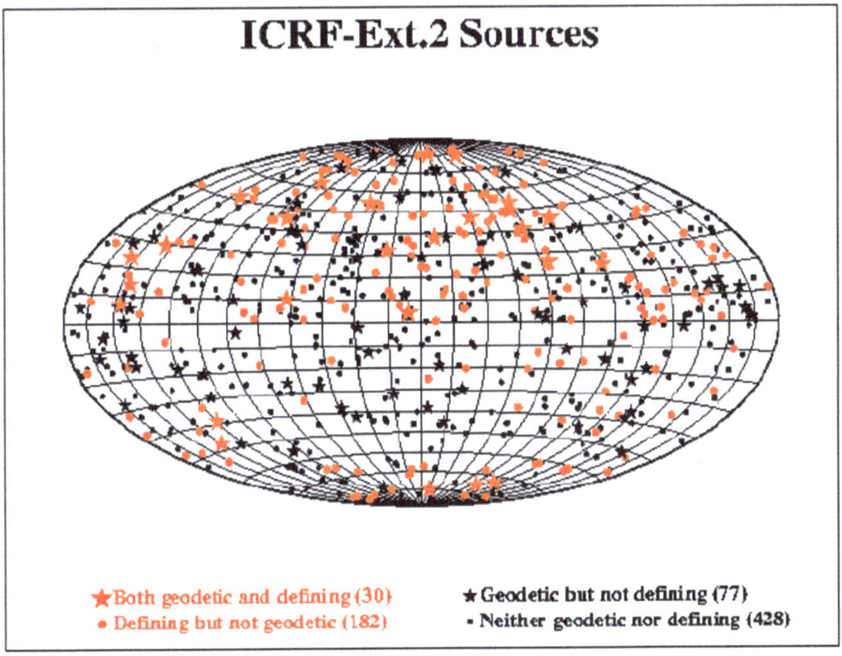

Fig. 9.5. Source locations of ICRF-Ext.2. The second extension of ICRF was computed based on VLBI data obtained between mid-1995 and the end of 2002 May and included an additional 109 new sources. From Fey et al. (2004). See also Box 1 in Section 2.2.

It is therefore essential that the proposed GGOS realizes a much more homogeneous coverage of the southern and northern hemispheres. This implies that about half of the 40 core sites of the future GGOS (equipped with VLBI telescopes) have to be located in the southern hemisphere. Because of rapid developments in communication technologies it should be possible to install such a network in the coming decade.

The ICRF is essential to geodesy as it is the frame for observations of Earth rotation and the celestial frame for the satellite orbits. The ICRF is also the basis for astrometry. The ICRF thus has different realizations at various wavelengths, the microwave VLBI realization being the most accurate at this time. The astrometric satellite GAIA is scheduled for launch in late 2011 and has the potential of generating an optical extragalactic realization with an order of magnitude better precision and two orders of magnitude more objects. Other space missions may refine the positions and proper motions of the brightest stars with a corresponding improvement of star tracking for satellite orientation. For most geodetic purposes, however, these improvements will not be usable because no correspondingly precise ground-based optical observing system exists. An accurate microwave realization for geodetic VLBI will still be needed.

9.8 GGOS data flow: from measurements to users

The official products generated by the technique-specific Services will be the basis for the products made available through the GGOS Portal (see Section 9.9). GGOS will thus rely on the data system infrastructure of the IAG Services.

The success of the IAG Services is partly due to the underlying support of their information and archive services. Each Service has a coordinating entity (coordinating center or central bureau) managing the daily operations of the Service. This function also facilitates communications and coordinates activities both within the Service and with a broad user community. A central coordinating function will be established for GGOS (see Chapter 10), providing coordination within GGOS and to the IAG Services.

The IAG Services' data centers are the central source for data for the analysis community and for products generated by the Services for the user community. GGOS will rely heavily on these data centers for service products and for input to the GGOS Portal.

9.8.1 Data centers and data flow

Each of the geometric IAG Services utilizes a similar structure (shown in Figure 9.6) for the flow of information, data, and products from the observing stations to the user community: Network Stations (track continuously, transmit data using predetermined schedules), Data Centers (interface to stations and users, perform data quality checks/conversion, archive data and products for analysis center and user access) and Analysis Centers (generate products). Participants in service activities

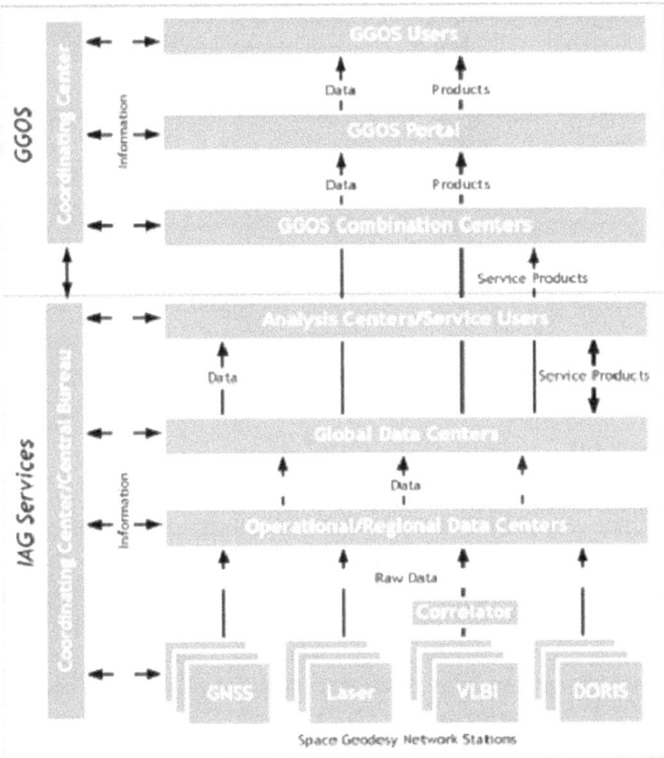

Fig. 9.6. Each of the geometric IAG Services operates with a similar component structure for data flow and archive. The service-specific global data centers will provide data and products to the GGOS combination centers; GGOS users will obtain data and products through the GGOS Portal or through direct access to the service data centers.

collaborate at all levels to ensure consistency and timely delivery of data and products.

Networks of tracking stations transmit data through various levels of data centers to ultimately reach the service analysis centers and user community. During the design phases of the IAG Services, it became clear that a distributed data flow and archive scheme would be vital for mission success. Thus, each Service established a hierarchy of data centers to distribute data from the network of tracking stations: operational, regional, and global data centers. This scheme provides efficient access to and storage of data, thus reducing traffic on the Internet, as well as a level of redundancy allowing for security of the data holdings. Operational data centers serve as the direct interface to the network stations (or correlators in the case of VLBI), connecting to the remote sites daily/hourly/sub-hourly, downloading the data, and archiving the raw station data. Regional data centers gather data from various operational data centers and maintain an archive for users interested in stations of a particular region. Furthermore, to reduce communication traffic, the regional data centers are used to collect data from several operational data centers before transmitting them to the global data centers. The global data centers are ideally the principal data source for the analysis centers and the general user community. Operational and regional data centers transmit data to these global data centers where they are then available on-line for ftp/web access. These data are utilized by the service analysis centers to create a range of products, which are then transmitted to the global data centers for public distribution. Multiple global data centers provide each Service with a level of redundancy, thus preventing a single point of failure should one data center become unavailable. Users can continue to reliably access data from one of the other available data centers. Furthermore, multiple, geographically-distributed global data centers reduce the network traffic that could occur to a single geographical location.

9.8.2 Synergies between observing techniques

Each of the four geometric IAG Services utilizes a similar flow of data, pioneered by the IGS and shown in Figure 9.6, from the measurement networks to the analysis centers. Standards, both technique-specific and cross-disciplinary, in data and product generation are adhered to throughout all levels in each of the Services. Each Service has developed its products using standard models and algorithms to ensure consistency over time. Data are currently archived in technique-specific formats (e.g., Receiver Independent Exchange Format (RINEX) for GNSS); however, products derived from the different techniques are moving towards common formats across data types (e.g., Software Independent Exchange Format (SINEX) for station positions, Standard Product 3 Orbit Format (SP3) for satellite orbits). All data are in American Standard Code for Information Interchange (ASCII), thus machine independent, and compressed for transmission and archiving. The Services are also evolving, as requirements change, by developing new formats and standards for the exchange of data and products.

9.8.3 Operating centers and communications

Operating (or operational data) centers are responsible for providing the communication infrastructure for network stations, downloading data on a routine basis, re-formatting and checking the downloaded data, maintaining these network stations, and archiving the raw data. Connections to the stations are typically provided through the Internet or dial-up methods with satellite communications used in more remote areas. Direct connections allow for rapid download, at least daily but often sub-daily (or even sub-hourly). Currently, GNSS and laser ranging stations are required to transmit data on a daily basis (although most stations send their data on an hourly basis), at a minimum, to these operating centers. VLBI data are sent from the network stations to a correlator on disk packs, and in some cases the data are electronically transferred via high-speed networks (e-transfer). As VLBI observations are organized in sessions, the data transmission follows a session schedule. DORIS stations uplink data to the DORIS receiver onboard the observed satellite, thus enabling installations in more remote areas. DORIS-equipped satellites then download these data to the DORIS control center for transmission to IDS data centers.

The future GGOS, striving for a much more homogeneous distribution of core sites and technique-specific observation networks, will have to rely heavily on satellite communication technologies, i.e., technologies that are accessible from remote areas of the globe. Communication links via satellites become cheaper every year, making this technology more and more attractive for GGOS.

In the case of satellite constellations observing the Earth in near real-time (e.g., for tsunami early warning systems using GNSS reflectometry), only inter-satellite communications or communications via geostationary satellites will ensure the data arrives at the data centers and analysis centers with minimum delay so that analysis can take place.

9.8.4 Future technologies and capabilities for data infrastructure

Several of the geometric IAG Services are moving into the era of real-time data streaming. Real-time and near real-time applications (e.g., weather forecasting, tsunami early warning systems) require low-latency data and product delivery. Real-time data transfer also allows operating centers and analysis centers to monitor station health and to provide rapid notification and correction of station problems. Standards, and protocols for real-time operations, liaisons with regional real-time networks, and technologies to broadcast products for real-time users are currently under development. Near real-time products derived from these data streams will be investigated. The development of the future VLBI system (VLBI2010) aim at real-time e-VLBI, where the transmission of station data is accomplished through high-speed network transfer to the correlator during an observing session and the data are correlated in real-time. Before full real-time capability, intermediate steps with e-transfer and correlation after the observing session will likely be necessary. GGOS will play a critical role in promoting standards by which real-time networks can operate and exchange data products on a global basis.

The data rates of observation of the space-geodetic techniques will dramatically increase in the decade to come. GNSS stations will observe more than 100, instead of 30 satellites, and the data rate may be as high as 50-100 Hz. Such data rates will enable not only, for example, the observation of seismic events with GNSS observations (site motion during the earthquake) and subsequent determination of rupture parameters, but also the monitoring of rapid scintillations in the ionosphere. The new generation of VLBI telescopes will record about ten times as much data as at present. The upcoming InSAR missions (TerraSAR-X and TanDEM-X) will collect data volumes of gthe order of petabytes (10^{15} bytes). The data infrastructure capable of handling such huge amounts of data has not yet been designed or developed, but should be part of the responsibility of GGOS in 2020.

9.9 GGOS User Interface: Database, Portal, and Clearinghouse

It is difficult to predict the development of web technologies, methodologies and approaches that will prevail in 2020. Considering developments over the last ten to fifteen years and extrapolating this into the future, suggests considerable changes in the nature of the interaction of human and web interfaces and the methodology for presenting information. Therefore, this section may be out-dated rather rapidly.

The GGOS User Interface will likely have three main elements:

(1) a GGOS database, which mainly contains general information, meta information and catalogues, and facilitates access to observations and products provided by the various IAG Services;
(2) the GGOS Portal, which will be the unique access point for all products and information made available through GGOS; and
(3) the GGOS Clearinghouse for geodesy, which will enable the search for information related to all aspects of geodesy.

The GGOS Portal will be the access point for all GGOS products. The Portal will also provide a route to the heterogeneous information systems of the IAG Services. The Portal will be linked with a GGOS database of relevant metadata and web services to enable searches for relevant data and products in the most effective way.

The IAG Services will provide very important and valuable data, information, and products, which are indispensable for Earth sciences and their applications. The GGOS Portal will give access to these data and products as well as general information about geodesy. The Portal will contribute to the GGOS objectives to promote and improve the visibility of geodetic research and to achieve maximum benefit for the scientific community and in society in general. Behind the GGOS Portal, each contributing Service will have its own visibility, and responsibility to maintain and manage its supporting data and information system.

The IAG Services produce products that are critical to the generation of GGOS products. These products and data are only available at the data centers of the individual components of GGOS. It is clear that for a future GGOS, all the relevant products for Earth sciences and applications will have to be made accessible through the GGOS Portal that leads the user – including the non-specialists working in different fields – to the individual products. The products and data themselves will be physically located at many different data and product centers and will be promoted by the individual IAG Services as well. For the benefit of new users who are not familiar with space geodesy, the initial web pages of the GGOS Portal will highlight the "burning questions" of geodesy and point the way to the relevant products, as well as their characteristics, location, availability, latency, accuracy, etc. General information about GGOS will also be available through the Portal, providing a valuable resource for both the external and internal GGOS communities.

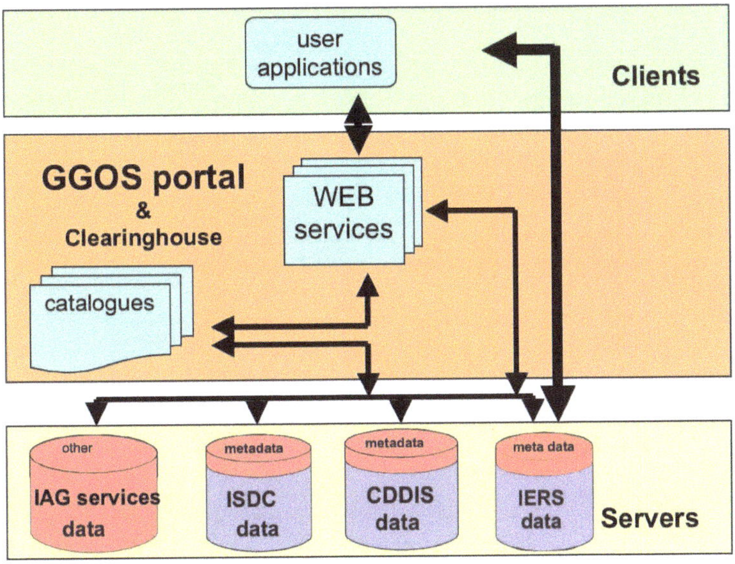

Fig. 9.7. GGOS portal architecture.

9.9.1 GGOS Portal architecture

The utility of the GGOS Portal will depend on data and information providers accepting and implementing a set of interoperability arrangements, including technical specifications for collecting, processing, storing, and disseminating shared data, metadata and products. GGOS interoperability will be based on non-proprietary standards, with preference given to formal international standards. The eXtensible Markup Language (XML) has become a quasi standard to facilitate the sharing of data across different information systems, particularly via the Internet. Moreover, web services for the support of interoperable machine-to-machine communication over a network are built on XML-based standards (e.g., Simple Object Access Protocol (SOAP), Web Services Description Language (WSDL)).

Data, products, and information from contributing IAG Services will becatalogued in a registry publicly accessible through the Clearinghouse. It is envisioned that this Clearinghouse is maintained collectively under the GGOS Portal. The catalogue will itself be subject to GGOS interoperability specifications, including the standard search and portrayal services.

The functions of the GGOS Portal (e.g., search capabilities for stations, satellites, data, products, institutions, data mining tools, visualization, web services, connec-

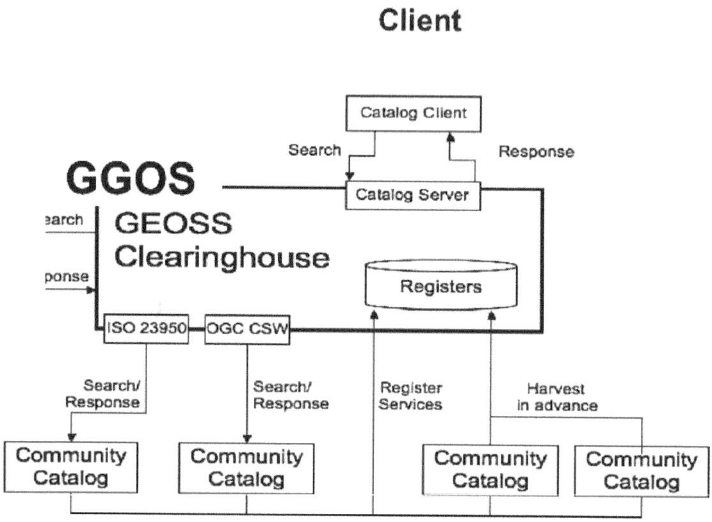

Fig. 9.8. GGOS Clearinghouse architecture – engineering viewpoint (according to D. Nebert). Here, the term clearinghouse is used in its modern meaning of a search-able registry, i.e., a set of catalogs.

tions to other catalogues, etc.) are supported by the GGOS Clearinghouse (Figure 9.7). The GGOS Clearinghouse will be a facility that collects and distributes information concerning the data catalogues and services. In a broader sense, the GGOS Clearinghouse will allow for a dialog between stakeholders on relevant issues in geodesy. The GGOS Portal will also provide access to a distributed network of catalogue services supporting the interoperability agreements of GGOS. Contributing IAG Services may nominate catalogues containing structured, standards-based metadata and other web services for access by the GGOS Clearinghouse. The Clearinghouse provides search capability across the catalogues and their registered resources by mapping these catalogues. The GGOS Portal will search the GGOS Clearinghouse but will also provide access to other GGOS resources e.g., calendar functions, forums, etc. Through the use of interoperability standards, additional portals may be established for national or professional communities to allow access to the GGOS Clearinghouse.

The metadata to be held by the Clearinghouse depends upon the search method. Two anticipated approaches to accessing remote catalogues (see Figure 9.8) include:

- Distributed search approach: search requests are sent in parallel to registered distributed catalogues of the IAG Services.
- Harvested approach: The Clearinghouse periodically harvests all metadata from registered distributed catalogues. A user search request is executed against the metadata harvested from the remote catalogues and the results are managed and visualized in the GGOS Clearinghouse.

9.9.2 GGOS Portal goals and objectives

The GGOS Portal will provide a web site which:

- represents a single web access point for all geodetic products relevant in the framework of GGOS;
- offers a set of tools for organized knowledge discovery including visualization to assist identification and selection of appropriate resources (information, data, products);
- accesses the GGOS Clearinghouse to search data catalogues, products and data sets generated by GGOS components;
- helps to answer the "burning questions of society" and points the way to the products, their characteristics, location, availability, latency, and accuracy;
- allows the searching of information and the retrieval of descriptive metadata from multiple, diverse target resources, databases, web pages, and library catalogues; and
- provides access to general information about GGOS.

Functions of the GGOS Portal include (but are not limited to):

- Basic functions of the GGOS web site such as hot spot information, news, tutorials, quick links, announcements, etc.
- A registry to host catalogues for metadata for all products of the IAG Services based on GGOS standards to ensure interoperability within the GGOS community and to other systems, in particular GEOSS.
- Search (temporal, spatial, multi-technique, keywords, etc.) of metadata, data, and product databases.
- Visualization of products (time series, maps, etc.).
- Information on and explanations of data, products, and geodetic techniques, with links to service-specific resources.

The GGOS Portal will likely be based on an open-source platform and web portal application allowing users to download, install and customize the portal services in their own environment. Based on modern architecture, standards and web services the GGOS Portal can be realized not only by single institutions but also by consortia with distributed server architecture. The Portal should be designed and implemented in such a way as to permit mirroring installation at alternate physical locations.

9.9.3 A GGOS clearinghouse mechanism for geodesy

A major function of the GGOS Clearinghouse will be to provide access to information on observations, products, and information relevant to GGOS, IAG, and geodesy in general. In a broader sense, the GGOS Portal, registry, and search engines should be complemented by a general clearinghouse mechanism (comparable to the clearinghouse mechanism for the Convention on Biodiversity, which links all stakeholders of the Convention; see http://www.cbd.int/chm/default.shtml). The mission of such a future GGOS clearinghouse mechanism would be to contribute to the implementation of GGOS, its standards, and its conventions, for the maximum benefit of its users. In particular, the GGOS clearinghouse mechanism should have three major goals:

- Promote and facilitate technical and scientific cooperation, among the IAG Services and Commissions, among GGOS components and other organizations, and within and between countries.
- Develop a global mechanism for exchanging and integrating information on geodesy.
- Develop the necessary human and technological networks.

Such an extended clearinghouse mechanism would have to be compatible with different levels of national/component capacity, driven by users' needs, and structurally decentralized. It would provide access to information, support decision-making, and have no vested interest in controlling the expertise or information. It would thus be created for the mutual benefit of all IAG Services and Commissions and other stakeholders.

Table 9.1. Parameter Space for a rigorous combination and integration of the geodetic observation techniques. Entry 1 defines the ICRF. Entries 2 to 5 related to the EOPs. Entries 6 and 7 together define the ITRF, while entries 7 to 10 are related to the gravity field. The atmosphere is covered by entries 11 and 12.

No.	Parameter	VLBI	GNSS	DORIS PRARE	SLR	LLR	Altimetry
1	Quasar Coordinates	X					
2	Nutation	X	(X)		(X)	X	
3	Polar Motion	X	X	X	X	X	
4	UT	X					
5	Length of Day		X	X	X	X	
6	Coordinates and Velocities	X	X	X	X	X	(X)
7	Geocenter		X	X	X		X
8	Gravity Field		X	X	X	(X)	X
9	Orbit		X	X	X	X	X
10	LEO		X	X	X		X
11	Ionosphere	X	X	X			X
12	Troposphere	X	X	X			X
13	Time/Frequency	(X)	X		(X)		

The activities of this clearinghouse mechanism would support GGOS' thematic and cross-cutting work programs by promoting cooperation, exchanging information and developing a network of partners. A first priority would be to ensure universal access to the GGOS Implementation Plan, including the underlying documents of the GGOS 2020 process, the GGOS standards, and conventions. The information provided would include case studies, national reports, and other relevant documentation. The mechanism would increase public awareness of the geodetic programs, issues, and products. It would be established as an Internet-based system to facilitate

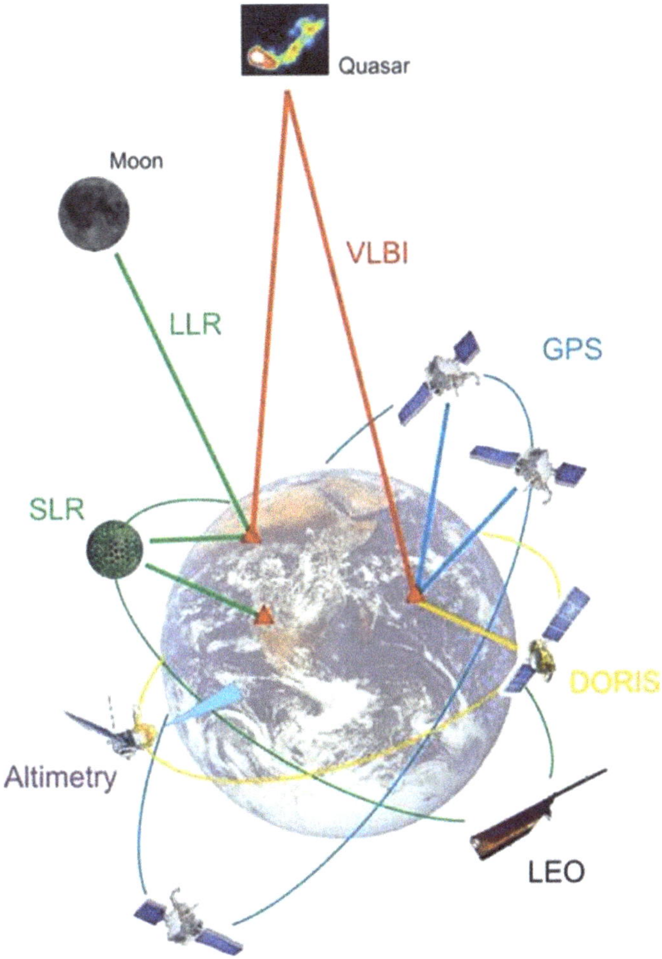

Fig. 9.9. Combination and integration of the geodetic observation techniques. The combined infrastructure allows the determination and maintenance of the global geodetic reference frames, and the determination of Earth's gravity field and rotation. The ground networks and navigation satellites (currently in particular GPS) are crucial in positioning, with applications to all SBAs. In particular, they allow the monitoring of volcanoes, earthquakes, tectonically active regions and landslide-prone areas. The Low Earth Orbit (LEO) satellites monitor sea level, ice sheets, water storage on land, atmospheric water content, high-resolution surface motion, and variations in the Earth' gravity field. The latter are cause, to a large extent, by regional and global mass transport in the hydrological cycle.

greater collaboration among the IAG Services and Commissions, the GGOS stakeholders, across national borders, through education and training projects, research cooperation, funding opportunities, and access to and transfer of technology.

This clearinghouse mechanism would be based on the philosophy that broad participation and easy information access must be a top priority. The underlying database can therefore be tapped through both traditional and electronic means of communication. Special efforts will have to be made to ensure the participation of organizations and institutions in developing countries.

9.10 Data analysis, combination, modeling, and products

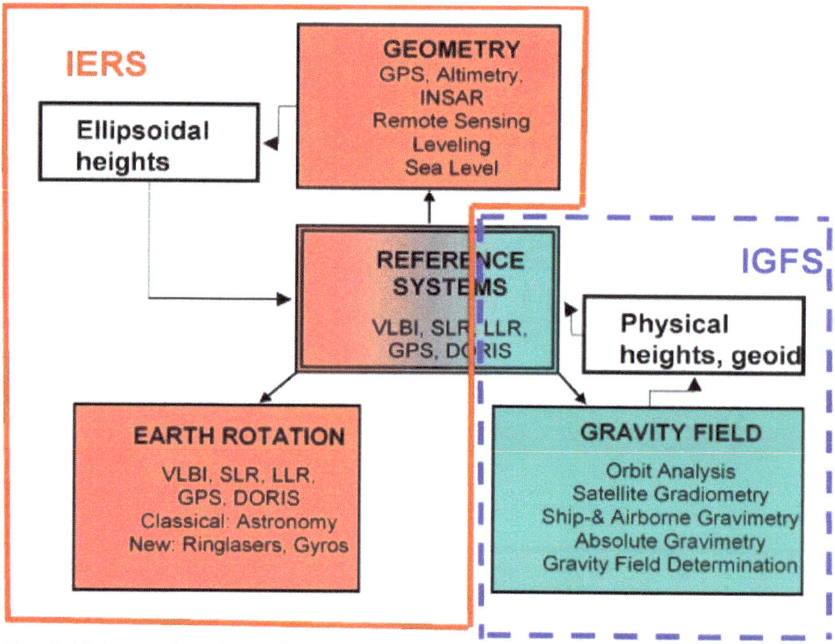

Fig. 9.10. Interactions in the Earth system centered around the three pillars of geodesy.

A major function of GGOS will be to facilitate the integration of the various levels of GGOS into a consistent observing system (Figure 9.9), delivering products and services as far as possible independent of the observing techniques and the processing. Considering the multi-technique, multi-component, and multi-parameter nature of GGOS, this will require consistency of processing strategies, models and standards across all components of GGOS. GGOS will facilitate communication and standardization between the analysis centers for the individual techniques and sensors, initiate intercomparison of products generated by the various components, and promote the study and modeling of technique-specific effects and other geodetic/geophysical signals. Redundancy and reliability will be achieved by having more than one analysis center for the major tasks and by developing full reprocessing capabilities for all data types.

Combination of the geometric products is currently achieved by the IERS. The International Gravity Field Service (IGFS) is developing the combination capabilities for the gravimetric products. The borderline between, and potential overlap of, IERS and IGFS (Figure 9.10) will require careful attention. GGOS will have to facilitate combination across the full parameter space (Table 9.1), and fully utilize synergies and advantages of the combination approach in partnership with the IERS and the IGFS.

The GGOS conventions will be a central issue for achieving consistency and highly accurate products. Currently, the conventions in the fields of geometry and rotation are taken care of by the IERS. In future, these conventions will have to be extended to cover the gravity field as well. They will have to address the geodetic, geophysical, geodynamic, etc., models to be used or, if not sufficient, to be developed. Coupling of models from oceanography, meteorology, geodesy, geophysics, glaciology, mass transport, energy budget, will have to be undertaken in order to achieve the GGOS accuracy goals. As pointed out in Chapter 8, 4-D Earth system modeling and the assimilation of diverse data into these 4-D Earth system models will have to be studied and eventually be covered by the conventions. The need for modeling and/or assimilation centers may thus arise. The importance of global geophysical fluids for validation will give a high weight to the Global Geophysical Fluid Center (GGFC) or an equivalent component of GGOS. However, the tools and methods for validation need more research and development.

GGOS as an observing system has to be more than just an Earth observing system collecting a tremendous volume of data. The observations have to be analyzed with state-of-the-art processing software and processing standards to generate time series of relevant geodetic, geodynamic, geophysical, hydrological and atmospheric parameters. To reach consistency between the different observation techniques the results of the individual techniques have to be rigorously combined and integrated using information on the local ties between the different instruments at co-location sites and satellites. Finally, the resulting products have to be validated and interpreted by making use of physical and geophysical models and modeling software packages, and by using additional observation data from other disciplines such as, for example, the meteorological, oceanographic, hydrological, etc. communities. This will require data analysis centers and centers combining the solutions from different analysis centers and different observation techniques and generating a series of GGOS products. Finally, the products have to be carefully validated.

With respect to the data analysis itself, software and modeling improvements are expected in form of, e.g., the development of new tropospheric mapping functions, gradient models, and atmospheric turbulence models, etc. Further areas of improvement are loading effects including mass loading models for hydrological variables, thermal and gravitational antenna deformations, and source structure effects. In addition, new analysis strategies will be investigated, in particular, the generation of consistent VLBI multi-purpose solutions for TRF, EOP, and CRF. Also, the "software noise" of solutions obtained from different software packages will need to be studied.

The processing and analysis procedures will have to progress towards fully automated processing in near real-time or even in real-time (particularly for early warning systems, GNSS seismology, atmosphere and ionosphere sounding). Full reprocessing capabilities for all data available will be crucial in order to provide long and consistent time series. A key development will be the combination of all data types at the observation level. This includes the combination of terrestrial data with LEO data (co-location, gravity, geocenter, atmosphere), as well as the combination with satellite altimetry data, and with InSAR and/or LIDAR. Finally, the combination of products from different analysis centers will increase redundancy, reliability, and accuracy.

The major outcome of GGOS in 2020 is expected to be a set of highly accurate, consistent and long-term stable products which will be the geodetic contribution to the observation and monitoring of the Earth system (i.e., to GEOSS and other international and regional initiatives). The high-level list of products is given in Section 7.5. It is anticipated that all GGOS product accuracies in 2020 will be of the order of about 10^{-9} relative to the absolute values of the measured quantities. However, in order to satisfy the goals mentioned above and in previous chapters, consistency between all GGOS products at the 10^{-9} level is also required.

Chapter 10
Towards GGOS in 2020

G. Beutler, M. Pearlman, H.-P. Plag, R. Neilan, M. Rothacher, R. Rummel

This chapter is concerned with the implementation of GGOS, addressing mainly the organizational aspects mentioned in Section 1.3. GGOS monitors and maintains the geometric and gravimetric reference frames, and provides the transformation between these systems with state-of-the-art observational tools. In doing so, GGOS provides the observational basis to, e.g., determine mass transport in the Earth system and thus serves the Earth science community with Earth observations and geodetic products relevant for many studies. The accuracy of, and the consistency between, all GGOS products shall be at the 10^{-9} level or better.

10.1 The GGOS high-level components

The tasks of GGOS outlined in Section 7.4 require the following components or entities:

1. **Terrestrial, technique-specific entities** coordinating the worldwide collection and primary analysis of the observations, and the generation of unique technique-specific products,
2. **Entities combining the technique-specific products** to develop technique-independent, combined product series.
3. An **entity proposing geodetic (and geodesy-related) space missions** in collaboration with the major space agencies (including geodetic missions to the Moon and other planets), resulting in an **uninterrupted series of geodesy-related space missions** to observe the time-varying gravity field, the time-varying sea-surface and ice-surface topography, and to maintain the geometric and gravimetric reference frames.
4. An **entity for communications and network coordination** to assist in design, and continuously improve the GGOS network.
5. **Bureau of Standards** to address the first five issues listed in Section 7.4, i.e., the conventions for the reference systems and frames, and the geodetic standards.

H.-P. Plag, M. Pearlman (eds.), *Global Geodetic Observing System*,
DOI 10.1007/978-3-642-02687-4_10, © Springer-Verlag Berlin Heidelberg 2009

6. **Coordination Office** to coordinate the activities of the GGOS entities and to provide the interface to the GGOS user community and the political decision-makers.

Many of the above entities, in particular the technique-specific networks and the center for combination of the geometry-related products and Earth rotation, already exist. These existing elements may need modification, fine-tuning, or improved co-ordination between themselves. It also may be necessary to put in place some of the existing elements on a new funding basis (see Section 10.3).

The GGOS elements of Levels 2 and 3 (see Section 9.1) related to geometry are also already in place (GNSS and laser ranging satellites, including the Moon with its laser reflectors). The same is not true for the space missions monitoring the Earth's gravity field and measuring sea and ice-sheet topography. Admittedly, there has been a long series of altimetry missions monitoring in the sea surface topography. In addition, with the satellite missions CHAMP, GRACE, and GOCE, an impressive series of gravity missions is currently active. There are, however, no concrete plans for monitoring the gravity field, and the fluid components, on a long term basis. GGOS shall develop a master plan for missions monitoring the Earth's gravity field and the geometry of the solid Earth, oceans and ice sheets. Therefore, GGOS shall have an entity for proposing geodetic space missions.

The proposed GGOS structure is shown in Figure 10.1. GGOS requires a decision-making body. For this purpose, GGOS shall have a Steering Committee in which all shareholders are represented, and in particular the IAG Services and Commissions. For day-to-day work, a smaller Executive Committee is proposed, which prepares proposals for the Steering Committee and oversees the work of the Coordination Office according to the decisions of the Steering Committee.

GGOS will have a Science Panel composed of experts in geodesy and (more generally) in Earth sciences to ensure the GGOS focus remains on the relevant scientific challenges and societal needs. This panel will be the main scientific advisory group for the GGOS Steering Committee.

This structure is defined in the GGOS Terms of Reference (ToR). The current version of the ToR was accepted by IAG during the IAG Executive Meeting in San Francisco in December 2008. The IAG By Laws as accepted during the IUGG meeting in Perugia, Italy, in July 2007 define GGOS as IAG's Observing System and an IAG component on the same (hierarchically highest) level as its Commissions and Services. Consequently, the GGOS Chair, appointed by the IAG, is also a member of the IAG Executive Committee.

10.2 Building on the heritage

10.2.1 Level 1: the terrestrial geodetic infrastructure

The terrestrial part of GGOS at present consists of

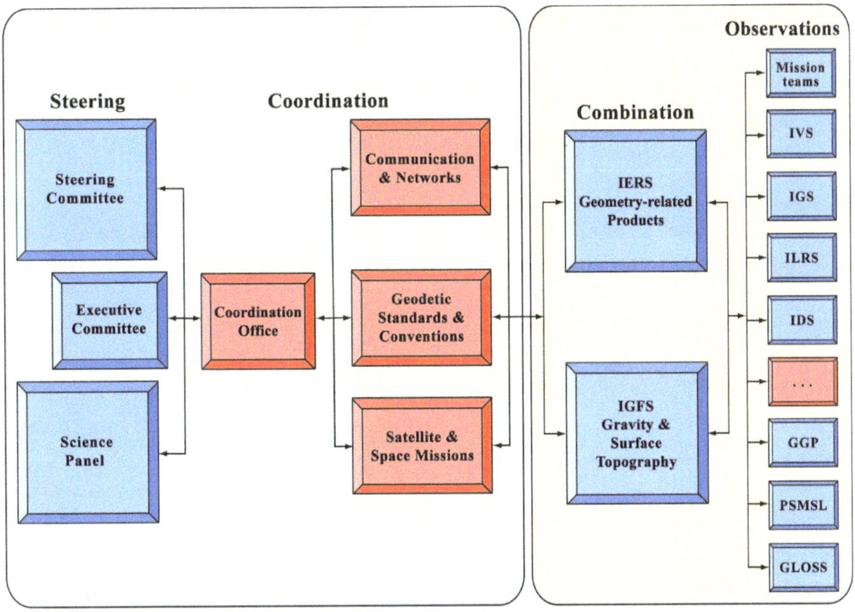

Fig. 10.1. Structure for the future GGOS. Lines and arrows indicate information flow and do not imply any hierarchical relations. Existing entities are indicated in blue, new entities currently being established in red.

- A global VLBI network of about 40 stations, coordinated by the IVS:
  - to maintain the celestial reference frame;
  - to guarantee that all Earth rotation parameters (in particular UT1-UTC, precession and nutation) may be derived in a combined analysis containing the observations of all space-geodetic techniques; and
  - to contribute to the terrestrial reference frame, including the scale.
- A global SLR/LLR network of about 40 observatories, coordinated by the ILRS:
  - to ensure that the tie between CM and the center of the station polyhedron may be observed with cm accuracy;
  - to contribute to the terrestrial reference frame and the determination of Earth rotation parameters, including the scale;
  - to calibrate/validate the GNSS-derived orbits (GPS, GLONASS, LEO);
  - to serve as back-up for LEO and MEO orbit determination.
- An international GNSS network of more than 300 sites, coordinated by the :
  - to provide highly accurate (cm-level) orbits and satellite clock corrections for all GNSS satellites as a prerequisite for precise (sub-cm) positioning and navigation;
  - to maintain the terrestrial reference frame (positions and velocities) in the required density;

  - to make the reference frame accessible anywhere on and near the surface of the Earth;
  - to provide a link between the GGOS technique-specific networks; and
  - to establish, together with VLBI and SLR, the series of Earth rotation parameters to the required quality and time resolution.

- A well-distributed network of more than 50 DORIS receivers for the primary purpose of determining the orbits of LEOs (e.g., TOPEX, JASON), but also contributing to the international terrestrial reference frame. DORIS is also used for calibrating space-geodetic techniques (e.g., GPS, SLR, DORIS using JASON). The DORIS operations are organized by the IDS.
- A well-distributed network of about 40 core sites with co-located SLR, GNSS, and VLBI and at least one other measurement technique (gravimetry or DORIS) at sites with acceptable weather and geological conditions. At least four of these sites should have full LLR capability.
- The IERS combining the products emerging from the technique-specific Services and publishing the products required for the maintenance terrestrial and celestial reference frame, including the series of Earth rotation parameters.
- A global network of absolute gravity stations and superconducting gravity stations co-located with space-geodetic reference stations, with the goal to tie the geometric reference frame with high long-term stability to the CM.
- A global network of tide gauges co-located with permanent GNSS sites.

In view of the impressive number of functioning IAG Services (Figure 10.1) with their fully operating networks and data processing elements, GGOS can be built on this very valuable heritage.

10.2.2 Level 2: the LEO satellite missions

The satellite gravity-related part of GGOS is not in as good a position as the geometry-related GGOS parts. It currently consists the IGFS, and the individual satellite mission teams, and it shall be augmented by a Satellite Mission entity. It should be stated that the IGFS has not (yet) assumed the role of comparing and combining gravity field results from all missions. Neither does it incorporate the results of the altimetry missions. The entity for satellite missions does not exist either. Attempts to establish a mission-independent *International Altimetry Service* (IAS) have not been successful so far. Likewise, GGOS should promote the establishment of a mission-independent global InSAR entity, preferably as a service. However, for both, satellite altimetry and InSAR, the question of access to the proprietary data is a major obstacle that GGOS needs to address in conjunction with the space agencies and, potentially, GEO.

10.2.3 Level 3: the GNSS and SLR satellites

A considerable number of "cannonball" satellites (LAGEOS 1 and 2, STARLETTE, etc.) were launched for geodetic purposes in the late 20th century. The continued observation of these satellites and the analysis of their observations are essential to derive, for example, the geocenter location with respect to the GGOS network and the (other) low degree and order coefficients of the Earth's potential field. The deployed SLR satellites potentially provide continuity over many decades.

In the near future there may be up to four fully operational GNSS based on the same principles of operation (GPS, GLONASS, GALILEO, and COMPASS). The observations of all available GNSS satellites (with all suitable techniques) will be important, from the point of view of the number of terrestrial observing sites. Moreover, many (if not most) Low Earth Orbiters (LEOs) deployed for Earth observation are or will be equipped with GNSS receivers and SLR reflectors to allow for an independent validation of the orbit quality derived from the microwave observations. It is therefore extremely important that all GNSS spacecrafts also be equipped with SLR reflectors. Co-location in space is the counterpart of co-location on the ground.

10.2.4 Level 4: lunar and planetary "geodesy" and missions

The Moon is in many respects "just another satellite". Thanks to its extremely low "cross section to mass ratio" (compared to the corresponding ratios for the artificial satellites) the lunar orbit offers in addition the unique opportunity to test the theories of gravitation. For monitoring changes in the distance between Earth and Moon and to test theories of gravitation, it is therefore essential that LLR to the reflectors deployed by the Apollo missions and Russian spacecrafts continues at least until transponders on the Moon and other planets become available. The lunar retro-reflectors provide long-term continuity for LLR.

When exploring the Moon, and the Solar System planets and their satellites, there are a number of issues which can only be addressed using geodetic techniques. In particular:

- establishment of a body-fixed reference frame (corresponding to the terrestrial reference frame);
- determination of the body's rotation (corresponding to Earth rotation); note that the celestial reference frame, established and maintained by the IVS, is a prerequisite for studying the rotation of such bodies;
- determination of the body's gravity field (via the trajectories of orbiters, including the gravity field of the Moon);
- evolution of the planets' satellite system (including the evolution of the Earth-Moon system); and
- mapping of the body's surface (via altimetry and InSAR).

10.2.5 Level 5: the extragalactic objects

The Quasars are the only objects observed by GGOS, which do not need to be deployed. It is a primary task of the IVS to define and maintain a list of quasars (with their coordinates) for geodetic purposes. The IVS was given this responsibility by the IAG (through International Union of Geodesy and Geophysics (IUGG)) and the IAU. Although the quasar distances are large enough such that Proper Motion can be neglected, the center of the microwave radiation may slightly vary. Therefore, the task of selecting such objects for geodetic purposes is not trivial.

10.3 Organizational considerations

10.3.1 History

The IAG has a long tradition of establishing scientific services. The first one, the ILS was created under the auspices of IAG at the end of the 19th century. Its creation is in a way exemplary for the motivation to create the GGOS. The creation of the ILS was motivated by the need to monitor Earth rotation, in particular polar motion (which at that time could by established by latitude observations with astronomical telescopes). The ILS, and then its successor, the International Polar Motion Service (IPMS), were remarkably stable. Polar motion determinations were generated by these Services for about eighty years. Thanks to these Services, today there is more than one hundred years of polar motion data. The ILS, Bureau International de l'Heure (BIH), and the IPMS were directly funded by government agencies for the declared purpose of monitoring the rotation of the Earth, which also included the definition and realization of universal time. Today geodesists are deeply indebted to these agencies for having initiated the geodetic study of the Earth, based on a worldwide collaboration of institutions and backed up by governmental commitments. With GGOS the IAG wishes to achieve a comparable system to serve the needs of the 21st century.

10.3.2 The revolution invoked by space geodesy

The funding situation in the field of geodesy and geodynamics changed dramatically in the second half of the 20thcentury. With the replacement of optical astrometry by new space-geodetic techniques, Earth monitoring for geodetic and geodynamics purposes was funded to a great extent by research and development funds. This step was also justified by the fact that modern space-geodetic methods opened the way to study not only aspects related to Earth rotation, but provided the metrological basis for a much wider range of applications (from global studies to regional and local ones). In the 1990s it also became evident that technique-specific services had to be created in order to exploit the full scientific potential of the new techniques. First the IGS, then the ILRS, the IVS, and eventually the IDS were created.

10.3.3 Current situation

The technique-specific IAG Services were established through Calls for Participation without offering funding for operations. Today these Services are therefore based on a voluntary collaboration of the contributing organizations. Naturally the funding situation differs from country to country (sometimes even from institution to institution). It may be puzzling for government representatives and science managers that despite considerable uncertainties the IAG Services have been so successful and stable. However, the question needs to be asked whether this model is sustainable in the long run.

GEO is building the GEOSS to a large extent on the same principle of voluntary commitment and best-effort contributions. The experience of the successful IAG Services shows that this principle requires a high degree of redundancy, and at the same time is problematic for providing a uniform global coverage of ground-based infrastructure. In particular for reference frame maintenance, large spatial gaps and temporal variations in the monitoring infrastructure (including changes in the polyhedron through new and disappearing stations) cause temporal inhomogeneities and degradations of accuracy.

In 2001 the IAG Council approved, upon recommendation of the IAG Executive, a new structure with the GGOS project as the IAG's flagship. It was the intention to view modern geodesy from the global perspective and to bundle the efforts of all branches of this science to serve one and the same goal. This book should be viewed as the principal result of the work of the GGOS planning committee (2003-2005) and the GGOS implementation committee (2005-2007).

10.3.4 Internal organization of GGOS

In Section 10.2 it was argued that GGOS must be built on the foundation of the IAG Services developed at the end of the 20th century. In Section 10.3.6 the two essential elements of the GGOS environment, namely the GEO (and its attempt to establish its GEOSS) and IGOS-P were introduced. The required GGOS structure was laid out in Section 10.1 (Figure 10.1). GGOS shall have:

- a Steering Committee as the plenary and decision-making component with representation of the shareholders;
- a Science Panel with broad expertise providing scientific advice to the Steering Committee;

- an Executive Committee developing GGOS according to the strategy and guidelines decided upon by the Steering Committee;
- a Coordination Office responsible for day-to-day operations;
- three entities (for satellite missions, geodetic standards, and networks and communications).

GGOS shall be based on the infrastructure provided by the IAG Services, i.e.,

- the two combination Services (IERS and IGFS);
- the technique-specific Services (IGS, ILRS, IVS, IDS, IGeS, GGP), the satellite and space mission teams, and future services such as an IAS and an InSAR services.

The missing parts of GGOS should be created as soon as possible. Their creation should be based on a Call for Participation issued by the IAG. A recommitment under the new boundary conditions (and an adapted list of duties and deliverables) shall be initiated for the established parts of GGOS. The conclusions and findings of this book shall serve as the basis for the establishment of the future GGOS.

10.3.5 Integration of relevant regional activities

GGOS will actively seek participation of regional geodetic programs to augment the GGOS global coverage in order to make the most economic use of available resources.

10.3.6 Integration of GGOS into global programs

It is encouraging to note that recently the necessity to preserve the infrastructure for global Earth observation was recognized on the ministerial level (see also Sections 1.3 and 5.1). In 2003, the *ad hoc* GEO was established as a result of a G8-meeting, and guided by a series of three ministerial-level Earth Observation Summits, GEO developed a plan for the implementation of GEOSS (see Chapter 5 for an overview or GEO, 2005a,b, for background information). In 2005, GEO was established permanently. Currently, GEO includes almost 80 member countries and more than 45 participating organizations. GEO is establishing the GEOSS with the vision *to realize a future wherein decisions and actions to the benefit of humankind are informed via coordinated, comprehensive and sustained Earth observations and information* (GEO, 2005a). GEOSS is building on and adding value to existing Earth observation systems by coordinating their efforts, addressing critical gaps, supporting their interoperability, sharing information, reaching a common understanding of user requirements, and improving delivery of information to users.

The IAG, represented in GEO by GGOS delegates, is one of the active participating organizations. GGOS clearly must be developed taking into account its role as a crucial part of the GEOSS. GGOS may be viewed as the metrological basis for many parts of the GEOSS.

The importance of geodesy, and in particular, GGOS, for Earth observations is increasingly acknowledged by GEO and comparable organization, as well as their users, such as United Nations authorities and scientific projects. Further progress in geodetic techniques and products is seen as an essential step towards a better understanding of the Earth system four the benefit of mankind. This has been recognized by GEO by including a task specifically addressing the geodetic reference frames in the GEO Work Plans 2007-2009 and 2009-2011.

GGOS is coordinated scientifically by IAG. On a higher organizational level GGOS must be associated with that international organization, which eventually will have all the other global observing systems under its auspices, while still remaining a major component of the IAG. Assuming a successful establishment of GEOSS, it is clear that the GGOS must be one of the systems in GEOSS. However, in order to improve the funding situation and to establish a firm link to an appropriate intergovernmental authority, GGOS should also be associated with a UN body. UNESCO appears to be an appropriate choice. It is hoped that this umbrella organization will sponsor elements of GGOS. IAG will provide (as it has done so far) the scientific oversight and expertise for GGOS and it will provide, through GGOS, the link to the umbrella organization.

Chapter 11
Recommendations

H.-P. Plag, G. Beutler, R. Gross, T. A. Herring, P. Poli, C. Rizos, M. Rothacher, R. Rummel, D. Sahagian, J. Zumberge

In this Chapter, we summarize the recommendations of the GGOS 2020 Writing Team. The recommendations are numbered, with the first number indicating the chapter from which a recommendation originates.

Recommendation 1.1 (Transition from research to operational):

Recognizing that

geodesy has a large potential to help meet the challenge in reaching sustainable development for a global society on a changing planet

it is recommended that

IAG and GGOS engage in improving the framework conditions for fully harnessing the potential of geodesy for Earth observation by actively promoting a transition of the geodetic observing system from research to operational, and facilitate the establishment of an operational core of GGOS with sufficient human resources for the sustained operation of this core.

Recommendation 1.2 (Global reference systems):

Recognizing that

the global geodetic reference frames are fundamental for all Earth observations

it is recommended that

IAG and GGOS facilitate, particularly in the frame of GEO, international agreement on a global geodetic reference system.

Recommendation 1.3 (Outreach and Eduction):

Recognizing that

society to a large extent is not aware of the vital role played by geodesy for realizing the principle of sustainable development, and that

educational aspects are extremely important (because they have the greatest impact on societal behavior) in order to prepare future generations to make use of the full benefits of geodesy

it is recommended that

IAG and GGOS make dedicated outreach efforts to science and society at large, with the goal to promote geodesy's role in reaching sustainable development, and to integrate this role of geodesy appropriately into education programs.

H.-P. Plag, M. Pearlman (eds.), *Global Geodetic Observing System*,
DOI 10.1007/978-3-642-02687-4_11, © Springer-Verlag Berlin Heidelberg 2009

Recommendation 2.1 (Adherence to conventions and standards):
Recognizing that
consistency within and across the "three pillars of geodesy" crucially depends on well-defined standards and widely respected conventions
it is recommended that
IAG continue to maintain geodetic standards and develop conventions for geodetic analyses and products, and that
every effort be made by the IAG Services and GGOS to adhere to relevant standards and conventions.
Recommendation 3.1 (Towards new reference systems):
Recognizing that
the combination of geodetic measurements will require new definitions of a terrestrial reference system and a consistent realization of this system
it is recommended that
particular attention be paid in GGOS to the development of these new reference systems.
Recommendation 4.1 (Promotion of ITRS and maintenance of ITRF):
Recognizing that
a stable and accurate geodetic reference frame must underpin Spatial Data Infrastructures (SDI), to ensure that all geo-referenced data used by a wide range of community groups and government agencies is unambiguously linked to the geodetic foundation
it is recommended that
ITRS be employed as the global geodetic reference system for SDI, and that
the ITRF be maintained and made accessible with an operational core ensuring ITRF with the accuracy, long-term stability, and the level of accessibility required by SDI applications.
Recommendation 4.2 (The link between science and applications):
Recognizing that
geodesy plays a vital role with respect to sustainable development, the provision of community services, support for many vital industries, security and emergency management, mapping and navigation, and others
it is recommended that
the link between "scientific geodesy" and "practical (or operational) geodesy" be strengthened, and made explicit so that national geodetic agencies are reminded of the mutual benefits of these two parts of geodesy, and of the fundamental contribution of geodesy to their mission.
Recommendation 4.3 (Links of IAG to other professional organizations):
Recognizing that
geodesy and GGOS are relevant to a number of international and national scientific and professional sister organizations of IAG, including, but not restricted to ISPRS, International Federation of Surveyors (FIG), International Association of Institutes of Navigation (IAIN), IEEE, and IUGG
it is recommended that

the IAG continue to work closely with these organizations by promoting the GGOS vision and its activities.

Recommendation 4.4 (Embracing new technologies):

Recognizing that

there is rapid technological development integrated into professional applications of geodesy

it is recommended that

GGOS embrace new geoinformation/geodetic technologies such as DInSAR, LIDAR, GNSS-RTK, structural monitoring systems, and multi-sensor precision navigation systems, in an integrated manner to address different spatial and temporal user requirements for high accuracy geometric information that is unambiguously tied to a rigorous geodetic framework.

Recommendation 5.1 (Representation in Earth Observation Committees):

Recognizing that

geodesy provides the metrological basis enabling Earth observation with the required accuracy and that

the global geodetic reference frame is a fundamental contribution to global Earth observation

it is recommended that

GGOS maintain a formal representation in existing Earth observing coordination committees (international bodies and commissions), and establish links to relevant committees as appropriate.

Recommendation 5.2 (Real-time access to data of GNSS tracking stations and promotion of occultation receivers):

Recognizing that

real-time or low-latency access to geodetic observations, in particular GNSS observations, is increasingly important for applications in numerical weather predictions, space weather predictions, early warning systems, and other societal applications

it is recommended that

existing and future ground-based GNSS sites installed by geodesists be connected in real-time to GNSS data and analysis centers thus enabling these non-geodetic applications, and that

future geodesy missions using GNSS receivers be radio occultation-compatible, whenever possible, and that

a framework be established to allow occultation data processing as a service to geodesy missions (i.e., outside the geodesy missions themselves).

Recommendation 5.3 (Gravity field and circulation models):

Recognizing that

detailed knowledge of the Earth's gravity field is important for atmosphere and ocean circulation models

it is recommended that

that GGOS establish proper contacts and interfacing to the Earth system modeling community with the goal to enable improvements of the gravity field representation in circulation models.

Recommendation 5.4 (GNSS and climate studies):
Recognizing that
GNSS observations are an important information source for climate studies related to water vapor
it is recommended that
GGOS (through its components) continue archiving GNSS observations and all necessary data for future reprocessing and use in climate studies, and that
long-term funding for this archiving be secured from relevant climate programs.
Recommendation 5.5 (GGOS and monitoring of the global water cycle):
Recognizing that
geodetic observations are fundamental for monitoring the global hydrological cycle on global to local scales
it is recommended that
GGOS encourage and support a global water cycle service that provides information on changes in the water storage on land, in ice sheets and in the oceans on a routine basis, potentially through assimilation of the geodetic observations in a Earth system model.
Recommendation 5.6 (GNSS seismology):
Recognizing that
GNSS can potentially contribute to the near real-time determination of the seismic magnitude and associated displacement field of large earthquakes
it is recommended that
GGOS promote the development of GNSS seismology, particularly for early warning and disaster assessment purposes.
Recommendation 6.1 (GGOS in support of planetary missions):
Recognizing that
planetary geodesy, radio science, interferometry (including imaging VLBI, astrometric VLBI, and Earth-Space VLBI), and interplanetary navigation all require an accurate geodetic foundation, and that
although the performance of GGOS is not a limiting factor in all of these applications, future requirements will be more demanding, especially those imposed by interplanetary navigation, and in some cases exceeding presentday capabilities of GGOS
it is recommended that
GGOS be developed in order to meet these future requirements, that in particular
GGOS generate real-time values of Earth orientation accurate to ≤ 3 mm, and that
GGOS enable calibrations of troposphere delay and ionosphere accurate to ≤ 3 mm and ≤ 2 TEC units, respectively.
Recommendation 7.1 (Threshold and target values for GGOS):
Recognizing that
it will not always be possible to implement the observing system meeting all observational requirements
it is recommended that

GGOS set up, together with relevant user groups, threshold and target values in terms of accuracy, spatial and temporal resolution, latency, and integrity, for the quantities to be observed or derived from geodetic observations.

Recommendation 7.2 (GGOS database of user needs and observational requirements):

Recognizing that

the user needs and observational requirements with respect to geodetic observations and products will evolve over time

it is recommended that

GGOS maintain a database of user needs and observation requirements, and a list of products to be provided by GGOS in order to meet these evolving requirements.

Recommendation 7.3 (Improved access to ITRF):

Recognizing that

the global geodetic reference frame and ready access to this frame plays a crucial role for many scientific, professional and societal applications

it is recommended that

GGOS focus on improved access to the global reference frame with low latency and high spatial resolution as well as the long-term stability of the frame.

Recommendation 8.1 (Future reference frame approach based on extended model):

Recognizing that

users in many applications increasingly require access to a geodetic reference frame with high spatial and temporal resolution in order to be able to detect "anomalous" motion of an object with respect to the reference frame

it is recommended that

GGOS encourage the development of a future reference frame approach based on a reference frame model with, in principle, infinite spatial and temporal resolution, and that

this reference frame be based on a dynamic Earth system model that assimilates observations and predicts the motion of all points on the surface of the Earth, as well as variations in the gravity field of the Earth system and the rotation of the solid Earth.

Recommendation 8.2 (Towards an integrated Earth system model):

Recognizing that

the future geodetic reference frame approach, in order to meet the demanding user requirements and to achieve the required spatial resolution, will have to be based on model prediction

it is recommended that

GGOS promote the development of an integrated Earth system model which can be used to predict the geodetic quantities in a self-consistent framework, and that

both forward-modeling and inversion methods be developed to predict geodetic quantities and to invert geodetic observations for the forcings, respectively.

Recommendation 9.1 (Augmentation of the current global geodetic infrastructure):

Recognizing that

the currently implemented global geodetic infrastructure is not sufficient to provide a monitoring of Earth's shape, gravity field and rotation meeting most of the users' needs, and to sustain the global geodetic reference frames required for many scientific and societal applications

it is recommended that

the global geodetic infrastructure not only be maintained at the current level but also be augmented, in order to close major spatial and technological gaps, with: (1) a global network of core sites on all continents, (2) absolute and superconducting gravimeters at a global network of reference sites, in particular the core sites, and (3) two additional dedicated SLR satellites, that

an operational core system be built up and maintained with the necessary infrastructure for an operational geodetic Earth system service providing quantitative information on changes in ice sheets, sea level, water cycle, and climate, as well as for hazards, disasters, and resource management application, and that

the operational core include at least: (i) the global geodetic networks for the determination and monitoring of the geodetic reference frames, including Earth rotation, (ii) continuous gravity satellites missions for the monitoring of mass transport, (iii) continuous satellite missions for the monitoring of ice sheets, sea surface height, and lake level variations, and (iv) continuous satellite missions for the imaging of the solid Earth's surface.

Recommendation 10.1 (Continuation of the IAG Services):

Recognizing that

the terrestrial technique-specific entities represented by the IAG Services are the basis of IAG's GGOS, and that their products are prerequisites for the realization GGOS

it is recommended that

the work of the technique-specific entities, based on the state-of-the-art observational and analysis tools, be continued, and that

funding for these technique-specific services be secured on a long-term basis.

Recommendation 10.2 (Uninterrupted sequence of satellite missions):

Recognizing that

uninterrupted geodesy-related satellite missions are required for the generation of the best possible time-varying gravity field, and the monitoring of sea and ice surface topographies, and that

today there is no consistent plan for deploying geodesy-related satellite missions

it is recommended that

GGOS, in close partnership with the space agencies and CEOS, develops a plan for an uninterrupted series of geodesy-related space missions based on scientific and societal needs, and that

GGOS have a specific entity developing these scenarios.

Recommendation 10.3 (Continuation of IERS):

Recognizing that

the results of the technique-specific entities (the IAG Services) are compared, validated, and combined to generate unique, technique-independent geodetic products (celestial, terrestrial, and, to a lesser extent, gravitational reference frames, constants, etc.), that

this work is undertaken by the IERS for the geometry-related products

it is recommended that

the work of the IERS be continued based on state-of-the-art validation and combination techniques, and that

funding for these activities be secured on a long-term basis.

Recommendation 10.4 (Plan for gravimetric mission-independent products):

Recognizing that

full utilization of the gravimetric satellite missions requires long time series based on all relevant techniques

it is recommended that

the IGFS develop a plan to generate mission-independent gravity products, which also include terrestrial and airborne data, that

an entity realizing this plan (inside or outside the IGFS) be identified, or, if not existing, be created, and that

long-term funding for this entity be secured.

Recommendation 10.5 (Establishment of an IAS):

Recognizing that

the geodetic products resulting from space missions (including SLR data, terrestrial and airborne gravity measurements) must be compared, validated, and combined into unique geodetic products, which have to be, moreover, consistent with the geometry-related products, and that

an international altimetry service could address one aspect of this problem, namely that of sea and ice surface topography based on the data of all altimetry missions available

it is recommended that

an IAS as a mission-independent altimetry service be established and incorporated into IAG and GGOS, and that

funding for this IAS be secured on a long-term basis.

Recommendation 10.6 (Establishment of an international InSAR Service):

Recognizing that

the InSAR observations are very versatile observations of Earth surface deformations and that

these observations serve a wide range of applications

it is recommended that

an international InSAR service be established and incorporated into IAG and GGOS, that

this service support the application of InSAR integrated with GNSS and make products related to Earth surface deformations routinely available, and that

funding for this service be secured on a long-term basis.

Recommendation 10.7 (Standards and conventions):

Recognizing that

in order to ensure consistency of observations, data processing, modeling and products across the "three pillars of geodesy" at a level of better than 10^{-9}, adher-

ence to geodetic standards and conventions is crucial
it is recommended that
a GGOS entity responsible for the geodetic standards and conventions be created (known as the "GGOS Bureau of Standards and Conventions"), that
this entity keep track of and make available a detailed and concise list of geodetic conventions, constants, and procedures, and that
this catalogue include the IERS conventions.

Recommendation 10.8 (Networks and communication):
Recognizing that
currently there is a large number of more or less independent technique-specific ground-tracking networks (and products) in GGOS, and that
coordination of these networks is not sufficient
it is recommended that
the IAG Services operating the technique-specific networks create, in cooperation with the IERS, a "GGOS Communications and Networks" entity with the objective of designing the networks (minimum number and distribution of core sites, co-location of techniques, etc.) and scoping the operation (communication and data flow between networks, and from stations to regional and global data centers) of the network as a whole.

Recommendation 10.9 (United Nations support for GGOS):
Recognizing that
the full implementation of GGOS, and particularly of an operational core system, requires broad international support for GGOS as organization
it is recommended that
IAG continue its active role in GEO and other relevant organizations, and that
IAG and GGOS continue the dialog on the association of GGOS with an appropriate United Nations agency (e.g., UNESCO).

Recommendation 10.10 (Establishment of a GGOS Coordinating Office):
Recognizing that
GGOS is based on a wide range of contributing organizations, institutions, space agencies, services, and systems, and that
GGOS has a wide range of users and stakeholders
it is recommended that
GGOS establish a central coordinating entity (known as the "GGOS Coordination Office") with the task to maintain an overview on GGOS contributors and users and their requirements as well as to support the GGOS decision-making entities on a day-to-day basis, and that
funding for this entity be secured on a long-term basis, preferably through the respective United Nations agency.

Recommendation AI.1 (GEO Resolution):
Recognizing that
the fundamental role for geodesy and the geodetic observation system for Earth observation in general, and GEO in particular, necessitates the continuous commitment of many national and regional institutions, organizations and governments to

GGOS
it is recommended that
the GEO Plenary consider a resolution recommending to the GEO member countries to maintain, and if necessary increase, their support of the operational infrastructure of GGOS at a level appropriate to meet the requirements of the SBAs addressed by GEO.

Recommendation AI.2 (GGOS Stakeholder Conference):
Recognizing that
the implementation of GGOS on the basis of the findings and recommendations resulting from the GGOS 2020 Process requires a continuous dialog engaging all stakeholders inside and outside of IAG
it is recommended that
a conference of the GGOS stakeholder organizations be organized to further develop the findings and recommendations of the GGOS 2020 book into key elements for the Implementation Plan of GGOS, and that
the GGOS 2020 book serve as the basis for discussion and decisions at this conference.

References

Ablain, M., Dorandeu, J., Le Traon, P.-Y., & Sladen, A., 2006. High resolution altimetry reveals new characteristics of the December 2004 Indian Ocean tsunami, *Geophys. Res. Lett.*, **33**, L21602, doi:10.1029/2006GL027533.

Abshire, J. B., Riris, H., Kawa, S. R., Sun, X., Chen, J., Mao, J., Stephens, M. A., Allan, G., Collatz, G. J., & Jian, P.-S., 2007. Laser sounder for global measurements of CO_2 concentrations in the troposphere from space, *Geophys. Res. Abst.*, **9**, EGU2007–A–10014.

Achard, F., Eva, H. D., Stibig, H. J., & et al., 2002. Determination of deforestation rates of the world's humid tropical forests, *Science*, **297**, 999–1002.

Achard, F., Eva, H. D., Mayaux, P., & et al., 2004. Improved estimates of net carbon emissions from land cover change in the tropics for the 1990s, *Global Biogeochemical Cycles*, **18**(2), Art. No. GB2008.

Alley, W. M., Healy, R. W., LaBaugh, J. W., & Reilly, V., 2002. Flow and storage in groundwater systems, *Science*, **296**, 1985–1990.

Alsdorf, D., Dunne, T., Hess, L., Melack, J. M., & Birkett, C., 2001a. Water level changes in a large Amazon lake measured with spaceborne radar interferometry and altimetry, *Geophys. Res. Let.*, **28**, 2671–2674.

Alsdorf, D. E., Melack, J. M., Dunne, T., Mertes, L. A. K., Hess, L. L., & Smith, L. C., 2000. Interferometric radar measurements of water level change: Amazon floodplain response to river stage, *Nature*, **404**, 174–177.

Alsdorf, D. E., Smith, L. C., & Melack, J. M., 2001b. Amazon water level changes measured with interferometric SIR-C radar, *IEEE Transactions on Geoscience and Remote Sensing*, **39**, 423–431.

Altamimi, Z., Sillard, P., & Boucher, C., 2002. A new release of the International Terrestrial Reference Frame for earth science applications, *J. Geophys. Res.*, **107**, 2214, doi: 10.1029/2001JB000561.

Altamimi, Z., Collileux, X., & Boucher, C., 2006. DORIS contribution to ITRF2005, *J. Geodesy*, **80**(8-11), 625–635, DOI 10.1007/s00190–006–0065–5.

Altamimi, Z., Collilieux, X., Legrand, J., Garayt, B., & Boucher, C., 2007. ITRF2005: A new release of the International Terrestrial Reference Frame based

H.-P. Plag, M.Pearlman (eds.), *Global Geodetic Observing System*,
DOI 10.1007/978-3-642-02687-4_BM2, © Springer-Verlag Berlin Heidelberg 2009

on time series of station positions and Earth Orientation Parameters, *J. Geophys. Res.*, **112**, B09401, doi:10.1029/2007JB004949.

Altamini, Z., Coulot, D., Berio, P., & Exertier, P., 2006. How can combination help to achieve consistency at the 0.1 ppb level?, Position paper presented at the GGOS 2006 Workshop 'Towards a Consistent Geodetic Foundation for Earth Observations, October 8-9, 2006,Munich, Germany. Available at http://www.iag-ggos.org/ggosws_2006/.

Andersen, O. B., Woodworth, P. L., & Flather, R. A., 1995. Intercomparison of recent global ocean tide models, *J. Geophys. Res.*, **100**, 25,261–25,282.

Anthes, R. A., Bernhardt, P. A., Chen, Y., Cucurull, L., Dymond, K. F., Ector, D., Healy, S. B., Ho, S.-P., Hunt, D. C., Kuo, Y.-H., Liu, H., Manning, K., McCormick, C., Meehan, T. K., Randel, W. J., Rocken, C., Schreiner, W. S., Sokolovskiy, S. V., Syndergaard, S., Thompson, D. C., Trenberth, K. E., Wee, T.-K., Yen, N. L., & Zeng, Z., 2008. The COSMIC/FORMOSAT-3 Mission: Early results, *Bull. Am. Meteoro. Soc.*, **89**(3), 313333, DOI: 10.1175/BAMS–89–3–313.

Asner, G. P., Knapp, D. E., Broadbent, E. N., & et al., 2005. Selective logging in the Brazilian Amazon, *Science*, **310**, 480–482.

Asner, G. P., Broadbent, E. N., Oliveira, P. J. C., & et al., 2006. Condition and fate of logged forests in the Brazilian Amazon, *Proc. National Academy of Sciences of the United States of America*, **103**(34), 12947–12950.

Assistant Secretary of Defence for Command, Control, Communication, and Intelligence, 2001. Global Positioning System Standard Positioning Service performance standards, Tech. rep., Assistant Secretary of Defence, Washington.

Banerjee, P., Pollitz, F. F., & Bürgmann, R., 2005. The size and duration of the Sumatra-Andaman earthquake from far-field static offsets, *Science*, **308**, 1769–1772.

Bao, L. F., Piatanesi, A., Lu, Y., Hus, H. T., & Zhou, X. H., 2005. Sumatra tsunami affects observations by GRACE satellites, *Eos, Trans. Am. Geophys. Union*, **86**, 353,356.

Barber, R. T., Sanderson, M. P., Lindley, S. T., Chai, F., Newton, J., Trees, C. C., Foley, D. G., & Chavez, F., 1996. Primary productivity and its regulation in the equatorial Pacific during and following the 1991-1992 El Niño, *Deep-Sea Research II*, **43**, 933–969.

ed. Battrick, B., 2006. *The Changing Earth: New Scientific Challanges for ESA's Living Planet Programme*, European Space Agency, ESTEC, Noordwijk, The Netherlands.

Becker, M., Zerbini, S., Baker, T., Bürki, B., Galanis, J., Garate, J., Georgiev, I., Kahle, H., Kotzev, V., Lobazov, V., Marson, I., Negusini, M., Richter, B., Veis, G., & Yuzefovich, P., 2002. Assessment of height variations by GPS at Mediterranean and Black Sea coast tide gauges from the SELF projects, *Global and Planetary Change*, **34**(1-2), 5–35.

Bernasconi, C. & others, ., 2005. *Climate Change and the European Water Dimension*, European Commission, EUR 21553.

eds Beutler, G., Hein, G. W., Melbourne, W. G., & G., S., 1996. *GPS Trends in Precise Terrestrial, Airborne, and Spaceborne Applications*, vol. 115 of **Proc. IAG symposium no.**, Springer-Verlag, New York.

Beutler, G., Rothacher, M., Schaer, S., Springer, T. A., Kouba, J., & Neilan, R. E., 1999. The International GPS Service (IGS): An interdisciplinary service in support of earth sciences, *Advances of Space Research*, **23**, 631–653.

Beutler, G., Mervart, L., & Verdun, A., 2004. *Methods of Celestial Mechanics: Volume I: Physical, Mathematical, and Numerical Principles*, Springer, Berlin etc.

Beutler, G., Drewes, H., Plag, H.-P., Reigber, C., Rothacher, M., & Rummel, R., 2005. IAG GGOS Implementation Plan, Tech. rep., GeoForschungZentrum Potsdam, Germany.

Beyerle, G. & Hocke, K., 2001. Observation and simulation of direct and reflected GPS signals in radio occultation experiments, *Geophys. Res. Lett.*, **28**(9), 1895–1898.

Bindoff, N. L., Willebrand, J., Artale, V., Cazenave, A., Gregory, J., Gulev, S., Hanawa, K., Le Qur, C., Levitus, S., Nojiri, Y., Shum, C., Talley, L., & Unnikrishnan, A., 2007. Observations: Oceanic climate change and sea level, in *Climate Change 2007: The Physical Science Basis. Contribution of Working Group I to the Fourth Assessment Report of the Intergovernmental Panel on Climate Change*, eds Solomon, S., Qin, D., Manning, M., Chen, Z., Marquis, M., Averyt, K. B., Tignor, M., & Miller, H. L., Cambridge University Press, Cambridge, United Kingdom and New York, NY, USA.

Bird, P., 2003. An updated digital model of plate boundaries, *Geochemistry Geophysics Geosystems*, **4**(3), 1027, doi:10.1029/2001GC000252.

Blewitt, G., 1993. Advances in Global Positioning System technology for geodynamics investigations: 1978–1992, in *Contributions of Space Geodesy to Geodynamics: Technology*, vol. 25 of **Geodynamics Series**, pp. 195–213, eds Smith, D. E. & Turcotte, D. L., American Geophysical Union.

Blewitt, G., 2003. Self-consistency in reference frames, geocenter definition, and surface loading of the solid earth, *J. Geophys. Res.*, **108**, 10.1029/2002JB002082.

Blewitt, G. & Clarke, P., 2003. Inversion of earth's changing shape to weigh sea level in static equilibrium with surface mass redistribution, *J. Geophys. Res.*, **107**, DOI: 10.1029/2002JB002290.

Blewitt, G., Lavallée, D., Clarke, P., & Nurutdinov, K., 2001. A new global mode of earth deformation: Seasonal cycle detected, *Science*, **294**, 2342–2345.

Blewitt, G., Coolbaugh, M. F., Sawatzky, D. L., Holt, W., Davis, J. L., & Bennett, R. A., 2003. Targeting of potential geothermal resources in the Great Basin from regional to basin-scale relationship between geodetic strain and geological structures, *Geothermal Resources Council Transactions*, **27**, 3–7.

Blewitt, G., Hammond, W. C., & Kreemer, C., 2005. Relating geothermal resources to Great Basin tectonics using GPS, *Geothermal Resources Council Transactions*, **29**, 331–335.

Blewitt, G., Altamimi, Z., Davis, J., Gross, R., Kuo, C., Lemoine, F., Neilan, R., Plag, H.-P., Rothacher, M., Shum, C. K., Sideris, M. G., Schöne, T., Tregoning,

P., & Zerbini, S., 2006a. Geodetic observations and global reference frame contributions to understanding sea level rise and variability, in *Understanding Sea-level Rise and Variability, A World Climate Research Programme Workshop and a WCRP contribution to the Global Earth Observation System of Systems, 6-9 June 2006, UNESCO, Paris*, pp. 127–143, WCRP, World Meterological Organization, Paris.

Blewitt, G., Kreemer, C., Hammond, W., Plag, H.-P., Stein, S., & Okal, E., 2006b. Rapid determination of earthquake magnitude using GPS for tsunami warning systems, *Geophys. Res. Lett.*, **33**, L11309, doi:10.1029/2006GL026145.

Board on Earth Sciences and Resources, 2003. *Living on an Active Earth: Perspectives on Earthquake Science*, The National Academies Press, Free online version available at http://darwin.nap.edu/books/0309065623/html.

eds Bock, Y. & Leppard, N., 1990. *Global Positioning System: An Overview*, vol. 102 of **Proc. IAG symposium**, Springer-Verlag, New York.

Bonaccorso, A., Calvari, S., Garfi, G., Lodato, L., & Patanè, D., 2003. Dynamics of the December 2002 flank failure and tsunami at Stromboli volcano inferred by volcanological and geophysical observations, *Geophys. Res. Lett.*, **20**, doi:10.1029/2003GL017702.

Bonforte, A. & Puglisi, G., 2003. Magma uprising and flank dynamics on Mount Etna volcano, studied using GPS data (1994–1995), *J. Geophys. Res.*, **108**(B3), 2153, doi:10.1029/2002JB001845.

Buffett, B. A., 1996a. Gravitational oscillations in the length of day, *Geophys. Res. Lett.*, **23**, 2279–2282.

Buffett, B. A., 1996b. A mechanism for decade fluctuations in the length of day, *Geophys. Res. Lett.*, **23**(25), doi: 10.1029/96GL03571.

Burris, J., Andrews, A., Riris, H., Abshire, J., Gates, A., Karinak, M., & Sun, X., 2006. Profiling CO_2 within the planetary boundary layer, Abstract, 23rd International Laser Radar Conference (ILRC23), July 2006 in Nara Japan.

Cardellach, E., Ruffini, G., Pino, D., Rius, A., Komjathy, A., & Garrison, J., 2003. Mediterranean balloon experiment: GPS reflection for wind speed retrieval from the stratosphere, *Rem. Sens. Env.*.

Cardellach, E., Ao, C. O., de la Torre Juarez, M., & Hajj, G. A., 2004. Carrier-phase delay altimetry with GPS-reflection/occultation interferometry from low-Earth orbiters, *Geophys. Res. Lett.*, **31**, L10402, doi:10.1029/2004GL019775.

Carlson, R. & 42 others, 2002. EarthScope scientific targets for the World's largest observatory pointed at the solid Earth, Workshop report, Geoscience Professional Services, Inc.

Chao, B. F., 1994. Man-made lakes and sea-level rise, *Nature*, **370**, 258.

Chao, B. F., 2003. Geodesy is not just for static measurements anymore, *Eos, Trans. Am. Geophys. Union*, **84**(16), 145–150.

Chao, B. F. & Gross, R., 1987. Changes in the Earth's rotation and low-degree gravitional field induced by earthquakes, *Geophys. J. Int.*, **91**, 569–596.

Chao, B. F. & Gross, R., 2005. Did the 26 December 2004 Sumatra, Indonesia, earthquake disrupt the Earth's rotation as the mass media have said?, *Eos, Trans. Am. Geophys. Union*, **86**(1), 1.

Chelton, D. B., Schlax, M. G., Freilich, M. H., & Milliff, R. F., 2004. Satellite measurements reveal persistent short-scale features in ocean winds, *Science*, **303**(5660), 978–983.

Chen, J., Devey, C., Fischer, C., Lin, J., & Whitmarsh, B., 2005. Ocean – abyss of time: Earth science for society, Tech. rep., International Year of Planet Earth, Earth Science for Society Foundation, Leiden, The Netherlands.

Chen, J. L. & Wilson, C. R., 2003. Low degree gravitational changes from earth rotation and geophysical models, *Geophys. Res. Lett.*, **30**(24), 2257, doi:10.1029/2003GL018688.

Cheng, M. & Tapley, B. D., 2004. Variations in the Earth's oblateness during the past 28 years, *J. Geophys. Res.*, **109**, B09402, doi:10.1029/2004JB003028.

Church, J. A., Gregory, J. M., Huybrechts, P., Kuhn, M., Lambeck, K., Nhuan, M. T., Qin, D., & Woodworth, P. L., 2001. Changes in sea level, in *Climate Change 2001: The Scientific Basis. Contribution of Working Group I to the Third Assessment Report of the Intergovernmental Panel on Climate Change*, pp. 639–693, eds Houghton, J. T., Ding, Y., Griggs, D. J., Noguer, M., van der Linden, P. J., Dai, X., Maskell, K., & Johnson, C. A., Cambridge University Press, Cambridge.

Clarke, P. J., Lavallée, D. A., Blewitt, G., van Dam, T., & Wahr, J. M., 2005. Effect of gravitational consistency and mass conservation on seasonal surface mass loading models, *Geophys. Res. Lett.*, **32**, doi: 10.1029/2005GL022441.

Commission on Geosciences, E. & Resources, 1997. *Satellite Gravity and the Geosphere: Contributions to the Study of the Solid Earth and Its Fluid Envelopes*, The National Academies Press, Free online version available at http://darwin.nap.edu/books/0309057922/html.

Commission on Physical Sciences, Mathematics, and Applications, 1990. *Geodesy in the Year 2000*, The National Academies Press, Free online version available at http://darwin.nap.edu/books/0309041457/html.

Crétaux, J. F., Soudarin, L., Cazenave, A., & Bouille, F., 1998. Present-day tectonic plate motions and crustal deformations from the DORIS system, *J. Geophys. Res.*, **103**(B12), 30167–30181, DOI 10.1029/98JB02239.

Crossley, D., Hinderer, J., Casula, G., Francis, O., Hsu, H.-T., Imanishi, Y., Jentzsch, G., Kaarianen, J., Merriam, J., Meurers, B., Neumeyer, J., Richter, B., Shibuya, K., Sato, T., & van Dam, T., 1999. Network of superconducting gravimeters benefits a number of disciplines, *Trans. Am. Geophys. U.*, **80**, 121–126.

Crowley, J. W., Mitrovica, J. X., Bailey, R. C., Tamisiea, M. E., & Davis, J. L., 2006. Land water storage within the Congo Basin inferred from GRACE satellite gravity data, *Geophys. Res. Lett.*, **33**, L19402, doi:10.1029/2006GL027070.

Dahl, A. L., 1998. IGOS from the perspective of the Global Observing Systems and their sponsors, in *Proceedings for the 27-th Int. Symp. on Remote Sensing of Environment: Information for Sustainability, June 8-12, 1998, Tromsø, Norway*, pp. 92–94, Norwegian Space Centre.

Davis, J. L., Elsegui, P., Mitrovica, J. X., & Tamisiea, M. E., 2004. Climate-driven deformation of the solid Earth from GRACE and GPS, *Geophys. Res. Lett.*, **31**, L24605, doi:10.1029/2004GL021435.

Day, J. W., Gunn, J. D., Folan, W. J., Yáñez-Arancibia, A., & Horton, B. P., 2007. Emergence of complex societies after sea level stabilized, *Eos, Trans. Am. Geophys. Union*, **88**(15), 169–170.

de Viron, O., Schwarzbaum, G., Lott, F., & Dehant, V., 2005. Diurnal and subdiurnal effects of the atmosphere on the Earth rotation and geocenter motion, *J. Geophys. Res.*, **110**, B11404, doi:10.1029/2005JB003761.

Defra, 2005. The threat posed by tsunamis to the uk, Tech. rep., British Geological Survey, Edinburgh, Study commissioned by Department for Environment, Food and Rural Affairs (Defra) Flood Management and produced by British Geological Survey, Proudman Oceanographic Laboratory, Met Office and HR Wallingford. Available at http://www.defra.gov.uk/environ/fcd/research.

DeMets, C., Gordon, R. G., Argus, D. F., & Stein, S., 1994. Effect of recent revisions to the geomagnetic reversal time scale on estimates of current plate motions, *Geophys. Res. Lett.*, **21**, 2191–2194.

DESDynI Writing Committee, 2007. Report of the July 16-19, 2007 Orlando, Florida Workshop to Assess the National Research Council Decadal Survey Recommendations for the DESDynI Radar/Lidar Space Mission, Available at http://science.hq.nasa.gov/earth-sun/index.html.

Dickey, J. O., Newhall, X. X., & Williams, J. G., 1985. Earth orientation from lunar laser ranging and an error analysis of polar motion services, *J. Geophys. Res.*, **90**, 9353–9362.

Dickey, J. O., Bender, P. L., Faller, J. E., Newhall, X. X., Ricklefs, R. L., Ries, J. G., Shelus, P. J., Veillet, C., Whipple, A. L., Wiant, J. R., Williams, J. G., & F., Y. C., 1994. Lunar laser ranging: A continuing legacy of the Apollo program, *Science*, **265**, 482–490.

Dixon, T. H., Amelung, F., Ferretti, A., Novali, F., Rocca, F., Dokka, R., Sella, G., Kim, S.-W., Wdowinski, S., & Whitman, D., 2006. Subsidence and flooding in New Orleans, *Nature*, **441**, 587–588.

Dong, D., Fang, P., Bock, Y., Cheng, M. K., & Miyazaki, S., 2002. Anatomy of apparent seasonal variations from GPS-derived site position time series, *J. Geophys. Res.*, **107**, 2075, doi:10.1029/2001JB000573.

Dooge, J. C. I., 2004. Bringing it all together, *Hydrol. Earth Syst. Sci. Discuss.*, **1**, 4173.

Drewes, H., 2006. The science rationale of the Global Geodetic Observing System GGOS, in *Dynamic Planet – Monitoring and Understanding a Dynamic Planet with Geodetic and Oceanographic Tools*, vol. 130 of **International Association of Geodesy Symposia**, pp. 703–710, Springer Verlag, Berlin.

Drinkwater, M. R., Floberghagen, R., Haagmans, R., Muzi, D., & Popescu, A., 2003. GOCE; ESA's first Earth Explorer Core mission, in *Earth Gravity Field from Space - from Sensor to Earth Sciences*, vol. 18 of **Space Science Series of ISSI**, pp. 419–432, eds Beutler, G. B., Drinkwater, M. R., Rummel, R., & von Steiger, R., Kluwer Academic Publishers, Dordrecht.

EEA, 1999. Sustainable water use in Europe, Part 1: Sectoral use of water, European Environment Agency, available at: http://themes.eea.eu.int/binary/e/enviasses01.pdf.

Elgered, G., Plag, H.-P., Marel, H., Barlag, S., & Nash, J., 2005. *COST Action 716 Exploitation of Ground-based GPS for Operational Numerical Weather Prediction and Climate Applications*, no. EUR 21639 in COST European cooperation in the field of scientific and technical research, European Commission.

Elósegui, P., Davis, J. L., Jaldehag, R. T. K., Johansson, J. M., Niell, A. E., & Shapiro, I. I., 1995. Geodesy using the Global Positioning System: the effects of signal scattering on estimates of site position, *J. Geophys. Res.*, **100**, 9921–9934.

Eltahir, E. A. B. & Yeh, P., 1999. On the asymmetric response of aquifer water level to droughts and floods in Illinois, *Wat. Resour. Res.*, **35**(4), 1199–1217.

Emanuel, K., 2003. Tropical cyclones, *Annual Review of Earth and Planetary Sciences*, **31**, 75–104.

ESA, 1999. Gravity field and steady-state ocean circulation mission, Reports for mission selection ESA SP-1233 (1), ESA publication divison, ESTEC, Noordwijk, The Netherlands.

Estefan, J. A. & Folkner, W. M., 1995. Sensitivity of planetary cruise navigation to Earth orientation calibration errors, Tech. Rep. TDA PR 42-123, July-September 1995, pp. 1-29, Jet Propulsion Laboratory, California Institute of Technology, Pasadena, CA, http://tmo.jpl.nasa.gov/progress_report/42-123/123E.pdf.

European Commission, 2004. Agreement on the promotion, provision and use of Galileo and GPS satellite-based navigation systems and related applications, Available at http://europe.eu.int/comm/dgs/energy_transport/galileo/documents% -/doc/ 2004_06_21_summit_2004_en.pdf.

European Commission and ESA, 2003. Global Monitoring for Environment and Security: Final report for the GMES initial period (2001-2003) version 3.5, Draft report, European Commission, Available at http://www.gmes.info/library/.

Fagard, H., 2006. Twenty years of evolution for the DORIS permanent network: from its initial deployment to its renovation, *J. Geodesy*, **80**(8-11), 429–456, DOI 10.1007/s00190–006–0084–2.

FAO, 2001. Global forest resources assessment 2000 – main report, Tech. Rep. FAO Forestry Paper No. 140, FAO, Rome, Italy, (also available at www.fao.org/forestry/site/7949/en).

Farrell, W. E., 1972. Deformation of the Earth by surface loads., *Rev. Geophys. Space Phys.*, **10**, 761–797.

Farrell, W. E. & Clark, J. A., 1976. On postglacial sea level, *Geophys. J. R. Astron. Soc.*, **46**, 647–667.

Fedrizzi, M., de Paula, E. R., Langley, R. B., Komjathy, A., Batista, I. S., & Kantor, I. J., 2005. Study of the March 31, 2001 magnetic storm effects on the ionosphere using GPS data, *Advances in Space Research*, **36**, 534–545.

Ferretti, A., Novali, F., Bürgmann, R., Hilley, G., & Prati, C., 2004. InSAR permanent scatterer analysis reveals ups and downs in San Francisco Bay area, *Eos, Trans. Am. Geophys. Union*, **85**(24), 317,324.

Fey, A. L., Ma, C., Arias, E. F., Charlot, P., Feissel-Vernier, M., Gontier, A.-M., Jacobs, C. S., Li, J., & MacMillan, D. S., 2004. The second extension of the International Celestial Reference Frame, *Astronomical J.*, **127**, 3587–3608.

Folkner, W. M., 1996. DSN station locations and uncertainties, Tech. Rep. TDA PR 42-128, October-December 1996, pp. 1-34, Jet Propulsion Laboratory, California Institute of Technology, Pasadena, CA, http://tmo.jpl.nasa.gov/progress_report/42-128/128F.pdf.

Folkner, W. M., Yoder, C. F., Yuan, D. N., Standish, E. M., & Preston, R. A., 1997. Interior structure and seasonal mass redistribution of Mars from radio tracking of Mars Pathfinder, *Science*, **278**, 1749–1752.

Fujii, Y. & Satake, K., 2007. Tsunami source of the 2004 SumatraAndaman earthquake inferred from tide gauge and satellite data, *Bull. Seismo. Soc. Am.*, **97**, S192 – S207.

Garcés, M., Caron, P., Hetzer, C., Le Pichon, A., Bass, H., & Bhattacharyya, J., 2005. Deep infrasound radiated by the Sumatra earthquake and tsunami, *Eos, Trans. Am. Geophys. Union*, **86**, 317, 320.

Garrison, J. L. & Katzberg, S. J., 2000. The application of reflected GPS signals to ocean remote sensing, *Remote Sens. Environ.*, **73**, 175–187.

Garrison, J. L., Katzberg, S. J., & Hill, M. I., 1998. Effect of sea roughness on bistatically scattered range coded signals from the Global Positioning System, *Geophys. Res. Lett.*, **25**(13), 2257–2260.

GEO, 2005a. The Global Earth Observing System of Systems (GEOSS) - 10-Year Implementation Plan, Available at http://earthobservations.org.

GEO, 2005b. Global Earth Observing System of Systems GEOSS - 10-Year Implementation Plan Reference Document - Draft, Tech. Rep. GEO 1000R/ESA SP 1284, ESA Publication Division, ESTEC, Noordwijk, The Netherlands, Avaliable at http://earthobservations.org.

Geoconnection, 2007. Canadian geospatial data infrastructure, http://www.geoconnections.org/CGDI.cfm/fuseaction/keyDocs.home/gcs.cfm.

Germain, O., Ruffini, G., Soulat, F., Caparrini, M., Chapron, B., & Silvestrin, P., 2004. The Eddy Experiment: GNSS-R speculometry for directional sea-roughness retrieval from low altitude aircraft, *Geophys. Res. Lett.*, **31**, L21307, doi:10.1029/2004GL020991.

Gewin, V., 2004. Mapping opportunities, *Nature*, **427**, 376–377.

Gleason, S., Hodgart, S., Sun, Y., Gommenginger, C., Mackin, S., Adjrad, M., & Unwin, M., 2005. Detection and processing of bistatically reflected GPS signals from low Earth orbit for the purpose of ocean remote sensing, *IEEE Trans. Geosci. Remote Sensing*, **43**(6), 1229–1241.

Gonzalez, F. I., Milburn, H. M., Bernard, E. N., & Newman, J. C., 1998. Deep-ocean assessment and reporting of tsunamis (DART): Brief overview and status report, in *Proceedings of the International Workshop on Tsunami Disaster Mitigation, 19-22 January 1998, Tokyo, Japan.*

Gower, J., 2005. Jason 1 detects the 26 December 2004 tsunami, *Eos, Trans. Am. Geophys. Union*, **86**(4), 37.

Grindlay, N. & Hearne, M., 2005. High risk of tsunami in the Northern Caribbean, *Eos, Trans. Am. Geophys. Union*, **86**, 121, 126.

Gross, R., 2006. Degree-2 harmonics of the Earth's mass load estimated from GRACE and Earth rotation data and models of surficial geophysical fluids, *Eos Trans. AGU*, **87**, Abstract G31A–03, West. Pac. Geophys. Meet. Suppl.

Gross, R. S., Blewitt, G., Clarke, P. J., & Lavallée, D., 2004. Degree-2 harmonics of the Earth's mass load estimated from GPS and Earth rotation data, *Geophys. Res. Lett.*, **31**, doi:10.1029/2004GL019589.

Haines, B. J., Bar-Sever, Y. E., Bertiger, W., Desai, S., & Willis, P., 2004. New strategies for the 1-cm precise orbit determination, *Marine Geodesy*, **27**(1-2), 299–318.

Hajj, G. A., Ao, C. O., C. O. Iijima, C. O., Kuang, D., Kursinski, E. R., Mannucci, A. J., Meehan, T. K., Romans, L. J., de la Torre Juarez, M., & Yunck, T. P., 2004. CHAMP and SAC-C atmospheric occultation results and intercomparisons, *J. Geophys. Res.*, **109**, D06109, doi:10.1029/2003JD003909.

Hansen, M. C. & DeFries, R. S., 2004. Detecting long-term global forest change using continuous fields of tree-cover maps from 8-km advanced very high resolution radiometer (AVHRR) data for the years 1982-99, *Ecosystems*, **7**(7), 695–716.

Hansen, M. C., DeFries, R. S., Townshend, J. R. G., & Sohlberg, R., 2000. Global land cover classification at 1 km spatial resolution using a classification tree approach, *Int. J. of Remote Sensing*, **21**(6 & 7), 13311364.

Hayashi, Y., 2008. Extracting the 2004 Indian Ocean tsunami signals from sea surface height data observed by satellite altimetry, *J. Geophys. Res.*, **113**, C01001, doi:10.1029/2007JC004177.

Heidbach, O., Barth, A., Connolly, P., Fuchs, K., Müller, B., Tingay, M., Reinecker, J., Sperner, B., & Wenzel, F., 2004. Stress maps in minutes: the 2004 world stress map release, *Eos, Trans. Am. Geophys. Union*, **85**, 521, 529.

Hein, G. W., Rodriguez, J. A. A., Wallner, S., Eisfeller, B., Pany, T., & Hartl, P., 2007. Envisioning a future GNSS system of systems, *Inside GNSS*, **2**(1), 58–67.

Hinderer, J. & Crossley, D., 2004. Scientific achievements from the first period (1997-2003) of the Global Geodynamics Project using a worldwide network of superconducting gravimeters, *J. Geodynamics*, **38**, 237–262.

Hirata, K., Satake, K., Tanioka, Y., Kuragano, T., Hasegawa, Y., Hayashi, Y., & Hamada, N., 2006. The 2004 Indian Ocean tsunami: Tsunami source model from satellite altimetry, *Earth Planets Space*, **58**(2), 195–201.

Hofmann-Wellenhof, B., Lichtenegger, H., & Collins, J., 1997. *Global Positioning System: Theory and Practice*, Springer-Verlag, New York.

Hogan, J., 2005. Warming debate highlights poor data, *Nature*, **436**, 896, doi:10.1038/436896a.

Hughes, C. W., Stepanov, V. N., Fu, L.-L., Barnier, B., & Hargreaves, G. W., 2007. Three forms of variability in Argentine Basin ocean bottom pressure, *J. Geophys. Res.*, **112**, C01011, doi:10.1029/2006JC003679.

IGOS-P Ocean Theme Team, 2001. An Ocean Theme for the IGOS Partnership, Tech. rep., IGOS Integrated Global Observing Strategy, Available at http://www.igospartners.org.

Ilk, K. H., Flury, J., Rummel, R., Schwintzer, P., Bosch, W., Haas, C., Schröter, J., Stammer, D., Zahel, W., Miller, H., Dietrich, R., Huybrechts, P., Schmeling, H.,

Wolf, D., Götze, H. J., Riegger, J., Bardossy, A., Günter, A., & Gruber, T., 2005. Mass transport and mass distribution in the earth system, Tech. rep., GOCE-Projectbüro Deutschland, Technische Universität München, GeoForschungsZentrum Potsdam.

InSAR Working Group, 2005. InSAR Workshop Summary Report, October 20-22, 2004, Oxnard, California, Tech. rep., NASA, Jet Propulsion Laboratory, California Institute of Technology, Pasadena, California.

International GPS Service, 2001. GPS Tide Gauge Benchmark Monitoring Pilot Project (TIGA - PP), Call for participation, International GPS Service, available at http://op.gfz-potsdam.de/staff/schoene/TIGA_CfP.pdf.

IOC, 1997. Global Sea Level Observing System (GLOSS) Implementation Plan - 1997, Tech. rep., Intergovernmental Oceanographic Commission, Technical Series, vol 50, 91 pp plus Annexes.

IOC, 2006. Manual on sea-level measurement and interpretation, Volume 4 - An update to 2006, Tech. Rep. 14, Intergovernmental Oceanographic Commission (IOC), Paris, 80 pages.

Jayles, C., Nhun-Fat, B., & Tourain, C., 2006. DORIS: system description and control of the signal integrity, *J. Geodesy*, **80**(8-11), 457–472, DOI 10.1007/s00190–006–0046–8.

Jerrett, D. & Nash, J., 2001. Potential uses of surface based GPS water vapour measurements for meteorological purposes, *Phys. Chem. Earth*, **26**, 457–462.

Johannessen, J. A., Balmino, G., Le Provost, C., Rummel, R., Sabadini, R., Sünkel, H., Tscherning, C. C., Visser, P., Woodworth, P., Hughes, C., LeGrand, P., Sneeuw, N., Perosanz, F., Aguirre-Martinez, M., Rebhan, H., & Drinkwater, M., 2003. The European Gravity Field and Steady-State Ocean Circulation Explorer satellite mission: its impact on geophysics, *Surveys in Geophys.*, **24**(4), 339–386.

Johnson, J. B., Lees, J. M., Gerst, A., Sahagian, D., & Varley, N., 2008. Long-period earthquakes and co-eruptive dome inflation seen with particle image velocimetry, *Nature*, **456**, 377–381, doi:10.1038/nature07429.

Jüttner, H.-U. & Plag, H.-P., 1999. On modelling of earth rotation variations in an integrated model and appropriate frame of reference, in *3. DFG Rundgespräch zum Thema Bezugssysteme*, vol. 5 of **Mitteilungen des Bundesamtes für Kartographie und Geodäsie**, pp. 59–62, Bundesamtes für Kartographie und Geodäsie.

Katzberg, S. J., Walker, R. A., Roles, J. R., Lynch, T., & Black, P. G., 2001. First GPS signals reflected from the interior of a tropical storm: Preliminary results from hurricane Michael, *Geophys. Res. Lett.*, **28**(10), 1981–1984.

Kaula, W. M., 1966. *Theory of Satellite Geodesy*, Blaisdell Publishing Co., Waltham, MA.

Kaula, W. M., 1970. The terrestrial environment: solid-earth and ocean physics, Tech. rep., Massachusetts Institute of Technology, Cambridge, Mass., Community 'Williamstown' Report prepared for National Aeronautics and Space Administration, Washington, D.C.

Kavaya, M. J., 1999. What is LIDAR?, Available at: http://-www.ghcc.msfc.nasa.gov/sparcle/sparcle_tutorial.html.

Kerr, R., 2005. Failure to gauge the quake crippled the warning effort, *Science*, **307**, 201.

Key, J. & the IGOS-Cryo Writing Team, 2004. Concept of an IGOS-Cryosphere Theme (IGOS-Cryo), Tech. rep., IGOS Integrated Global Observing Strategy, Version 2.2, October 7, 2004, available at http://www.igospartners.org.

Khan, S. A., Wahr, J., Stearns, L. A., Hamilton, G. S., van Dam, T., Larson, K. M., & Francis, O., 2007. Elastic uplift in southeast Greenland due to rapid ice mass loss, *Geophys. Res. Lett.*, **34**(L21701, doi:10.1029/2007GL031468).

Kierulf, H. P., Bockmann, L., Kristiansen, O., & Plag, H.-P., 2002. Foot-print of the space-geodetic observatory, Ny-Ålesund, Svalbard, in *Proceedings of the Second IVS General Meeting, Tsukuba, Japan, 4-6 February 2002*, pp. 86–90, NASA Goddard Space Flight Center, Greenbelt, MD.

Komjathy, A., Zavorotny, V., Axelrad, P., Born, G., & Garrison, J., 2000. GPS signal scattering from sea surface: wind speed retrieval using experimental data and theoretical model, *Rem. Sens. Environ.*, **73**, 162–174.

Komjathy, A., Sparks, L., Wilson, B., & Mannucci, A. J., 2005. Automated daily processing of more than 1000 ground-based GPS receivers to study intense ionospheric storms, *Radio Science*, **40**, RS6006 doi:10.1029/2005RS003279.

Konopliv, A. S., Yoder, C. F., Standish, E. M., Yuan, D. N., & Sjogren, W. L., 2006. A global solution for the Mars static and seasonal gravity, Mars orientation, Phobos and Deimos masses, and Mars ephemeris, *Icarus*, **182**, 23–50.

Kovalevsky, J., 1997. Connection of the HIPPARCOS catalogue to the extragalactic reference frame, in *Proceedings from the Hipparcos Venice '97 symposium Presenting The Hipparcos and Tycho Catalogues and first astrophysical results of the Hipparcos astrometry mission*, ESTEC, The Netherlands, available at http://astro.estec.esa.nl/SA-general/Projects/Hipparcos/hipp_venice.html.

Kreemer, C. & Holt, W. E., 2001. A no-net-rotation model of present-day surface motions, *Geophys. Res. Lett.*, **28**, 4407–4410.

Kreemer, C., Holt, W. E., & Haines, A. J., 2003. An integrated global model of present-day plate motion and plate boundary deformation, *Geophys. J. Int.*, **154**, 8–34.

Kreemer, C., Blewitt, G., & Hammond, W. C., 2006a. Using geodesy to explore correlations between crustal deformation characteristics and geothermal resources, *Geothermal Resources Council Transactions*, **30**, 441–446.

Kreemer, C., Blewitt, G., Hammond, W. C., & Plag, H.-P., 2006b. Global deformations from the great 2004 Sumatra-Andaman earthquake observed by GPS: implications for rupture process and global reference frame, *Earth Planets Space*, **58**, 141–148.

Kunze, E. & Smith, S. G. L., 2004. The role of small-scale topography in turbulent mixing of the global ocean, *Oceanography*, **17**(1-2), 55–64.

Kursinski, E., Hajj, G., Hardy, K., Romans, L., & Schofield, J., 1995. Observing tropospheric water vapor by radio occultation using the Global Positioning System, *Geophys. Res. Lett.*, **22**(17), 2365–2368.

Kusche, J. & Schrama, E. J. O., 2005. Surface mass redistribution inversion from global GPS deformation and Gravity Recovery and Climate Experiment (GRACE) gravity data, *J. Geophys. Res.*, **110**, B09409, doi:10.1029/2004JB003556.

La Rocca, M., Galluzzo, D., Saccorotti, G., Tinti, S., Cimini, G. B., & Del Pezzo, E., 2004. Seismic signals associated with landslides and with a tsunami at Stromboli volcano, Italy, *Bull. Seismological Soc. America*, **94**, 1850–1867.

Lambeck, K., 1988. *Geophysical Geodesy - The Slow Deformations of the Earth*, Oxford Science Publications.

Lambert, S. B., Bizouard, C., & Dehant, V., 2006. Rapid variations in polar motion during the 2005-2006 winter season, *Geophys. Res. Lett.*, **33**, L13303, doi:10.1029/2006GL026422.

Lanari, R., De Natale, G., Berardino, P., Sansosti, E., Ricciardi, G. P., Borgstrom, S., Capuano, P., Pingue, F., & Troise, C., 2002. Evidence for a peculiar style of ground deformation inferred at Vesuvius volcano, *Geophys. Res. Lett.*, **29**(9), doi:10.1029/2001GL014571.

Langley, R. B., King, R. W., & Shapiro, I. I., 1981. Earth rotation from lunar laser ranging, *J. Geophys. Res.*, **86**, 11913–11918.

Larson, K. M., Bodin, P., & Gomberg, J., 2003. Using 1-hz GPS data to measure deformations caused by the Denali Fault Earthquake, *Science*, **300**, 1421–1424.

Latychev, K., Mitrovica, J. X., Tromp, J., Tamisiea, M. E., Komatitsch, D., & Christara, C. C., 2005. Glacial isostatic adjustement on 3-D Earth models: a finite-volume formalism, *Geophys. J. Int.*, **161**, 421–444.

Lawford, R. & the Water Theme Team, 2004. A Global Water Cycle Theme for the IGOS Partnership, Tech. rep., IGOS Integrated Global Observing Strategy, Report of the Global Water Cycle Theme Team, April 2004, available at http://www.igospartners.org.

Le Provost, C., Le Grand, P., Dombrowsky, E., Le Traon, P. Y., Losch, M., Ponchaut, F., Schröter, J., Sloyan, B., & Sneeuw, N., 1999. Impact of GOCE for ocean circulation studies, Final report, ESA study contract 13175/98/NL/GD.

Leick, A., 2003. *GPS Satellite Surveying*, J. Wiley and Sons, Inc., New York.

Lemoine, F., Kenyon, S., Factor, J., Trimmer, R., Pavlis, N., Chinn, D., Cox, C., Klosko, S., Luthcke, S., Torrence, M., Wang, Y., Williamson, R., Pavlis, E., Rapp, R., & Olsen, T., 1998. The development of the joint NASA GSFC and the National Imagery and Mapping Agency (NIMA) geopotential model EGM96, Tech. rep., NASA Goddard Space Flight Center.

Lemoine, J. M. & Capdeville, H., 2006. A corrective model for Jason-1 DORIS Doppler data in relation to the South Atlantic anomaly, *J. Geodesy*, **80**(8-11), 507–523, DOI 10.1007/s00190–006–0068–2.

Lesack, L. F. W. & Melack, J. M., 1995. Flooding hydrology and mixture dynamics of lake water derived from multiple sources in an Amazon floodplain lake, *Water Resourc. Res.*, **31**, 329–345.

Loveland, T. R., Reed, B. C., Brown, J. F., Ohlen, D. O., Zhu, Z., Yang, L., & Merchant, J. W., 2000. Development of a global land cover characteristics database and IGBP DISCover from 1 km AVHRR data, *Int. J. of Remote Sensing*, **21**, 1303–1330.

Lovelock, J., 1979. *GAIA: A new look at life on Earth*, Oxford University Press, 157pp.

Lowe, S. T., LaBrecque, J. L., Zuffada, C., Romans, L. J., Young, L. E., & Hajj, G. A., 2002. First spaceborne observation of an Earth-reflected GPS signal, *Radio Sci.*, **37**, doi:10.1029/2000RS002539.

Lu, Z., Wicks, C., Dzurisin, D., Thatcher, W., Freymueller, J., McNutt, S., & Mann, D., 2000. Aseismic inflation of Westdahl volcano, Alaska, revealed by satellite radar interferometry, *Geophys. Res. Lett.*, **27**, 1567–1570.

Luthcke, S. B., Rowlands, D. D., Lemoine, F. G., Klosko, S. M., & Chinn, D., 2006. Monthly spherical harmonic gravity field solutions determined from GRACE inter-satellite range-rate data alone, *Geophys. Res. Lett.*, **33**, L02402, doi.10.1029/2005GL024846.

Ma, C., Arias, E. F., Eubanks, T. M., Fey, A. L., Gontier, A. M., Jacobs, C., Sovers, O. J., & Archinal, B. A. Charlot, P., 1998. The International Celestial Reference Frame as realized by very long baseline interferometry, *Astronomical J.*, **116**, 516–546.

Mandrake, L., Wilson, B., Wang, C., Hajj, G., Mannucci, A., & Pi, X., 2005. A performance evaluation of the operational Jet Propulsion Laboratory/University of Southern California Global Assimilation Ionospheric Model (JPL/USC GAIM), *J. Geophys. Res.*, **110**, A12306, doi:10.1029/2005JA011170.

Mannucci, A. J., Wilson, B. D., Yuan, D. N., Ho, C. H., Lindqwister, U. J., & Runge, T. F., 1998. A global mapping technique for GPS-derived ionospheric total electron content measurements, *Radio Science*, **33**, 565–582.

Mannucci, A. J., Iijima, B. A., Lindqwister, U. J., Pi, X., Sparks, L., & Wilson, B. D., 1999. GPS and ionosphere, in *URSI Reviews of Radio Science, 1996-1999*, Oxford University Press.

Marsh, S. & the Geohazards Theme Team, 2004. Geohazards Theme Report, Tech. rep., IGOS Integrated Global Observing Strategy, Available at http://www.igospartners.org.

Marson, I., 2000. Regional and local sea-level variations, gravimetry, in *Sea level change and coastal processes - Implications for Europe*, pp. 91–93, European Commission, Directorate-General for Research, EUR 19337.

Martin-Mur, T. J., Abraham, D. S., Berry, D., Bhaskaran, S., Cesarone, R. J., & Wood, L. J., 2006. The JPL roadmap for deep space navigation (AAS 06-223), in *Proceedings of the AAS/AIAA 2006 Space Flight Mechanics Meeting*, vol. 124 of **Advances in the Astronautical Sciences**, pp. 1925–1932.

Martin-Neira, M., 1993. A passive reflectometry and interferometry system (PARIS): Application to ocean altimetry, *ESA Journal*, **17**, 331–355.

Massonet, D., Rossi, M., Carmosa, C., Adragna, F., Peltzer, G., Feigl, K., & Rabaute, T., 1993. The displacement field of the Landers earthquake mapped by radar interferometery, *Nature*, **364**, 138–142.

Mayaux, P., Holmgren, P., Achard, F., Hugh, E., Stibig, H.-J., & Branthomme, A., 2005. Tropical forest cover change in the 1990s and options for future monitoring, *Royal Society, Philosophical Transactions: Biological Sciences*, **360**(1454), 373–384.

McCabe, M. F. & Wood, E. F., 2006. Scale influences on the remote estimation of evapotranspiration using multiple satellite sensors, *Remote Sensing of the Environment*, **105**(4), 271–285.

McCarthy, D. D. & Petit, G., 2004. *IERS Conventions 2003*, IERS Technical Note 32, International Earth Rotation Service, Also available at http://www.iers.org.

Meier, M. F., Dyurgerov, M. B., Rick, U. K., O'Neel, S., Pfeffer, W. T., Anderson, R. S., & Glazovsky, A. F., 2007. Glaciers dominated eustatic sea-level rise in the 21st century, *Science*, **317**, 1064–1067.

Menke, W. & Levin, V., 2005. A strategy to rapidly determine the magnitude of great earthquakes, *Eos, Trans. Am. Geophys. Union*, **86**, 185,189.

Milburn, H. B., Nakamura, A. I., & Gonzalez, F. I., 1996. Real-time tsunami reporting from the deep ocean, in *Proceedings of the Oceans 96 MTS/IEEE Conference, 23-26 September 1996, Fort Lauderdale, FL*, pp. 390–394.

Milly, P. C. D., Cazenave, A., & Gennero, M. C., 2003. Contribution of climate-driven change in continental water storage to recent sea-level rise, *Proceedings of the National Academy of Sciences*, **100**(23), 13158–13161.

Milne, G. A., Mitrovica, J. X., & Davis, J. L., 1999. Near-field hydro-isostasy: the implementation of a revised sea-level equation, *Geophys. J. Int.*, **139**, 464–482.

Mitrovica, J. X. & Milne, G. A., 1998. Glaciation-induced perturbations in the Earth's rotation: A new appraisal, *J. Geophys. Res.*, **103**(B1), 985–1005.

Mitrovica, J. X. & Milne, G. A., 2003. On post-glacial sea level: I. General theory, *Geophys. J. Int.*, **154**, 253–267.

Mitrovica, J. X., Davis, J. L., & Shapiro, I. I., 1994. A spectral formalism for computing three-dimensional deformations due to surface loads 2.Present-day glacial isostatic adjustment, *J. Geophys. Res.*, **99**, 7075–7101.

Mitrovica, J. X., Tamisiea, M. E., Davis, J. L., & Milne, G. A., 2001. Recent mass balance of polar ice sheets inferred from patterns of global sea-level change, *Nature*, **409**, 1026–1028.

Mitrovica, J. X., Gomez, N., & Clark, P. U., 2009. The sea-level fingerprint of West Antarctic Collapse, *Science*, **323**(5915), 753–756, DOI: 10.1126/science.1166510.

Miyazaki, S., Larson, K., Choi, K., Hikima, K., Koketsu, K., Bodin, P., Haase, J., Emore, G., & Yamagiwa, A., 2004. Modeling the rupture process of the 2003 September 25 Tokachi-Oki (Hokkaido) earthquake using 1-Hz GPS data, *Geophys. Res. Lett.*, **31**, doi:10.1029/2004GL021457.

Morel, L. & Willis, P., 2005. Terrestrial reference frame effects on mean sea level determined by TOPEX/Poseidon, *Adv. Space Res.*, **36**, 358–368.

Mound, J. E. & Buffett, B. A., 2005. Mechanisms of core-mantle angular momentum exchange and the observed spectral properties of torsional oscillations, *J. Geophys. Res.*, **110**(B8), doi: 10.1029/2004JB003555.

eds Mueller, I. & Zerbini, S., 1989. *The Interdisciplinary Role of Space Geodesy*, vol. 22 of **Lecture Notes in Earth Sciences**, Springer-Verlag Berlin etc.

Mueller, I. I., Rajal, B. S., & Zhu, Y. S., 1982. Comparison of polar motion data from the 1980 project MERIT short campaign, in *High-Precision Earth Rotation and*

Earth-Moon Dynamics, vol. 94 of **Astrophysics and Space Science Library**, pp. 141–146, D. Reidel Publishing Company, Dordrecht, Holland.

Mulholland, J. D., 1980. Scientific advances from ten years of lunar laser ranging, *Rev. Geophys. Space Phys.*, **18**, 549–564.

NASA, 1991a. Solid earth sciences in the 1990s. Volume 1 – Program Plan, NASA Technical Memorandum 4256, NASA Office of Space Science and Applications, Washington D.C.

NASA, 1991b. Solid earth sciences in the 1990s. Volume 2 – Panel Reports, NASA Technical Memorandum 4256, NASA Office of Space Science and Applications, Washington D.C.

NASA, 1991c. Solid earth sciences in the 1990s. Volume 3 – Measurement Techniques and Technology, NASA Technical Memorandum 4256, NASA Office of Space Science and Applications, Washington D.C.

National Research Council, 2004. *Groundwater Fluxes Across Interfaces*, The National Acadamies Press, Washington, DC, Report prepared by the Committee on Hydrological Science, Water Science and Technology Board, Board on Atmospheric Sciences and Climate, Division on Earth Life Studies.

National Research Council, 2007. *Earth Science and Applications from Space: National Imperatives for the Next Decade and Beyond - Committee on Earth Science and Applications from Space: A community Assessment and Strategy for the Future*, National Library, available at http://books.nap.edu/catalog/11820.html.

Nawa, K., Suda, N., Satake, K., Fujii, Y., Sato, T., Doi, K., Kanao, M., & Shibuya, K., 2007. Loading and gravitational effects of the 2004 Indian Ocean Tsunami at Syowa Station, Antarctica, *Bull. Seismo. Soc. Am.*, **97**, S271–S278.

Niebauer, T. M., Sasagawa, G. S., Faller, J. E., Hilt, R., & Klopping, F., 1995. A new generation of absolute gravimeters, *Metrologia*, **32**(3), 159–180.

Niell, A., Whitney, A., Petrachenko, B., Schlüter, W., Vandenberg, N., Hase, H., Koyama, Y., Ma., C., Schuh, H., & Tuccari, G., 2006. VLBI 2010: Current and future requirements for Geodetic VLBI Systems, in *International VLBI Service for Geodesy and Astronomy 2005 Annual Report*, pp. 13–40, NASA/TP-2006-214136, available at http://ivscc.gsfc.nasa.gov/publications/ar2005/spcl-vlbi2010/.

Nocquet, J.-M., Calais, E., Alamimi, Z., Sillard, P., & Boucher, C., 2001. Intraplate deformation in western Europe deduced from an analysis of the International Terrestrial Reference Frame 1997 (ITRF) velocity field, *J. Geophys. Res.*, **106**, 11,239–11,257.

Occhipinti, G., Lognonné, P., Alam Kherani, E., & Hébert, H., 2006. Three-dimensional waveform modeling of ionospheric signature induced by the 2004 Sumatra tsunami, *Geophys. Res. Lett.*, **33**, L20104, doi:10.1029/2006GL026865.

Okada, Y., 1992. Internal deformation due to shear and tensile faults in a half-space, *Bull. Seismo. Soc. America*, **82**, 1018–1040.

Oki, T. & Kanae, S., 2006. Global hydrological cycles and world water resources, *Science*, **313**, 1068–1072, DOI: 10.1126/science.1128845.

Pavelyev, A. G., Volkov, A. V., Zakharov, A. I., Krutikh, S. A., & Kucherjavenkov, A. I., 1996. Bistatic radar as a tool for Earth investigation using small satellites, *Acta Astronautica*, **39**(9-12), 721–730.

Peltier, W. R., 2004. Global glacial isostasy and the surface of the ice-age Earth: The ICE-5G(VM2) model and GRACE, *Ann. Rev. Earth Planet Sci.*, **32**, 111–149.

Pfeffer, W. T., Harper, J. T., & O'Neel, S., 2008. Kinematic constraints on glacier contributions to 21st-century sea-level rise, *Science*, **321**(5894), 1340–1343, DOI: 10.1126/science.1159099.

Pielke Jr., R. & Landsea, C. W., 1998. Normalized hurricane damage in the United States: 1925-95, *Weather and Forecasting*, **13**, 621–631.

Plag, H.-P., 2005. The GGOS as the backbone for global observing and local monitoring: a user driven perspective, *J. Geodynamics*, **40**, 479–486, doi:10.1016/j.jog.2005.06.012.

Plag, H.-P., 2006a. National geodetic infrastructure: current status and future requirements - the example of Norway, Bulletin 112, Nevada Bureau of Mines and Geology, University of Nevada, Reno, 97 pages.

Plag, H.-P., 2006b. GGOS and it user requirements, linkage and outreach, in *Dynamic Planet – Monitoring and Understanding a Dynamic Planet with Geodetic and Oceanographic Tools*, vol. 130 of **International Association of Geodesy Symposia**, pp. 711–718, Springer Verlag, Berlin.

Plag, H.-P. & Jüttner, H.-U., 2001. Inversion of global tide gauge data for present-day ice load changes, in *Proceed. Second Int. Symp. on Environmental research in the Arctic and Fifth Ny-Ålesund Scientific Seminar*, no. Special Issue, No. 54 in Memoirs of the National Institute of Polar Research, pp. 301–317.

Plag, H.-P., Ambrosius, B., Baker, T. F., Beutler, G., Bianco, G., Blewitt, G., Boucher, C., Davis, J. L., Degnan, J. J., Johansson, J. M., Kahle, H.-G., Kumkova, I., Marson, I., Mueller, S., Pavlis, E. C., Pearlman, M. R., Richter, B., Spakman, W., Tatevian, S. K., Tomasi, P., Wilson, P., & Zerbini, S., 1998a. Scientific objectives of current and future WEGENER activities, *Tectonophysics*, **294**, 177–223.

Plag, H.-P., Engen, B., Clark, T. A., Degnan, J. J., & Richter, B., 1998b. Post-glacial rebound and present-day three-dimensional deformations, *J. Geodynamics*, **25**, 263–301.

eds Plag, H.-P., Axe, P., Knudsen, P., Richter, B., & Verstraeten, J., 2000a. *European Sea Level Observing System (EOSS): Status and future developments*, vol. EUR 19682, Office for Official Publication of the European Communities, Luxembourg, 72 pages.

Plag, H.-P., Romagnoli, C., & Zerbini, S., 2000b. Regional and local sea level variations - introduction, in *Sea level change and coastal processes - Implications for Europe*, pp. 81–84, European Commission, Directorate-General for Research.

Plag, H.-P., Gross, R., Chao, B. F., & Van Dam, T., 2005. Forcing of polar motion in the Chandler frequency band: An opportunity to evaluate interannual climate variations, *Eos, Trans. Am. Geophys. Union*, **86**(3), 26.

Plag, H.-P., Beutler, G., Forsberg, R., Ma, C., Neilan, R., Pearlman, M., Richter, B., & Zerbini, S., 2006a. Linking the Global Geodetic Observing System (GGOS) to the Integrated Global Observing Strategy Partnership (IGOS-P) through the

Theme 'Earth System Dynamics', in *Dynamic Planet – Monitoring and Understanding a Dynamic Planet with Geodetic and Oceanographic Tools*, vol. 130 of **International Association of Geodesy Symposia**, pp. 727–734, Springer Verlag, Berlin.

Plag, H.-P., Blewitt, G., Kreemer, C., & Hammond, W. C., 2006b. Solid Earth deformations induced by the Sumatra earthquakes of 2004-2005, in *Dynamic Planet – Monitoring and Understanding a Dynamic Planet with Geodetic and Oceanographic Tools*, vol. 130 of **International Association of Geodesy Symposia**, pp. 549–556, Springer Verlag, Berlin.

Plag, H.-P., Blewitt, G., & Herring, T. A., 2007a. Towards a consistent conventional treatment of surface-load induced deformations, Position Paper presented at the IERS Workshop on Conventions, September 20-21, 2007, Sevres, France. Available at http://www.bipm.org/en/events/iers/papers.html.

Plag, H.-P., Hammond, W., Kreemer, C., & Blewitt, G., 2007b. Integrating point and image geodesy: mutual benefits and requirements, Poster presented at the DESDynI Workshop, July 17-19, 2007, Orlando, Florida, USA.

Plag, H.-P., Kreemer, C., & Hammond, W., 2007c. Combination of GPS-observed vertical motion with absolute gravity changes constrain the tie between reference frame origin and Earth center of mass, Poster presented at the National EarthScope Meeting, Monterey, California, 26-30 March 2007.

Plag, H.-P., Kreemer, C., & Hammond, W., 2009. Global absolute gravity measurements help to tie the geodetic reference frame to Earth's center of mass, *Geophys. Res. Lett.*, p. In review.

Plumb, J., Larson, K., White, J., & Powers, E., 2005. Absolute calibration of a geodetic time transfer system, *IEEE Trans. Ultrason. Ferroelectr. Freq. Control*, **52**(11), 1904–1911.

Poli, P., Moll, P., Rabier, F., Desroziers, G., Chapnik, B., Berre, L., Healy, S. B., Andersson, E., & El Guelai, F.-Z., 2007. Forecast impact studies of zenith total delay data from European near real-time GPS stations in Météo France 4DVAR, *J. Geophys. Res.*, **112**, D06114, doi:10.1029/2006JD007430.

Pollitz, F. F., 1996. Coseismic deformation from earthquake faulting on a layered spherical Earth, *Geophys. J. Int.*, **125**, 1–14.

Pollitz, F. F., 1997. Gravitational-viscoelastic postseismic relaxation on a layered spherical Earth, *J. Geophys. Res.*, **102**, 17,921–17,941.

Pongratz, J., Bounoua, L., DeFries, R. S., & et al., 2006. The impact of land cover change on surface energy and water balance in Mato Grosso, Brazil, *Earth Interactions*, **10**, Art. No. 19.

Pritchard, M. E. & Simons, M., 2002. A satellite geodetic survey of large-scale deformation of volcanic centres in the central Andes, *Nature*, **418**, 167–171.

Ramankutty, N., 2004. Croplands in West Africa: A geographically explicit dataset for use in models, *Earth Interactions*, **8**, 1–22.

Ray, J. & Senior, K., 2003. IGS/BIPM pilot project: GPS carrier phase for time/frequency transfer and time scale formation, *Metrologia*, **40**, S270–288.

Ray, J., Dong, D., & Altamimi, Z., 2004. IGS reference frames: Status and future improvements, *GPS Solutions*, pp. DOI:10.1007/s10291–004–0110–x.

Ray, J. R. & Senior, K., 2001. Temperature sensitivity of timing measurements using Dorne Margolin antennas, *GPS Solutions*, **5**, 24–30.

Ray, J. R. & Senior, K., 2005. Geodetic techniques for time and frequency comparisons using GPS phase and code measurements, *Metrologia*, **42**, 215–232.

Raymond, C., Burbanks, D., Chao, B. F., Donnellan, A., Gillespie, A., Henyey, T. L., Herring, T., Jordam, T., Turcotte, D. L., & Zoback, M. L., 2003. Global Earthquake Satellite System (GESS) - A 20 year plan to enable earthquake prediction, Tech. Rep. JPL 400-1069 03/03, NASA JPL, California Institute of Technology, Pasadena, California.

Reigber, C., Schwintzer, P., & Lühr, H., 1999. The CHAMP geopotential mission, *Boll. Geof. Teor. Appl.*, **40**, 285–289.

Reigber, C., Schwintzer, P., Neumayer, K.-H., Barthelmes, F., König, R., Förste, C., Balmino, G., Biancale, R., Lemoine, J.-M., Loyer, S., Bruinsma, S., Perosanz, F., & Fayard, T., 2003. The CHAMP-only EIGEN-2 Earth Gravity Field Model, *Adv. Space Res.*, **31**(8), 1883–1888.

Reigber, C., Schmidt, R., Flechtner, F., König, R., Meyer, U., Neumayer, K.-H., Schwintzer, P., & Zhu, S. Y., 2005. An Earth gravity field model complete to degree and order 150 from GRACE: EIGEN-GRACE02S, *J. Geodynamics*, **39**, 1–10, doi: 10.1016/j.jog.2004.07.001.

Richter, B., Zerbini, S., Matonti, F., & Simon, D., 2004. Long-term crustal deformation monitored by gravity and space techniques at Medicina, Italy and Wettzell, Germany, *J. Geodynamics*, **38**, 281–292, doi:10.1016/j.jog.2004.07.013.

Rieck, C., Jarlemark, P., Jaldehag, K., & Johansson, J., 2003. Thermal influence on the receiver chain of GPS carrier phase equipment for time and frequency transfer, in *Proc. 2003 IEEE Int. Frequency Control Symp.and PDA Exhibition Jointly with the 17th European Frequency and Time Forum (Tampa, Florida, 5-8 May 2003)*, pp. 326–331.

Rizos, C. & Drane, C. R., 2004. The role of telegeoinformatics in ITS, in *Telegeoinformatics: Location-Based Computing and Services*, eds Karimi, H. A. & Hammad, A., CRC Press LLC, ISBN 0-415-36976-2,377 pages.

Robertson, D. S., 1991. Geophysical applications of very-long-baseline interferometry, *Rev. Mod. Phys*, **63**(4), 899–918.

Rodell, M. & Famiglietti, J. S., 2001. An analysis of terrestrial water storage variations in Illinois with implications for the Gravity Recovery and Climate Experiment (GRACE), *Wat. Resour. Res.*, **37**, 1327–1340.

Rodell, M. & Famiglietti, J. S., 2002. The potential for satellite-based monitoring of groundwater storage changes using GRACE: The High Plains aquifer, central U.S., *J. Hydrol.*, **263**, 245–256.

Rodell, M., Houser, P. R., Jambor, U., Gottschalck, J., Mitchell, K., Meng, C.-J., Arsenault, K., Cosgrove, B., Radakovich, J., Bosilovich, M., Entin, J. K., Walker, J. P., Lohmann, D., & Toll, D., 2004. The Global Land Data Assimilation System, *Bull. Amer. Meteor. Soc.*, **85**(3), 381–394.

Rodell, M., Chen, J., Kato, H., Famiglietti, J., Nigro, J., & Wilson, C., 2006. Estimating ground water storage changes in the Mississippi River basin (USA) using GRACE, *Hydrogeology Journal*, pp. doi:10.1007/s10040–006–0103–7.

Roeloffs, E. A., 1988. Fault stability changes induced beneath a reservoir with cyclic variations in water level, *J. Geophys. Res.*, **93B**, 2107–2124.

Roemmich, D., Willis, J., Gilson, J., Stammer, D., Koehl, A., Yemenis, T., Chambers, D. P., Landerer, F., Marotzke, J., Gregory, J., Suzuki, T., Church, J., White, N., Domingues, C., Cazenave, A., & Letraon, P.-Y., 2006. Global ocean warming and sea level rise, in *Understanding Sea-level Rise and Variability, A World Climate Research Programme Workshop and a WCRP contribution to the Global Earth Observation System of Systems, 6-9 June 2006, UNESCO, Paris*, pp. 33–79, WCRP, World Meterological Organization, Paris.

Rothacher, M., Beutler, G., Weber, R., & Hefty, J., 2001. High-frequency variations in Earth rotation from Global Positioning System data, *J. Geophys. Res.*, **106**(B7), 13,711–13,738.

Rowley, R. J., Kostelnick, J. C., Braaten, D., Li, X., & Meisel, J., 2007. Risk of rising sea level to population and land area, *Eos, Trans. Am. Geophys. Union*, **88**(9), 105, 107.

Rummel, R., 2000. Global Integrated Geodetic and Geodynamic Observing System (GIGGOS), in *Towards an Integrated Global Geodetic Observing System*, vol. 120 of **International Association of Geodesy Symposia**, pp. 253–260, Springer, Berlin.

eds Rummel, R., Drewes, H., Bosch, W., & Hornik, H., 2000. *Towards an Integrated Global Geodetic Observing System*, vol. 120 of **International Association of Geodesy Symposia**, Springer, Berlin.

Rummel, R., Drewes, H., & Beutler, G., 2002. Integrated Global Geodetic Observing System (IGGOS): A candidate IAG project, in *Vistas for Geodesy in the New Millennium*, vol. 125 of **International Association of Geodesy Symposia**, pp. 609–614, Springer, Berlin.

Rummel, R., Rothacher, M., & Beutler, G., 2005. Global Geodetic Observing System (GGOS): Science rationale, *J. Geodynamics*, **40**, 357–362.

Sahagian, D., 2000. Global physical effects of anthropogenic hydrological alterations sea level and water redistribution, *Global Planet. Change*, **25**(1-2), 39–48.

Sahagian, D. L., Schwartz, F. W., & Jacobs, D. K., 1994. Direct anthropogenic contribution to sea level rise in the twentieth century, *Nature*, **367**, 54–57.

Salichon, J., LeCozannet, G., Hosford, S., Missotten, R., & McManus, K., 2007. Igos-g theme report, Tech. rep., BRGM, Orlean, France.

Salstein, D. A., Viron, O., Yseboodt, M., & Dehant, V., 2001. High-frequency geophysical fluid modeling necessary to understand earth rotation variability, *EOS*, **82**, 237–238.

Schaer, S., Gurtner, W., & Feltens, J., 1998. IONEX: The IONosphere Map EXchange format version 1, in *Proceedings of the 1998 IGS Analysis Centers Workshop, ESOC, Darmstadt, Germany, February 9-11, 1998.*

eds Schellnhuber, H.-J. & Wenzel, V., 1998. *Earth System Analysis - Integrating Science for Sustainable Development*, Springer, Berlin and others.

Schierless, L., Schunk, R. W., Sojka, J. J., & Thompson, D. C., 2004. Development of a physics-based reduced state Kalman filter for the ionosphere, *J. Geophys. Res.*, pp. RS1S04, doi:10.1029/2002RS002797.

Seitz, F., Stuck, J., & Thomas, M., 2005. White noise Chandler wobble excitation, in *Forcing of polar motion in the Chandler frequency band: A contribution to understanding interannual climate variations*, vol. 24, p. This issue, Cahiers du Centre Européen de Géodynamic et de Séismologie.

Sella, G. F., Doxon, T. H., & Mao, A. L., 2002. Revel: A model for recent plate velocities from space geodesy, *J. Geophys. Res.*, **107**, 10.1029/2000JB000033.

Seneviratne, S., Viterbo, P., Lüthi, D., & Schär, C., 2004. Inferring changes to terrestrial water storage using ERA-40 reanalysis data: The Mississippi River basin, *J. Climate*, **17**, 2039–2057.

Shedlock, K. M., Giardini, D., Grünthal, G., & Zhang, P., 2000. The GSHAP Global Seismic Hazard Map, *Seismological Res. Let.*, **71**, 679–689.

Shelus, P. J., 2001. Lunar laser ranging: Glorious past and a bright future, *Surv. Geophysics*, **22**, 517–535.

Sherwood, R., Mannucci, A., Zuffada, C., & Heeg, C., 2006. GPS system of systems for science study, JPL Document D-34579, Jet Propulsion Laboratory, Pasadena, California, USA.

Shiklomanov, I. A., 1993. World fresh water resources, in *Water in Crisis*, ed. Gleick, P. H., Oxford University Press, New York.

Sideris, M., 2007. IAG closing address, Presentation at the IUGG Meeting, Perugia, Italy, July, 2007.

Skole, D. & Tucker, C., 1993. Tropical deforestation and habitat fracmentation in the Amazon – satellite data from 1978 to 1988, *Science*, **261**, 1104–1104.

Skole, D. L., Chomentowski, W. H., Salas, W. A., & et al., 1994. Physical and human dimensions of deforestation in Amazonia, *Bioscience*, **44**(5), 314–322.

Smith, D. B., 1998. The emerging IGOS partnership, in *Proceedings for the 27-th Int. Symp. on Remote Sensing of Environment: Information for Sustainability, June 8-12, 1998, Tromsø, Norway*, pp. 99–102, Norwegian Space Centre.

eds Smith, D. E. & Turcotte, D. L., 1993a. *Contributions of Space Geodesy to Geodynamics: Crustal Dynamics*, vol. 23 of **American Geophysical Union Geodynamics Series**, American Geophysical Union, Washington, D. C.

eds Smith, D. E. & Turcotte, D. L., 1993b. *Contributions of Space Geodesy to Geodynamics: Earth Dynamics*, vol. 24 of **American Geophysical Union Geodynamics Series**, American Geophysical Union, Washington, D. C.

eds Smith, D. E. & Turcotte, D. L., 1993c. *Contributions of Space Geodesy to Geodynamics: Technology*, vol. 25 of **American Geophysical Union Geodynamics Series**, American Geophysical Union, Washington, D. C.

Smith, D. E., Christodoulidis, D. C., Kolenkiewicz, R., Dunn, P. J., Klosko, S. M., Torrence, M. H., Fricke, S., & Blackwell, S., 1985. A global geodetic reference frame from LAGEOS ranging (SL5.1AP), *J. Geophys. Res.*, **90**, 9221–9233.

Smith, D. E., Kolenkiewicz, R., Dunn, P. J., Robbins, J. W., Torrence, M. H., Klosko, S. M., Williamson, R. G., Pavlis, E. C., Douglas, N. B., & K., F. S., 1990. Tectonic motion and deformation from satellite laser ranging to LAGEOS, *J. Geophys. Res.*, **95**, 22013–22041.

Smith, D. E., Kolenkiewicz, R., Nerem, R. S., Dunn, P. J., Torrence, M. H., Robbins, J. W., Klosko, S. M., Williamson, R. G., & Pavlis, E. C., 1994. Contemporary global horizontal crustal motion, *Geophys. J. Int.*, **119**, 511–520.

Smith, D. E., Zuber, M. T., Frey, H. V., Garvin, J. B., Head, J. W., Muhleman, D. O., Pettengill, G. H., Phillips, R. J., Solomon, S. C., Zwally, H. J., & Banerdt, W. B., 1998. Topography of the northern hemisphere of Mars from the Mars orbiter laser altimeter, *Science*, **279**, 1686–1692.

Smith, D. E., Zuber, M. T., Sun, X., Neumann, G. A., Cavanaugh, J. F., McGarry, J. F., & Zagwodzki, T. W., 2006. Two-way laser link over interplanetary distance, *Science*, **311**(5757), 53, DOI: 10.1126/science.1120091.

Smith, R. D., Maltrud, M. E., Bryan, F., & Hecht, M. W., 2000. Numerical simulation of the North Atlantic Ocean at 1/10°, *J. Phys. Oceanogr.*, **30**, 1532–1561.

Smith, W. H. F. & Sandwell, D. T., 2004. Conventional bathymetry, bathymetry from space, and geodetic altimetry, *Oceanography*, **17**(1-2), 8–23.

Smrekar, S. E., McGill, G. E., Raymond, C. A., & Dimitriou, A. M., 2004. Geologic evolution of the Martian dichotomy in the Ismenius area of Mars and implications for plains magnetization, *J. Geophys. Res.*, **109**, E11002.

eds Solomon, S., Qin, D., Manning, M., Chen, Z., Marquis, M., Averyt, K. B., Tignor, M., & Miller, H. L., 2007. *Climate Change 2007: The Physical Science Basis. Contribution of Working Group I to the Fourth Assessment Report of the Intergovernmental Panel on Climate Change*, Cambridge University Press, Cambridge, United Kingdom and New York, NY, USA.

Solomon, S. C. & the Solid Earth Science Working Group, 2002. *Living on a restless planet*, NASA, Jet Propulsion Laboratory, Pasadena, California, also available at http://solidearth.jpl.nasa.gov.

Song, Y. T., Ji, C., Fu, L.-L., Zlotnicki, V., Shum, C. K., Yi, Y., & Hjorleifsdottir, V., 2005. The 26 December 2004 tsunami source estimated from satellite radar altimetry and seismic waves, *Geophys. Res. Lett.*, **32**, L20601, doi:10.1029/2005GL023683.

Soudarin, L. & Crétaux, J. F., 2006. A model of present-day tectonic plate motions from 12 years of DORIS measurements, *J. Geodesy*, **80**(8-11), 609–624, DOI 10.1007/s00190–006–0090–4.

Soudarin, L., Crétaux, J.-F., & Cazenave, A., 1999. Vertical crustal motions from the DORIS space-geodesy system, *Geophys. Res. Lett.*, **26**(9), 1207–1210.

Sovers, O. J., Fanselow, J. L., & Jacobs, C. S., 1998. Astrometry and geodesy with radio interferometry: Experiments, models, results, *Rev. Mod. Phys.*, **70**(4), 1393–1454.

Space Studies Board, 2005. *Earth science and applications from space: urgent needs and opportunities to serve the Nation*, Committee on Earth Science and Applications from Space: A Community Assessment and Strategy for the Future, Space Studies Board, National Research Council, Washington, D.C., USA, Available at http://www.nap.edu/catalog/11281.html.

Spencer, P. S. J., Robertson, D. S., & Mader, G. L., 2004. Ionospheric data assimilation methods for geodetic applications, in *Proceedings of IEEE PLANS 2004, Monterey, California, April 26-29, 2004*, pp. 510–517.

Spencer, R. & Vassie, J. M., 1997. The evolution of deep ocean pressure measurements in the U.K., *Progress in Oceanography*, **40**, 423–435.

Spiess, F. N., 1990. Seafloor geodesy by the year 2000, in *Geodesy in the Year 2000*, pp. 100–113, ed. Commission on Physical Sciences, Mathematics, and Applications, The National Academies Press, Free online version available at http://darwin.nap.edu/books/0309041457/html.

Steffen, K., Thomas, R., Rignot, E., Cogley, G., Dyurgerov, M., Raper, S., & Huybrechts, P., 2006. Cryospheric contributions to sea-level rise and variability, in *Understanding Sea-level Rise and Variability, A World Climate Research Programme Workshop and a WCRP contribution to the Global Earth Observation System of Systems, 6-9 June 2006, UNESCO, Paris*, pp. 79–107, WCRP, World Meterological Organization, Paris.

Stolz, A., Bender, P. L., Faller, J. E., Silverberg, E. C., Mulholland, J. D., Shellus, P. J., Williams, J. G., Carter, W. E., Currie, D. G., & Kaula, W. M., 1976. Earth rotation measured by lunar laser ranging, *Science*, **193**, 997–999.

Stroeve, J., Serreze, M., Gearheard, S., Holland, M., Maslanik, J., Meier, W., & Scambos, T., 2008. Arctic Sea ice extent plummets in 2007, *Eos, Trans. Am. Geophys. Union*, **89**(2), 13–14.

Strozzi, T., Tosi, L., Wegmüller, U., Teatini, P., Carbognin, L., & Rosselli, R., 2002. Thematic and land subsidence maps of the Lagoon of Venice from ERS SAR Interferometry, *Geophys. Res. Lett.*, **29**(13), 345–347.

Sun, W. & Okubo, S., 1998. Surface potential and gravity changes due to internal dislocations in a spherical Earth - II. Application to a finite fault, *Geophys. J. Int.*, **132**, 79–88.

Sun, W., Okubo, S., & Fu, G. Y., 2006. Green's functions of coseismic strain changes and investigation of effects of Earth's spherical curvature and radial heterogeneity, *Geophys. J. Int.*, **167**, 1273–1291.

Talwani, P. & Acree, S., 1985. Pore pressure diffusion and the mechanism of reservoir-induced seismicity, *Pure Appl. Geophys.*, **122**, 947–965.

Tapley, B. D., Schutz, B. E., & Eanes, R. J., 1985. Station coordinates, baselines, and Earth rotation from LAGEOS laser ranging: 1976-1984, *J. Geophys. Res.*, **90**, 9235–9248.

Tapley, B. D., Schutz, B. E., Eanes, R. J., Ries, J. C., & Watkins, M. M., 1993. Lageos laser ranging contributions to geodynamics, geodesy, and orbital dynamics, in *Contributions of Space Geodesy to Geodynamics: Earth Dynamics*, vol. 24 of **American Geophysical Union Geodynamics Series**, pp. 147–173, eds Smith, D. E. & Turcotte, D. L., American Geophysical Union, Washington, D. C.

Tapley, B. D., Bettadpur, S., Ries, J. C., Thompson, P. F., & Watkins, M. M., 2004a. GRACE measurements of mass variability in the Earth system, *Science*, **305**, 503–505.

Tapley, B. D., Bettadpur, S., Watkins, M., & Reigber, C., 2004b. The gravity recovery and climate experiment: mission overview and early results, *Geophys. Res. Lett.*, **31**, L09607, doi: 10.1029/2004GL019920.

Tardin, A. T. & da Cunha, R. P., 1990. Report inpe-5015-rpe/609, Tech. rep., Instituto de Pesquisas Espaciais, Sao Jose dos Campos, Brazil.

Tavernier, G. & et al., 2006. The International DORIS Service: genesis and early achievements, *J. Geodesy*, **80**(8-11), 403–417, DOI 10.1007/s00190–006–0082–4.

Tedesco, M., 2007. A new record in 2007 for melting in Greenland, *Eos, Trans. Am. Geophys. Union*, **88**(39), 383.

Thomas, M., Dobslaw, H., Stuck, J., & Seitz, F., 2005. The ocean's contribution to polar motion excitation - as many solutions as numerical models?, in *Forcing of polar motion in the Chandler frequency band: A contribution to understanding interannual climate variations*, vol. 24, p. This issue, Cahiers du Centre Européen de Géodynamic et de Séismologie.

Thomas, R., Rignot, E., Casassa, G., Kanagaratnam, P., Acuna, C., Akins, T., Brecher, H., Frederick, E., Gogineni, P., Kabill, W., Manizade, S., Ramamoorthy, H., Rivera, A., Russell, R., Sonntag, J., Swift, R., Yungle, J., & Zwally, J., 2004. Accelerated sea-level rise from West Antarctica, *Science*, **306**, 255–258.

Thornton, C. L. & Border, J. S., 2000. Radiometric tracking techniques for deep space navigation, Monograph 1, Jet Propulsion Laboratory, California Institute of Technology, Pasadena, CA, http://descanso.jpl.nasa.gov/Monograph/series1/Descanso1_all.pdf.

Titov, V., Gonzáalez, F. I., Bernard, E. N., Eble, M. C., Mofjeld, H. O., Newman, J. C., & Venturato, A. J., 2005a. Real-time tsunami forecasting: Challenges and solution, *Natural Hazards*, **35**, 41–48.

Titov, V., Rabinovich, A. B., Mofjeld, H. O., Thomson, R. E., & González, F. I., 2005b. The global reach of the 26 December 2004 Sumatra tsunami, *Science*, **309**, 2045–2048.

Townshend, J. R. & the IGOL Writing Team, 2004. Integrated Global Observations of the Land: A proposed theme to the IGOS Partnership - Version 2, Tech. rep., IGOS Integrated Global Observing Strategy, Proposal prepared by the IGOL Proposal Team, May 2004, available at http://www.igospartners.org.

Townshend, J. R. G. & Justice, C. O., 1995. Spatial variability of images and the monitoring of changes in the normalized difference vegetation index, *Intern. J. Remote Sensing*, **16**(12), 2187–2195.

Treuhaft, R. N., Lowe, S. T., Zuffada, C., & Chao, Y., 2001. 2-cm GPS altimetry over Crater lake, *Geophys. Res. Lett.*, **22**(23), 4343–4346.

eds Turner II, B. L., Clark, W. C., Kates, R. W., Richards, J. F., Mathews, J. T., & Meyer, W. B., 1990. *The Earth as Transformed by Human Action: Global and Regional Changes in the Biosphere Over the Past 300 Years*, University Press, Cambridge, 713 pages.

UNAVCO, 1998. Supporting research into Earth processes and hazards via high-precision geodesy using the Global Positioning System, Available online at http://www.unavco.org/pubs_reports/brochures/1998_UNAVCO/1998_UNAVCO.html.

United Nations, 2006. *Water - a shared responsibility. The United Nations Water Development Report 2*, UNESCO, Paris, and Berghahn Books, New York.

U.S. Climate Change Research Program, 2007. *Our Changing Planet – The U.S. Climate Change Science Program for Fiscal Year 2008*, U.S. Climate

Change Research Program, Washington, D.C., A report by the U.S. Climate Change Research Program and the Subcommittee on Global Change Research, and a Supplement to the President's Budget for Fiscal Year 2008. Available at http://www.usgcrp.gov/usgcrp/Library/ocp2008/default.htm.

Van Camp, M., Williams, S. D. P., & Francis, O., 2005. Uncertainty of absolute gravity measurements, *J. Geophys. Res.*, **110**, B05406, doi:10.1029/2004JB003497.

Van Dam, T., Plag, H.-P., Francis, O., & Gegout, P., 2003. GGFC Special Bureau for Loading: current status and plans, in *Proceedings of the IERS Global Geophysical Fluid Center Workshop, Munich, November 20-21, 2002*, no. 30 in International Earth Rotation and Reference System Service, IERS Technical Note, pp. 180–198.

Vecchi, G. A., Wittenberg, A. T., Held, I. M., Leetmaa, A., & Harrison, M. J., 2006. Weakening of tropical pacific atmospheric circulation due to anthropogenic forcing, *Nature*, **441**, 73–76, doi:10.1038/nature04744.

Velicogna, I. & Wahr, J., 2005. Greenland mass balance from GRACE, *Geophys. Res. Lett.*, **32**, L18505, doi:10.1029/2005GL023955.

Velicogna, I. & Wahr, J., 2006. Measurements of time-variable gravity show mass loss in Antarctica, *Science*, **311**, 1754–1756.

Vigny, C., Simons, W. J. F., Abu, S., Bamphenyu, R., Satirapod, C., Choosakul, N., Subarya, C., Socquet, A., Omar, K., Abidin, H. Z., & Ambrosius, B. A. C., 2005. Insight into the 2004 Sumatra-Andaman earthquake from GPS measurements in southeast Asia, *Nature*, **436**, 201–206.

Wahr, J., Swenson, S., Zlotnicki, V., & Velicogna, I., 2004. Time-variable gravity from GRACE: First results, *Geophys. Res. Lett.*, **31**, L11501, doi:10.1029/2004GL019779.

Wahr, J., Swenson, S., & Velicogna, I., 2006. The accuracy of GRACE mass estimates, *Geophys. Res. Lett.*, **33**, L06401, doi:10.1029/2005GL025305.

Wahr, J. M., 1981. A normal mode expansion for the forced response of a rotating earth, *Geophys. J. R. Astron. Soc.*, **64**, 651–675.

Wahr, J. M., Dazhong, H., & Trupin, A., 1995. Prediction of vertical uplift caused by changing polar ice volumes on visco-elastic Earth, *Geophys. Res. Lett.*, **22**, 977–980.

Wang, Z. & Ormsbee, L., 2005. Comparison between probabilistic seismic hazard analysis and flood frequency analysis, *Eos, Trans. Am. Geophys. Union*, **86**, 45,51–52.

Williams, D. & Townshend, J. R. G., 1998. The concept of an Integrated Global Observing Strategy, in *Proceedings for the 27-th Int. Symp. on Remote Sensing of Environment: Information for Sustainability, June 8-12, 1998, Tromsø, Norway*, pp. 95–98, Norwegian Space Centre.

Williams, D., Rank, D., Kijek, R., Cole, S., & Pagiatakis, S., 2005. National geodetic infrastructure requirements study, Study report NRCan 03 - 0628, Natural Resources Canada, Final unpublished Report, prepared by BearingPoint; available on request from NRC.

Williams, J. G., Newhall, X. X., & Dickey, J. O., 1993. Lunar laser ranging: Geophysical results and reference frames, in *Contributions of Space Geodesy to Geodynamics: Earth Dynamics*, vol. 24 of **American Geophysical Union Geodynamics Series**, pp. 83–88, eds Smith, D. E. & Turcotte, D. L., American Geophysical Union, Washington, D. C.

Willis, P. & Heflin, M. B., 2004. External validation of the GRACE GGM01C gravity field using GPS and DORIS positioning results, *Geophys. Res. Lett.*, **31**(13), L13616, doi:10.1029/2004GL020038.

Willis, P., Haines, B., Berthias, J. P., Sengenes, P., & Le Mouel, J. L., 2004. Behaviour of the DORIS/Jason oscillator over the South Atlantic Anomaly, *C. R. Geosci.*, **336**(9), 839–846. DOI 10.1016/j.crte.204.01.004.

Willis, P., Jayles, C., & Bar-Sever, Y., 2006. DORIS: From orbit determination for altimeter missions to geodesy, *C. R. Geoscience*, p. doi:10.1016/j.crte.2005.11.013.

Woodworth, P. & Player, R., 2003. The Permanent Service for Mean Sea Level: an update to the 21st century, *J. Coastal Research*, **19**, 287–295.

Woodworth, P. L., Aarup, T., Merrifield, M., Mitchum, G. T., & Le Provost, C., 2003. Measuring progress of the Global Sea Level Observing System, *Eos, Trans. Am. Geophys. Union*, **84**(50), 565, 10.1029/2003EO500009.

Woppelmann, G., Zerbini, S., & Marcos, M., 2006. Tide gauges and geodesy: a secular synergy illustrated by three present-day case studies, *C.R. Geoscience*, p. doi:10.1016/j.crte.2006.07.006.

Wu, X., Bar-Sever, Y. E., Folkner, W. M., Williams, J. G., & Zumberge, J. F., 2001. Probing Europa's hidden ocean from tidal effects on orbital dynamics, *Geophys. Res. Lett.*, **28**(11), 2245–2248.

Wu, X., Argus, D. F., Heflin, M. B., Ivins, E. R., & Webb, F. H., 2002. Site distribution and aliasing effects in the inversion for load coefficients and geocenter motion from GPS data, *Geophys. Res. Lett.*, **29**, 2210, doi: 10.1029/2002GL016324.

Wu, X., Heflin, M. B., Ivins, E. R., Argus, D. F., & Webb, F. H., 2003. Large-scale global surface mass variations inferred from GPS measurements of load-induced deformation, *Geophys. Res. Lett.*, **30**, 1742, doi: 10.1029/2003GL017546.

Wu, X., Heflin, M. B., Ivins, E. R., & Fukumori, I., 2006. Seasonal and interannual global surface mass variations from multisatellite geodetic data, *J. Geophys. Res.*, **111**, B09401, doi:10.1029/2005JB004100.

Yoder, C. F., Konopliv, A. S., Yuan, D. N., Standish, E. M., & Folkner, W. M., 2003. Fluid core size of Mars from detection of the solar tide, *Science*, **300**, 299–303.

ed. Zebker, H., 2005. *InSAR Workshop Summary Report, October 20-22, 2004, Oxnard, California*, NASA, Jet Probulsion Laboratory, Pasadena, California, USA.

Zerbini, S., Plag, H.-P., Baker, T., Becker, M., Billiris, H., Bürki, B., Kahle, H.-G., Marson, I., Pezzoli, L., Richter, B., Romangoli, C., Sztobryn, M., Tomasi, P., Tsimplis, M., Veis, G., & Verrone, G., 1996. Sea level in the Mediterranean: a first step towards separation of crustal movements and absolute sea-level variations, *Global and Planetary Change*, **14**, 1–48.

Zerbini, S., Negusini, M., Romagnoli, C., Domenichini, F., Richter, B., & Simon, D., 2002. Multi-parameter continuous observations to detect ground deformation and to study environmental variability impacts, *Global Planet. Change*, **34**(1-2), 37–58.

Zerbini, S., Richter, B., Rocca, F., van Dam, T., & Matonti, F., 2006. A combination of space and terrestrial geodetic techniques to monitor land subsidence: Case Study, the Southeastern Po Plain, Italy, *J. Geophys. Res.*, Submitted.

Zuffada, C., Elfouhaily, T., & Lowe, S., 2003. Sensitivity analysis of wind vector measurements from ocean reflected GPS signals, *Rem. Sens. Environ.*, **88**, 341–350, doi:10.1016/S0034–4257(03)00175–5.

Zwally, H. J., Abdalati, W., Herring, T., Larson, K., Saba, J., & Steffen, K., 2002. Surface melt-induced acceleration of Greenland ice-sheet flow, *Science*, **297**, 218–222.

Zwally, H. J., Giovinetto, M. B., Li, J., Cornejo, H. G., Beckley, M. A., Brenner, A. C., Saba, J. L., & Yi, D., 2005. Mass changes of the Greenland and Antarctic ice sheets and shelves and contributions to sea-level rise: 1992-2002, *J. Glaciology*, **51**, 509–527.

Acronyms and abbreviations

ACES Atomic Clock Ensemble in Space
AG Absolute gravimeter
AIRS Atmospheric Infra-Red Sounder
AMSU Advanced Microwave Sounding Unit
ANZLIC Australia and New Zealand Land Information Council
ASCII American Standard Code for Information Interchange
ASDD Australian Spatial Data Directory
ASDI Australian Spatial Data Infrastructure
ATC Air Traffic Control
AVHRR Advanced Very High Resolution Radiometer
BIH Bureau International de l'Heure
BIPM Bureau International des Poids et Mesures
BPR Bottom Pressure Recorder
CAM Core Angular Momentum
CCRS Conventional Celestial Reference System
CDP Crustal Dynamics Project
CE Center of Mass of the Solid Earth
CEOS Committee for Earth Observation Satellites
CEP Celestial Ephemeris Pole
CF Center of Figure of the solid Earth
CGDI Canadian Geospatial Data Infrastructure
CHAMP Challenging Minisatellite Payload
CM Center of Mass of the whole Earth system
CMB Core-Mantle Boundary
CNES Centre National d'Etudes Spatiales
COP Community of Practice
COPES Coordinated Observation and Prediction of the Earth System
CORS Continuously Operating GPS Stations
CRS Celestial Reference System
CTF Control track farming
CTRS Conventional Terrestrial Reference System

DART Deep-ocean Assessment and Reporting of Tsunamis
DEM Digital Elevation Model
DIAL DIfferential Absorption LIDAR
DIMERS Dynamical Integrated Modular Earth Rotation System
DORIS Doppler Orbitography and Radiopositioning Integrated by Satellites
DSM Digital Surface Model
DSN Deep Space Network
DST Dynamic Sea Surface Topography
DyMEG Dynamic Model for the Earth Rotation and Gravity
EC European Commission
ECEF Earth-centered, Earth-fixed
EGG97 European Gravimetric Geoid 1997
EGM Earth Gravity Model
EGNOS European Geostationary Navigation Overlay Service
ENSO El Niño Southern Oscillation
EOP Earth Orientation Parameters
EOS Earth Observation Summit
EOS Earth Observation Satellite
ERS Earth Remote Sensing
ESA European Space Agency
ESSP Earth System Science Pathfinder
EU European Union
FAGS Federation of Astronomical and Geophysical Data Analysis Services
FANS Future Air Navigation System
FAO Food and Agriculture Organization
FGDC Federal Geographic Data Committee
FIG International Federation of Surveyors
FOC Full Operational Capability
G3OS Global Three Observing Systems
GA Geoscience Australia
GAGAN GPS Aided GEO Augmented Navigation
GAIM Global Assimilative Ionospheric Model
GCOS Global Climate Observing System
GEO Group on Earth Observations
GEOSS Global Earth Observation System of Systems
GFO Geosat-Follow-On
GGFC Global Geophysical Fluid Center
GGOS Global Geodetic Observing System
GGP Global Geodynamics Project
GIA Glacial Isostatic Adjustment
GIM Global Ionospheric Mapping
GIS Geographic Information Systems
GLDAS Global Land Data Assimilation System
GLONASS Global Navigation Satellite System
GLOSS Global Sea Level Observing System

GMES Global Monitoring of Environment and Security
GNSS Global Navigation Satellite System
GNSSS Global Navigation Satellite System of Systems
GOCE Gravity field and steady-state Ocean Circulation Explorer
GOES Geostationary Satellite Server
GOOS Global Ocean Observing System
GPM Global Precipitation Measurement
GPS Global Positioning System
GRACE Gravity Recovery and Climate Experiment
GRGS Groupe de Recherche de Géodésie Spatiale
GRS Geodetic Reference System
GSHM Global Seismic Hazard Map
GSO GeoStationary Orbit
GSWP Global Soil Wetness Project
GTHM Global Tsunami Hazard Map
HF High Frequencies
HRSC High Resolution Stereo Camera
HSB Humidity Sounder for Brasil
IAG International Association of Geodesy
IAIN International Association of Institutes of Navigation
IAU International Astronomical Union (the International Astronomical Union)
ICAO International Civil Aviation Organisation
ICRF International Celestial Reference Frame
ICRS International Celestial Reference System
ICSM Intergovernmental Committee on Surveying and Mapping
ICSU International Council for Science
ICT Information and Communication Technology
IDS International DORIS Service
IERS International Earth Rotation and Reference Systems Service
IGFS International Gravity Field Service
IGOS Integrated Global Observing Strategy
IGOS-P Integrated Global Observing Strategy Partnership
IGN Institut Géographique Nationale
IGS International GNSS Service
ILRS International Laser Ranging Service
ILS International Latitude Service
IMO International Maritime Organisation
INS Inertial Navigation Systems
InSAR Interferometric Synthetic Aperture Radar
INSPIRE Infrastructure for Spatial Information in Europe
IOC International Oceanographic Commission
IPCC Intergovernmental Panel on Climate Change
IPMS International Polar Motion Service
IRNSS Indian Regional Navigation Satellite System
ISIS International SAR Information System

IRS Incoherent Scatter Radars
IT Information Technology
ITRF International Terrestrial Reference Frame
ITRS International Terrestrial Reference System
IUGG International Union of Geodesy and Geophysics
IVS International VLBI Service for Geodesy and Astrometry
IYPE International Year Of Planet Earth
LAGEOS LAser GEOdynamics Satellite
LBS Location-Based Service
LEO Low Earth Orbiters
LIDAR LIght Detection And Ranging
LLR Lunar Laser Ranging
LOD Length of Day
LOLA Lunar Orbiter Laser Altimeter
LRO Lunar Reconnaissance Orbiter
LSL Local Sea Level
MEO Medium Earth Orbit
MERIT Monitoring Earth Rotation and Inter-comparison of Techniques
MGS Mars Global Surveyor
MLA Mercury Laser Altimeter
MODIS Moderate-Resolution Imaging Spectroradiometer
MOLA Mars Orbiting Laser Altimeter
MSAS Multifunctional Transport Satellite Space-based Augmentation System
MSS Mean-Squared Slope
MYRTLE Multi Year Return Time Level Equipment
NGRS Australian National Geospatial Reference System
NGS National Geodetic Survey
NGSLR Next Generation SLR
NIGCOMSAT Nigerian Communication Satellite
NOAA National Oceanic and Atmospheric Administration
NRT near-real time
NSDI National Spatial Data Infrastructure
NSRS National Spatial Reference System
OMB Office of Management and Budget
OSSE Observing System Simulation Experiments
PARIS Passive Reflectometry and Interferometry System
PBO Plate Boundary Observatory
PF Precision Farming
PGR Post-Glacial Rebound
PNT Positioning, Navigation and Timing
ppb parts per billion
POD Precision Orbit Determination
POL Proudman Oceanographic Laboratory
PSMSL Permanent Service for Mean Sea Level
QZSS Quasi-Zenith Satellite System

RF Radio-Frequency
RFID Radio-Frequency Identification
RFO Reference Frame Origin
RINEX Receiver Independent Exchange Format
RLE Revised Local Reference
RTK Real-Time Kinematic
SAR Synthetic Aperture Radar
SBA Societal Benefit Area
SBAS Satellite-Based Augmentation System
SDI Spatial Data Infrastructure
SG Superconducting Gravimeter
SI International System of Units
SINEX Software Independent Exchange Format
SiS Signal in Space
SLR Satellite Laser Ranging
SMOS Soil Moisture and Ocean Salinity
SOAP Simple Object Access Protocol
SP3 Standard Product 3 Orbit Format
SRTM Shuttle Radar Topography Mission
SSM/I Special Sensor Microwave/Imager
SST Satellite-to-Satellite Tracking
TAI International Atomic Time
TCG Geocentric Coordinate Time
TDR Time Domain Reflectometry
TEC Total Electron Content
TIGA TIde GAuge Pilot Project
TRMM Tropical Rainfall Measuring Mission
TRS Terrestrial Reference System
TT Terrestrial Time
TYIP Ten-Year Implementation Plan
UAV Unmanned Aerial Vehicle
UK-DMC United Kindom's Desaster Monitoring Constellation
UN United Nations
UNESCO United Nations Educational, Scientific and Cultural Organization
UNFCCC United Nations Framework Convention on Climate Change
UR User Requirement
USO Ultra-Stable Oscillator
UT Universal Time
UT1 Universal Time 1
UTC Coordinated Universal Time
UV Ultraviolet
UWB UWB
VHF Very High Frequencies
VLBA Very Long Baseline Array
VLBI Very Long Baseline Interferometry

VSI VLBI Standard Interface
WAAS Wide Area Augmentation System
WCRP World Climate Research Programme
WEGENER Working group of European Geoscientists for the Establishment of Networks for Earth-science Research
WGS 84 World Geodetic System 1984
WMO World Meteorological Organisation
WOCE World Ocean Circulation Experiment
WSDL Web Services Description Language
WSSD World Summit on Sustainable Development
WWW World Wide Web
XML eXtensible Markup Language

Index